NANO-CMOS
DESIGN FOR
MANUFACTURABILILTY

NANO-CMOS DESIGN FOR MANUFACTURABILILTY

Robust Circuit and Physical Design for Sub-65 nm Technology Nodes

Ban Wong

Franz Zach

Victor Moroz

Anurag Mittal

Greg Starr

Andrew Kahng

WILEY

A JOHN WILEY & SONS, INC., PUBLICATION

Published by John Wiley & Sons, Inc., Hoboken, New Jersey
Published simultaneously in Canada

For general information on our other products and services or for technical support, please contact our Customer Care Department within the United States at (800) 762-2974, outside the United States at (317) 572-3993 or fax (317) 572-4002.

Wiley also publishes its books in a variety of electronic formats. Some content that appears in print may not be available in electronic formats. For more information about Wiley products, visit our web site at www.wiley.com.

Library of Congress Cataloging-in-Publication Data:

Nano-CMOS design for manufacturability : robust circuit and physical design for sub-65 nm technology nodes / by Ban Wong ... [et al.].
    p. cm.
  Includes bibliographical references and index.
  ISBN 978-0-470-11280-9 (cloth)
  1. Metal oxide semiconductors, Complementary–Design and construction. 2. Integrated circuits–Design and construction. 3. Nanoelectronics. I. Wong, Ban P., 1953-
    TK7871.99.M44N362 2008
    621.39′5–dc22

                                           200807589

Printed in the United States of America

10 9 8 7 6 5 4 3 2 1

CONTENTS

III THE ROAD TO DFM   333

7 Nano-CMOS Design Tools: Beyond Model-Based
Analysis and Correction   335

PREFACE

This book is the sequel to our first book, *Nano-CMOS Circuit and Physical Design* and takes the design to technology nodes beyond the 65-nm geometries. In the first part we address the challenges that arise as a result of the new process techniques to keep pace with Moore's law beyond the 65-nm technology node, as well as the inherently higher performance level of designs in those technology nodes. Embedded SiGe, contact etch stop layer, and other process-straining techniques require layout constraints to minimize process variability. Subwavelength optical lithography will dictate layout disciplines as well. Pattern-distortion-feature-limited yield is on the increase and left unchecked would be a major impediment to successful product development at the 65-nm node and beyond. Device scaling is lending itself to critical parameters approaching the atomic scale and is hence subject to atomic uncertainties. These are of concern at the 45-nm mode and will become critical at the 22-nm technology node and below. Signal and power integrity will become even more significant for these designs.

Designers who are focused on cell size and local performance have little understanding of pattern printability. As a result, pattern-distortion features limited yield, and unpredictable circuit performance on silicon is on the rise. Catastrophic failures due to the difficulty in applying proper optical proximity correction (OPC) or failure of the OPC/RET algorithm induced by layout designed with no knowledge of its impact on patterning can be a result of current design methodology. Until very recently, DFM was a mere buzzword. Driven by the ever-rising trend toward pattern-distortion-feature-limited yield and unpredictable silicon performance, as well as leaving much of the process entitlement on the table, resulting in a much lower performance product than could be achieved with full entitlement, many are turning to DFM (design for manufacturability) to alleviate this yield loss mechanism and to harness the full process entitlement. The classical approach through design rules (causing explosive design rule growth) is intractable by itself, yet cannot fully describe the parametric changes as a result of lithographic distortion as well as stress proximity differences. This stiffles design productivity through long design rule checks, yet a clean design using design rule checking does not result in a problem-free design. Restrictive design rules have been touted as the panacea for all these ills, but are they?

This drives up design cost as well as mask cost, which raises the barrier for entry, especially for startup companies. Without the increased design starts contributed by the many startups, the industry would no longer be healthy and the future would be bleak. This book provides the needed bridge from physical and circuit design to fabrication processing, as well as harnessing systematic variability instead of having to apply design margins that would further reduce process entitlement available to designers.

In **Part I** we introduce the newly exacerbated effects that will require designers' attention. We begin with a discussion of the lithography-related aspects of DFM. As the semiconductor industry enters the subwavelength domain, the traditional wisdom that chip yields are ultimately dominated by defects no longer holds. Lithography and design-driven factors contribute increasingly to the ultimate chip yield. This notion is enforced by the experience of many fabless companies whose product plans were severely delayed by manufacturing issues, particularly at the beginning of a new technology node. Various fabless companies have experienced costly mask reworking and delays in their schedules due to printing-related issues. Subwavelength lithography faces a variety of process variations: dose, focus, mask CD (critical dimension) variations, lens aberrations, and across-field nonuniformities. With reduced k_1 factors the process variabilities create increased CD variations as well.

Stress engineering is a new and hot topic in the semiconductor industry. It was introduced by Intel in 2003 and is currently being embraced by the rest of the industry as the main vehicle to keep up with Moore's law for the next decade. The most popular stress engineering techniques for bulk CMOS transistors are the strained CESL (contact etch stop layer) and eSiGe (embedded SiGe source–drain). A combination of these two techniques has been demonstrated to improve transistor drive current by up to 85% or to suppress leakage current by two orders of magnitude. Such a dramatic reduction in the leakage current is used to prolong the battery life of mobile applications, and a boost in the drive current is used for high-performance applications. In contrast to the obvious benefits, stress engineering introduces a number of side effects that can disturb or even ruin circuit performance if left unattended. Chapter 3 begins with a review of the physics of stress effects in silicon and a list of all possible stress sources in semiconductor manufacturing. Then methods of introducing beneficial stresses into a transistor are described, and how each stress component depends on the specific layout of the transistor and its neighborhood is shown. The final part of the chapter covers how to improve a layout to maximize stress-enhanced transistor performance and how to minimize layout-related performance variability.

In **Part II** we describe design solutions to mitigate the impact of process effects as discussed in Part I. In this part we discuss methodology to make subwavelength patterning technology work in manufacturing. We also discuss design solutions to deal with signal and power integrity, both of which will be very significant in designs at the 65-nm nodes and beyond.

Successful design of circuits in nano-CMOS processes requires a change in design, parasitic extraction, and verification methodologies to ensure robust, reliable designs and physical implementations. Given the ever-increasing complexity of the advanced processes (multiple oxide thicknesses and various threshold voltage options, where second-order effects are becoming first-order effects), analog and mixed-signal designers are faced with a new barrage of issues and options that must be considered when designing these circuits. This problem is exacerbated by the ever-increasing integration of functions on a single chip that is transforming more and more analog and mixed-signal designers into system-on-a-chip companies, which now face designing processes unfamiliar to them. In Chapter 5 we provide basic guidelines for designing these circuits, with manufacturability and yield in nano-CMOS processes as the primary focus. The reader will become familiar with the various aspects of the design that can affect both manufacturability and yield, including such topics as design methodologies for minimizing the effects of process variation, design rule modifications to improve manufacturability, and device selection criteria for analog blocks.

Even in digital designs, we can no longer ignore subwavelength patterning distortion, the ever-increasing parametric variability induced by these advanced technology nodes, and the fine critical dimensions that amplify the effect of CD variation on circuit performance and functionality. We address the necessary paradigm shift in the circuit design and physical Implementation methodology to arrive at a high-yielding and scalable design that also harnesses full process entitlement. Power integrity, recently recognized as a lever for yet better performance, is also described in Part II.

We conclude the book in **Part III** by dealing with new tools needed to support DFM efforts. Extraction technology must take the subwavelength distortions into consideration. New extraction tools must be based on printed contours to include the size changes and distortion effects on parasitics. There is also a need for a static timer that is aware of the effects on timing due to wire and device distortions as a result of chemical–mechanical polishing as well as subwavelength patterning distortions. Simulation tools will be needed to capture the effects of transistor drive strength due to layout. This tool will also guide physical designers so that the physical design will result in the least device drive strength variability. We have to learn to live with variability due to the nano-CMOS technology nodes. We also introduce an auto-correction tool that is capable of fixing the layout of cells with multiple optimization goals if provided with the priority of each of the various objectives.

BAN WONG
FRANZ ZACH
VICTOR MOROZ
ANURAG MITTAL
GREG STARR
ANDREW KAHNG

ACKNOWLEDGMENTS

The book draws heavily on the knowledge base built by many individuals and conferences as well as the authors' direct experience in each of the areas presented. As such, we are indebted to the many persons who have contributed in one way or another to the completion of this book.

First, we thank Don Draper of Rambus for his contributions to Chapter 4. We thank Professor Asenov of the University of Glasgow, UK, for allowing us to reproduce the data created by his research team reproduced in Table 6.2. We appreciate the contributions of Etsuya Morita and Maarten Berkins of Takumi in providing detailed descriptions of the Takumi auto-correction tool, and Bob Gleason for the inverse lithographic concept used in this book. We thank Roy Prasad, CEO of Invarium, for supplying some of the figures used in the book as well as useful discussions with his technical teams. Special thanks to Nitin Deo of Clearshape for providing access to their technical teams for discussions on some of the concepts on contour-based extraction that are used in the book.

We acknowledge with thanks Professor Kahng's students, postdoctoral scholars, and visiting scholars at the UCSD VLSI CAD Laboratory, as well as colleagues at Blaze DFM, Inc., for interesting and fruitful discussions and collaborations. Particular thanks are due to Puneet Gupta, Chul-Hong Park, Kwangok Jeong, Sam Nakagawa, Puneet Sharma, and Swamy Muddu.

We are also indebted to Chris Progler, CTO of Photronics, for reviewing the text and providing many valuable suggestions as well as providing several figures.

1

INTRODUCTION

1.1 VALUE OF DESIGN FOR MANUFACTURABILITY

Many designers are still unclear as to the value of having design for manufacturability (DFM) inserted into their design flow, simply because DFM requires additional resources (tool cost and design resources), design delays, and so on. No designer is willing to sacrifice schedule while increasing the resources required—just to achieve a better yield. Designers are always seeking ways to improve performance, power, and die size while minimizing design margins and eliminating the need to rework the design as a result of circuit bugs. To be attractive to designers, DFM must offer avenues to achieve these goals as well.

The ultimate reward for using DFM is an economic one. A design with higher performance, lower power, and smaller die size translates to a higher average selling price (ASP) and lower manufacturing costs. An improved and predictable yield during manufacturing results in a reduced time to market, higher profits, and a longer product lifetime. Nowak [1] describes this economic concept, which bridges the return on investment (ROI) gap between design and manufacturing. The concept is well illustrated in Figure 1.1, where the dashed line shows a possible life cycle of a design in an advanced technology node without deploying DFM, and the solid line shows a life cycle of similar design incorporating DFM. Figure 1.1 suggests that a design with DFM will result in a faster, more predictable yield ramp, thus less time to market, higher profits, and a longer product life cycle.

Nano-CMOS Design for Manufacturability: Robust Circuit and Physical Design for Sub-65 nm Technology Nodes
By Ban Wong, Franz Zach, Victor Moroz, Anurag Mittal, Greg Starr, and Andrew Kahng
Copyright © 2009 John Wiley & Sons, Inc.

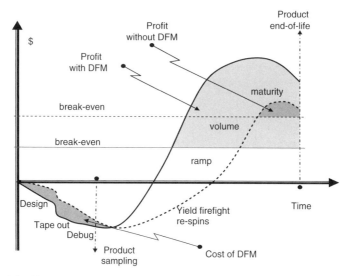

Figure 1.1 Economics of DFM. (Adapted with permission from Nowak/Radojcic, ref. 1.)

As the technology advances, the critical dimensions (CDs) of many critical layers are well into the subwavelength regime, and film thickness has approached atomic layer dimensions. The result is greater variability. For example, the gate dielectric of a typical 65-nm node is on the order of four atomic layers thick. It would be impossible for any process to place four atomic layers precisely on the gate oxide of every transistor on an entire wafer. Also, with gate length CDs in the sub-40-nm range, 4 nm of variability would represent a 10% change in the CDs. This is a meager CD budget, yet would have to be shared among the various processes: lithography, etching, optical and etching proximity effects, mask error, and mask error enhancement factor (MEEF) [2, Chap. 3]. Without designer intervention, it would be impossible to achieve the CD budget. A large polysilicon ("poly") CD variation would result in a large spread in product performance, power, and leakage.

Dopant fluctuation is unavoidable at these dimensions, where the number of dopants for a transistor in a typical standard cell in the 65-nm node is less than 100 dopant atoms, resulting in a large $\sigma\Delta V_t$ (threshold voltage) value. It has been determined that the dopant location is also important at these dimensions [3]. Figure 1.2 shows that the V_t of two transistors with the same number of dopant atoms can still be very different, depending on the location of the dopants. The dopant location is a result of the difference in energy that each dopant acquired as it was being propelled toward the silicon wafer. No one has yet been able to cut the energy tail of the implanters. Process variability has greater impact now, and if not designed for, reduces parametric yields [3–5].

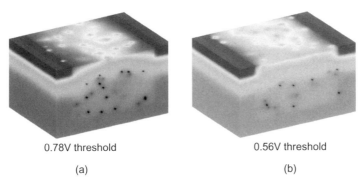

0.78V threshold   0.56V threshold

(a)   (b)

Figure 1.2   Random dopant location-induced V_t fluctuation (a) 0.78-V threshold; (b) 0.56-V threshold. Both devices have 170 dopants in the channel depletion region. (From ref. 3.)

1.2 DEFICIENCIES IN BOOLEAN-BASED DESIGN RULES IN THE SUBWAVELENGTH REGIME [6]

With older technology, Boolean-based design rules worked well and have been the design sign-off to guarantee a manufacturable design. As technology scales, problems arise. Figure 1.3a shows a failure under certain process conditions for a design that meets design rules. In the subwavelength regime, the limitations of Boolean-based design rules are beginning to show. What is more critical is the case shown in Figure 1.3b, where the lines are not yet shorted but are already so close that they pose a reliability hazard. How many such reliability hazards are lurking in the design? That will be difficult to quantify or detect without a model-based tool. Figure 1.4 shows another typical failure as a result of pattern distortion due to the optical proximity effect in the subwavelength regime. The two landing pads are equidistant from the line above. The only difference within this context is the tab on the horizontal line that is above the right landing pad (pad B). This difference in proximity caused the right landing pad to be pulled toward the horizontal line above, resulting in shorts under certain process conditions. Incidentally, the short occurred between structures that are not minimally spaced relative to each other. In the absence of the proximity effect as in landing pad A, no short is seen.

For the 90-nm node, the first-order proximity effect is to the structures immediately adjoinly the polygon of concern. At the 45-nm node the proximity effect influence is as far as a few structures away from the polygon. When the proximity effects are so far reaching, it is very difficult to code Boolean-based rules to describe this effect so that designers can design for it. At these advanced nodes a model-based approach would be inevitable to fully describe this effect to designers so that it can be avoided in their design.

For each technology generation, manufacturers attempt to deal with the problem by resolving the densest pattern that the design rules allow. However, this does not mean that they can print the full chip of any design. Most

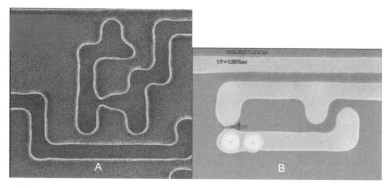

Figure 1.3   Failure at a process corner for a design that met design rules.

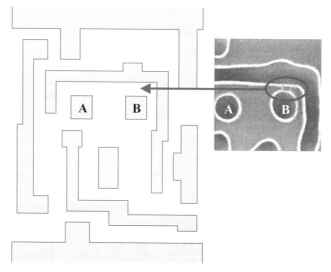

Figure 1.4   Example of the need for a model-based solution.

problems are the result of two-dimensional patterns, such as line-end patterns, where the diffraction patterns are much more complicated than the simple dense line–space pattern that the manufacturers have developed for the lithography process. The line ends are also most susceptible to etch distortion and proximity effects, as well as a different response to poly CD trim. These types of issues will get worse as the technology advances, resulting in lots of opportunities for designer intervention on designs of the future. Model-based tools will be required to help find and fix these hotspots. Boolean-based design rules alone will no longer be sufficient. Without model-based tools, the Boolean-based design rules will have to be very complex as well as there being too many to be practical. Even then, they will still not be well enough developed to address all the new and newly exacerbated design–process interactions.

1.3 IMPACT OF VARIABILITY ON YIELD AND PERFORMANCE

At the process and device levels, we are seeing line edge roughness (LER) contributing more significantly to channel-length variability at the CDs of nano-CMOS devices. This results in higher device OFF-current (I_{off}) variability, as shown in Figure 1.5. Chips designed for used in hand held devices will be affected by this higher I_{off} variation and must be designed for, or this will present a yield issue that delays product introduction (see in Figure 1.1.)

Poly CD control is getting difficult, but the criticality for many circuits is not abating, as shown in Figure 1.6, where 6% lower poly CDs result in an unusable product. For microprocessors the speed versus average selling price is nonlinear; in fact, it increases exponentially with speed. There is a huge financial motivation to make the poly CDs as narrow as possible and still have a good product yield. The lower the poly CDs, the higher the speed of the part will be. The better the CD control, the lower the CDs can be pushed. Figure 1.7 shows that the margins drop very quickly with reduced poly CDs. There are a lot of opportunities for a designer to participate in improving the poly CD control to produce the best-performing part with the least yield loss. Figure 1.8 shows the V_t response to poly CDs. When the CDs are larger than the target values by 10%, we see that the V_t spread improves, whereas a −10% value for the poly CDs results in a greater change in V_t from the target as well as a larger CD σ value. Many of the devices have V_t values below zero, which means they will never shut off, resulting in massive current leakage. Design opportunities to improve poly CD control are covered in more detail in Chapter 6.

Figure 1.9 shows the delay distribution of two similar designs with different layout styles. It is clear from these data that the design style can have a large effect on the σ value of the delay distribution.

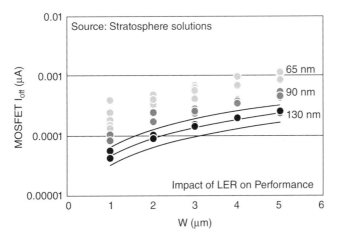

Figure 1.5 I_{off} variability as a result of line-edge roughness. (Courtesy of Stratosphere Solutions.)

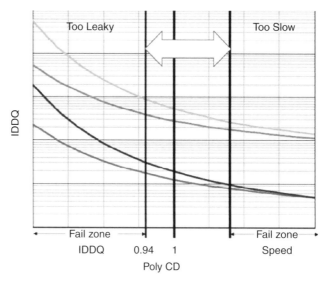

Figure 1.6 IDDQ response to poly CD.

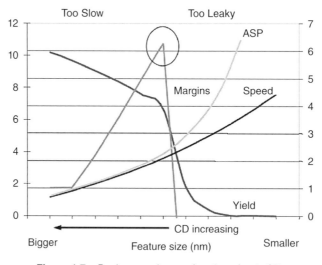

Figure 1.7 Design margin as a function of poly CD.

Patterning reproducibility will be a challenge, as shown in Figure 1.10. The two poly lines shown in the figure are supposed to be identical. As a result of patterning-reproducibility limitations as well as LER, they look different under high magnifications. Transistor matching will be difficult at the nano-CMOS nodes.

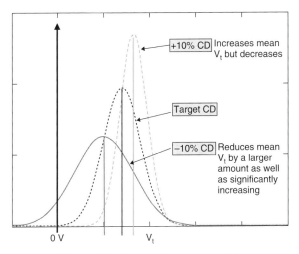

Figure 1.8 V_t spread versus poly CD.

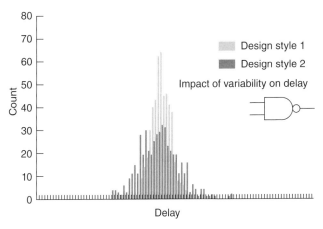

Figure 1.9 Layout style and design context affect variability. (Courtesy of Stratosphere Solutions.)

Process-Related Variability

- Line-edge roughness, random dopant fluctuations and location, poly CD control, and so on: have greater impact on electrical properties
- Mask error and mask error enhancement factor in a low-k_1 lithography process: affect electrical properties and device matching
- OPC and patterning reproducibility: a matching issue for analog, memory, and clock distribution designs

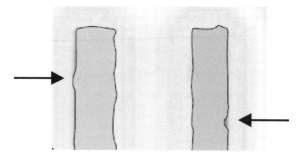

Figure 1.10   Challenge of patterning reproduceability at nano-CMOS dimensions. (From ref. 4.)

Device-Level Variability

- V_t, device ON current (I_{on}), and I_{off} variability: increase
- Resistance and capacitance variability: increases for vias, contacts, and metal
- Transistor, resistor, and capacitor: mismatch increases
- NBTI and other V_t shift mechanisms
- Stress and strain variability: film thickness and proximity
- Lithography distortions: OPC, optical proximity, and lithographic condition dependent

Cell- or Circuit-Level Variability

- Delay distributions: much broader
- Cell (block) placement and rotation: induce variability
- Layout- and context-dependent stress and strain variability: results in I_{on} variability
- Well proximity effect
- Layout effect on mobility

Chip-Level Variability

- Increase in *di/dt*: can result in timing and functional failures
- Signal integrity: has a major impact on timing and functionality
- Greater spread in chip standby power
- Voltage and temperature gradient: affects device performance
- Substrate noise (triple well isolation may not work for radio frequencies)
- Jitter

1.4 INDUSTRY CHALLENGE: THE DISAPPEARING PROCESS WINDOW

Since the 65-nm technology node was developed, manufacturers have been working below a robust manufacturing level. Even when the process has been extended by all known techniques, with the numerical aperture (NA) pushed as high as feasible using the latest scanners, manufacturers are unable to keep the process above the robust manufacturability line (Figure 1.11). The result is that the process window is shrinking as the optical lithography process is falling behind Moore's law. The illumination wavelength has been stuck at 193 nm, due to difficulties in bringing extreme ultraviolet scanners on line, and to materials and integration concerns with the 157-nm wavelength (Table 1.1) [5–7]. The knobs that are available are NA and immersion lithography to make greater than unity NA work with the depth of focus (DOF) margin in these

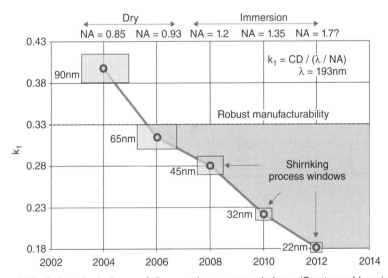

Figure 1.11   Industry's challenge of disappearing process windows. (Courtesy of Invarium.)

Table 1.1   Illumination sources available for optical lithography

Wavelength (nm)	Illumination Source	Year Introduced
436	g-line	1980
365	i-line	1989
248	KrF laser	1995
193	ArF laser	2002
193i	ArF laser	2006
157	F_2 laser	Failed
13.4	EUV	?

processes. Aggressive resolution enhancement techniques (RETs), phase-shift masking (PSM), and illumination optimization are the other knobs available. Even when all these techniques are deployed to the fullest, the process window is still shrinking [7].

Therefore, designers must now do their part to bridge the process–design gap to make their designs manufacturable within these technology nodes. In Chapter 2 we go into more detail on the challenges of the lithographic process and the solutions available to manufacturers as well as designers.

1.5 MOBILITY ENHANCEMENT TECHNIQUES: A NEW SOURCE OF VARIABILITY INDUCED BY DESIGN–PROCESS INTERACTION

To remain competitive, all foundries have resorted to some sort of stress engineering using stress memory techniques (SMTs), contact etch stop layer (CESL) stressing film, or recessed source–drains using embedded silicon–germanium (eSiGe). CESL tensile and compressive film is used to enhance nMOS and pMOS drive current, respectively (more details in succeeding chapters).

Figure 1.12 shows all the layout parameters that affect transistor carrier mobility. The mobility is affected by the stress applied to the transistor channel by the stressing film or recessed eSiGe source–drain and shallow trench isolation (STI). The stress applied to the channel by the CESL film is proportional to the volume and proximity of the film to the channel. The volume and proximity of the film to the channel are modulated by poly pitch, contact pitch, contact-to-poly (channel) space, and contact critical dimensions (CDs). The number of contacts on a transistor affect the degree to which the CESL stressing film is perforated, which relaxes the film, resulting in reduced stressing film

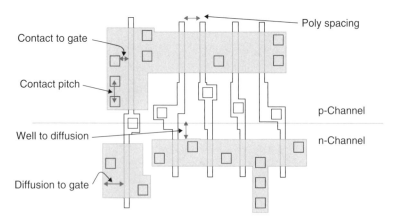

Figure 1.12 Effect of layout parameters on carrier mobility. (Courtesy of Synopsys.)

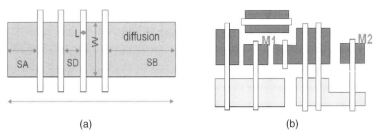

Figure 1.13 Stress proximity–device drive strength modeling issue. (a) BSIM4 LOD model is indifferent to context (the surrounding layout patterns). (b) M1 and M2 have identical W/L and LOD, but different stress and drive strength. (Courtesy of Synopsys.)

effectiveness. This results in lower carrier mobility, hence lower transistor drive.

STI in the 90- and 65-nm technologies exerts a compressive stress on transistors. As the distance of the edge of the diffusion (where STI begins) to the transistor channel is increased, the effect of the STI stress as seen by the transistor channel diminishes. Therefore, varying that distance will change the transistor drive current as the stress level changes. Since STI exerts compressive stress, it degrades nMOS drive current but improves pMOS drive.

Figure 1.13 illustrates modeling issues with the current BSIM length of diffusion (LOD) model. The parameters available to the current LOD model include SA, SB, and SD. These parameters cannot differentiate transistors of differing context and can seriously misrepresent the stress effects experienced by a similar transistor with a different context. For example, M1 and M2 in Figure 1.13b have similar LODs but very different contexts and so will experience significantly different channel stress, which will not be captured by the current BSIM4 LOD model. The result is that a designer using the BSIM4 LOD model will find that his or her simulations will defer from the behavior of the circuit on silicon. We discuss this issue and its solutions more in subsequent chapters.

Figure 1.14 shows the effect of eSiGe recess depth on the stress in the channel—hence the drive strength. The greater the recess depth, the greater is the stress on the channel. However, at minimum poly-to-poly spacing we see a steep change in the stress level with a very slight change in the poly-to-poly spacing. Therefore, at minimumal poly-to-poly spacing, the drive strength of the device changes a lot if the process varies the spacing. Poly-to-poly spacing variation can be a result of poly CD variation. Poly CD variation in such a context affects the transistor drive by varying the transistor threshold voltage as well as the mobility of the carriers as the stress-level changes with the poly-to-poly spacing. Therefore, it is prudent not to design critical circuits with minimum poly-to-poly spacing to move into the flatter portion of the curve in Figure 1.14.

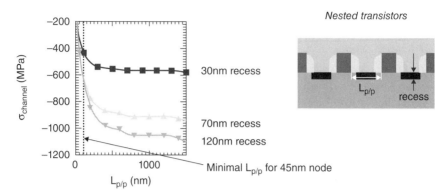

Figure 1.14 Layout process interaction. (Courtesy of Synopsys.)

1.6 DESIGN DEPENDENCY OF CHIP SURFACE TOPOLOGY

Ever since the industry switched from aluminum (Al) to copper (Cu) interconnects, the planarization and copper clearing technique used is chemical–mechanical polishing (CMP). Due to the differences in the materials' (copper and dielectric) resistance to wear during the CMP process, the topology of the chip is affected by differing copper density on the chip, which in turn is dependent on the design. Manufacturers use various techniques to normalize the copper density, but these techniques are not perfect and on some designs it may be difficult to normalize the density without changing the design characteristics and intent. The result is that the topology of most chips is far from being flat (Figure 1.15). This can cause catastrophic failures, due to the inability of CMP to clear the copper on the higher levels of interconnect, as shown in Figure 1.16. Figure 1.17 shows that additional interconnect layers compound the topological impact, which eats into the already narrow DOF margin in the lithographic process and can create lithographic hotspots as well.

Chip surface undulation also results in parametric variation, as it causes varying copper loss on interconnects at different locations. The interconnect resistivity and capacitance will vary so that as-drawn extraction will not reflect reality and will result in overdesign to allow for the unmodeled variations. There is therefore a need for a tool to help model these effects and to correct for them as much as possible.

1.7 NEWLY EXACERBATED NARROW WIDTH EFFECT IN NANO-CMOS NODES

It has been observed that the effective channel length of devices at the STI edge will be longer than the channel length away from the STI edge [8,9] (Figures 1.18 and 3.3). The result is that narrow devices will have longer

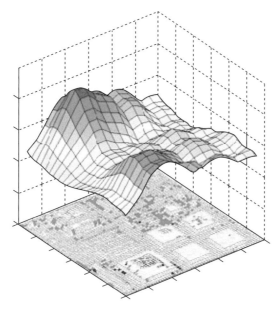

Figure 1.15   Design-dependent undulating chip surface. (Courtesy of Praesagus.)

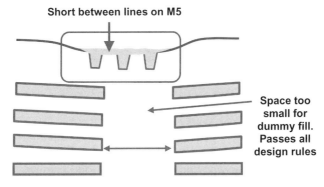

Figure 1.16   Catastrophic CMP failure.

channel lengths, hence lower drive current and higher V_t. Devices with different widths will exhibit varying V_t and drive. This can have a significant effect on memory bitcell designs where the pull-down and pass transistors are of different widths. If manufacturers cannot eliminate this effect, it must be reflected in the model, so that design engineers can design for it. As shown in Figure 1.18, the longer channel length persists even 100 nm from the STI edge, which means that a 200-nm device will be affected by this newly exacerbated phenomenon. Although all known techniques will be utilized to minimize this effect, it remains to be seen if all manufacturers can eliminate it. Therefore, when designing circuits that incorporate devices with narrow widths, as in

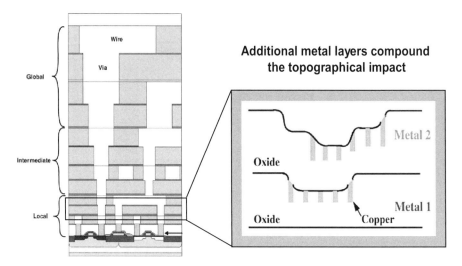

Figure 1.17   Growing problem with more metal layers. (Courtesy of Praesagus.)

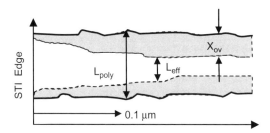

Figure 1.18   Narrow width effect on L_{eff}. X_{ov}, extension overlap; L_{eff}, effective channel length; solid lines, gate edge; dashed lines, junction edge.

memory bitcells, keeper devices, and other designs that use minimum geometry devices, it is important to keep this effect in mind.

1.8   WELL PROXIMITY EFFECT

This effect has been observed to affect circuit performance and functionality since the advent of the 130-nm node. Designers must exercise caution when placing critical circuit components, particularly transistors, in the scatterbands of a well (Figure 1.19). For about $2\,\mu m$ from the edge of the well, we will see dopants scattered by the implant mask, resulting in higher well doping in the 2-μm band around the edge of the well. Inner corners are much worse than outer corners, as shown by the well scatter dose map in Figure 1.19. Designs that require matched transistors, self-time circuits, and hold-time delay

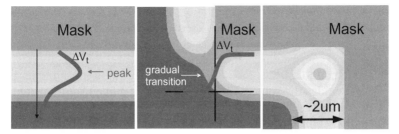

Figure 1.19   Well scatter dose map.

elements should preferably be located away from the well scatterbands. In later chapters we discuss design techniques to minimize the effect of well proximity effects on circuits that must be placed in the scatterbands as a result of space constraints, due to pitch-matching layouts.

1.9 NEED FOR MODEL-BASED DFM SOLUTIONS BEYOND 65 NM

Although the lithographic process is lagging behind aggressive dimensional scaling, we are seeing the circle of influence of proximity being enlarged proportionately. Furthermore, chip surface undulation modulated by the design effect on the CMP step adds another level of abstraction that affects the lithographic process, making it even more difficult to be described by rules alone. Problematic proximity interactions cannot be accomplished easily when using rules. Even if one could code the design rule checking (DRC) deck with the ability to find such interactions, it would be unmanageably complex, with a disproportionately large number of rules, to the extend that DRC runs would take so long that real-time data for the chip tapeout could not be provided. Even if the run-time problem could be surmounted using expensive supercomputers, the rules would be so extensive and complex that design productivity would be an issue. The misinterpretation of the rules and errors in the implementation would also cause delays in tapeouts and result in missing the product market window. Figure 1.3b is a classic case that can only be uncovered by a model-based solution.

The effects of lithographic distortions modulated by chip surfaces can be captured accurately only by using a physics-based model calibrated to the process [10]. The benefits include the ability to push rules to achieve a smaller chip area guided by the model. This will result in a smaller chip than in the case of blind adoption of the restricted design rules. Contour-based extraction also relies on a calibrated model which can be used to turn systematic variation into deterministic effects that can be simulated. Therefore, no extra margins need to be set aside for systematic variations, reducing overdesign.

The key to successful implementation of a DFM flow is the ability to provide early design feedback when the design is still fluid, to avoid major restructuring of the design, resulting in delayed tapeout. There is also a need for correct by construction DFM and timing-aware routing, to minimize reroutes and lead to a faster time to market. This requires a fast model-based analysis that guides the router. The first-generation guided router would be a hybrid of a rule-based guide and a fast model-based solution to deal with constructs that cannot be described by rules alone and to confirm the manufacturability of a particular construct. The model-based solution also guides the repair of a high-priority hotspot.

1.10 SUMMARY

Although we have not listed possible design process interactions exhaustively, the list in this chapter serves to illustrate the point that down the road of technology scaling we will see these effects becoming more predominant. We will also find new effects that we have not seen in earlier technology nodes. Process variability is also here to stay and is getting worse as dimensional scaling pushes toward atomic dimensions, where, for example, gate dielectric thickness has already approached a thickness of four atomic layers. Hence, we have to learn to deal with these effects and the increase in variability so that our circuits will be functional on the first silicon representation for cost and time-to-market economics. As a result, the need for model-based solutions will abound to help designers know when they can use minimum rules for a smaller chip area and how to avoid process-sensitive layout constructs for a predictable yield ramp for volume production. Model-based solutions also provide the basis to turn systematic variations into deterministic events, thus eliminating the need for additional design margins, which results in over-designing. This book combines the expertise of circuit engineers, process technologists, lithography experts, computer-aided design developers, and academic researchers to provide practical solutions and silicon-proven methodology to deal with the problems described.

REFERENCES

1. M. Nowak, Bridging the ROI gap between design and manufacturing, SNUG 2006, Santa Clara, CA, 2006.
2. B. P. Wong et al., *Nano-CMOS Circuit and Physical Design*, Wiley, Hoboken, NJ, 2004.
3. A. Asenov, Random dopant induced threshold voltage lowering and fluctuations in sub-0.1 μm MOSFET's: a 3-D "atomistic" simulation study, *IEEE Trans. Electron Devices*, vol. 45, no. 12, pp. 2505–2513, Dec. 1998.

4. M. Pelgrom, Nanometer CMOS: an analog challenge, IEEE Distinguished Lectures, Fort Collins, CO, May 11, 2006.

5. B. P. Wong, Design in the nano-CMOS regime, DFM tutorial, Asia and South Pacific Design Automation Conference, Yokohama, Japan, Jan. 23, 2007.

6. B. P. Wong, Bridging the gap between dreams and nano-scale reality, DFM tutorial, Design Automation Conference, San Francisco, CA, July 28, 2006.

7. Y. Borodovsky, Marching to the beat of Moore's law, SPIE Microlithography Conference, San Jose, CA, 2006.

8. V. Moroz et al., Physical modeling of defects, dopant activation and diffusion in aggressively scaled bulk and SOI devices: atomistic and continuum approaches, Materials Research Society Conference, Boston, MA, Apr. 19, 2006.

9. H. Fukutome et al., Direct measurement of effects of shallow-trench isolation on carrier profiles in sub-50 nm N-MOSFETs, Symposium on VLSI Technology, Kyoto, Japan, 2005.

10. D. Perry et al., Model-based approach for design verification and co-optimization of catastrophic and parametric-related defects due to systematic manufacturing variations, SPIE Microlithography Conference, San Jose, CA, 2007.

I

NEWLY EXACERBATED EFFECTS

2

LITHOGRAPHY-RELATED ASPECTS OF DFM

2.1 ECONOMIC MOTIVATIONS FOR DFM

The cost of entering the 65-nm node, and even more so the 45-nm node, are quite high: so high, in fact, that the number of players on both the manufacturing and design sides is dwindling. On the manufacturing side the price tag for a wafer fabrication line has increased exponentially over the years (Figure 2.1). Whereas in 1970 the price of a wafer fabrication facility was about $10 million, it reached $100 million in the mid-1990s. In 2007, the cost of a 45-nm line has reached a staggering $3 to 4 billion. The exponential growth in the cost of a wafer fabrication line is fueled by several factors: an increase in the average fabrication output, a rise in equipment cost due to larger wafer sizes, and tighter process control requirements for advanced nodes. Lithographic equipment is a significant contributor to rising the cost of the wafer fabrication line (see Figure 2.1). The price increase for exposure tools has largely been in line with the increase in capital investment necessary for a wafer fabrication line. In addition, in recent years the percentage of the total fabrication cost attributed to lithographic tools has risen from 14% in 1997 to 18% in 2007 [1].

In addition to rising manufacturing costs, the R&D investment to develop the process for a new technology node is reaching staggering numbers. For example, R&D expenditure for the development of a logic 65-nm-node process is about $1.5 billion. It increased to $2.5 billion for a 45-nm process and is expected to reach $3 billion for the 32-nm node. The revenue necessary to

Nano-CMOS Design for Manufacturability: Robust Circuit and Physical Design
for Sub-65 nm Technology Nodes
By Ban Wong, Franz Zach, Victor Moroz, Anurag Mittal, Greg Starr, and Andrew Kahng
Copyright © 2009 John Wiley & Sons, Inc.

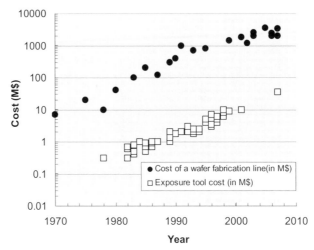

Figure 2.1 Historical cost trend of a wafer fabrication line and exposure tool cost from 1970 to 2007.

support the R&D investment for the 65 nm node is approximately $8 billion and exceeds the revenue of all but the biggest CMOS logic semiconductor companies. As a result, the number of companies able to participate in the most advanced processes is dwindling. At the 130-nm node, about 50 companies were building fabrication lines, between 10 and 20 companies built 65-nm fabrication lines, and it is expected that the number of companies building 32-nm logic fabrication lines will be only about 5 to 10.

The price of a mask set is another lithography-related contributor to the increased cost of a technology. The price tag for a complete mask set in the early manufacturing phase of the technology was at about $1 million for the 90-nm node, $1.5 million at 65 nm, and is expected to reach $2 million at 45 nm and about $3 million for 32 nm. Primary contributions to mask costs are the data-handling steps prior to mask building, mask writing, development, and etching steps, as well as mask inspection. Figure 2.2 provides an overview of the relative contributions of these steps to the overall mask cost. The biggest contributor is the mask writing step, followed closely by mask inspection and material costs. The dominating effect of mask writing on mask cost is due to the high cost of the advanced mask writer (about $20 million) combined with a low tool throughput; writing times of 10 to 24 hours are not unusual [2]. The dramatic increase in the total number of shapes driven by the complexities of optical proximity correction (OPC) in the deep-subwavelength domain is the primary reason for the low throughput. Mask inspection costs are driven by similar factors: very high tool costs (up to $30 million) and relatively low throughputs (on the order of hours). Furthermore, more stringent requirements for mask blanks contribute to the rising material costs.

Costs have increased not only on the manufacturing side but on the design side as well. Figure 2.3 provides an overview of the trend in design costs, starting at about $2 million for the 250-nm node and rising to an estimated $75

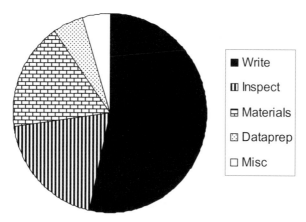

Figure 2.2   Approximate breakdown of mask costs: percentage contributions from data preparation, mask writing, mask inspection, and materials.

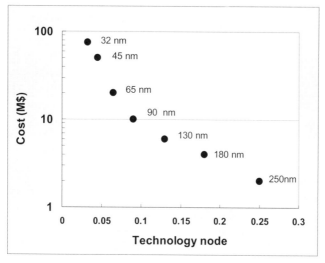

Figure 2.3   Design costs as a function of technology node. Data for 65, 45, and 32 nm are estimates.

million for the 32-nm node. Studies indicate that the increased complexities of design verification in the nanotechnology domain are a main contributor to the cost explosion. However, despite the considerable effort put into verification tools, an increasing percentage of designs fail on the first tapeout. Whereas first-time success in silicon designs was the norm in 0.18 technology (about 80%), it began declining in the 130-nm node (60%), and this trend has continued downward since then. For advanced node, design respin is now quite common to address initial yield issues. Another key trend being observed is the fact that the mechanisms affecting yield have clearly shifted through the

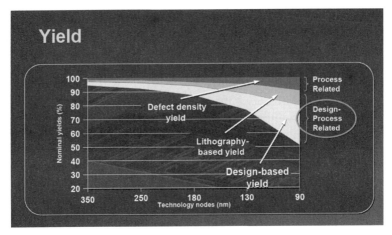

Figure 2.4   Trend chart on the development of defect density–driven, lithography-based, and design-based yield degradations.

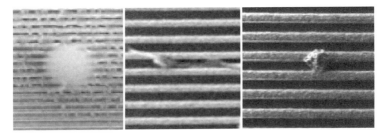

Figure 2.5   Yield loss factors due to random defectivity.

technology nodes. Lithography-based yield detractors and parametric yield issues are affecting yields increasingly at advanced nodes (Figure 2.4). Addressing these issues may require either a mask redesign with a modified version of the optical proximity correction, lengthy process optimization to reduce variabilities, or a modified design. We show brief examples for common yield issues. Figure 2.5 shows examples of random defects that affect product yields. A large variety of these defects, generated by lithography, etching, deposition, cleaning, and CMP equipment in the fabrication contribute to yield losses at all patterning steps. This type of defectivity was the predominant cause for yield loss up to about the 130-nm node. Beginning with the 90-nm node, systematic yield loss mechanisms became more prevalent. These mechanisms include:

- Impact of micro and macro loading on via and line etch rates
- Optical proximity effects and impact on via opens
- Misalignment issues on metal line ends and via coverage

• Topography effects due to pattern-density-dependent CMP removal rate

In these mechanisms the probability of failure is strongly linked to specific layout attributes. Figure 2.6 shows a via failure mechanism in which the failure rate depends on the distance between contacts, due primarily to the decreasing lithographic process window. Other mechanisms that contribute to layout-dependent contact and via failure rates are etch loading effects. Those etch rates depend on global and local pattern densities. Another example of a lithography-related systematic failure is shown in Figure 2.7. In this case, due

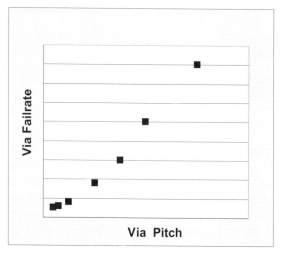

Figure 2.6 Example of a layout dependent systematic yield loss mechanism: via fail rate as a function of via pitch.

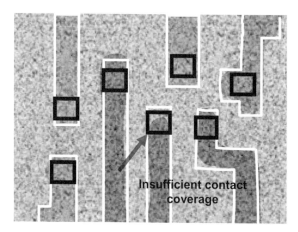

Figure 2.7 Example of a lithography-related systematic failure. Due to the particular configuration of neighboring lines, an insufficient hammerhead was generated, which in turn leads to insufficient contact coverage.

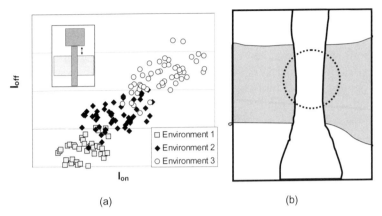

(a)   (b)

Figure 2.8   Example of a variability-driven parametric yield loss mechanism. (a) The I_{on} vs. I_{off} characteristic of devices with identical transistor width and length. (b) The on-wafer shape shown deviates from the ideal case. This deviation depends on varying optical environments. As a result three different design environments for the same transistor lead to systematic layout dependent variations.

to layout constraints, the hammerhead at the line end is too small, resulting in contact opens. The extent to which the line end fails to cover the via properly may vary depending on specific details of the layout as well as details of the OPC algorithm. The line-end foreshortening may be so severe as to result in catastrophic failures (i.e., contact opens even under nominal lithographic conditions). In less severe cases, yield degradation or yield excursion due to the smaller lithographic process window for dose, focus, and overlay is observed.

Parametric failures, the last category that we discuss, are increasingly taking a front stage in 65- and 45-nm designs. Parametric failures are failures where chip performance parameters (e.g., overall leakage) do not meet specifications. Those may be attributed to the increased variability encountered in advanced processes. As an example, we show layout-driven variability in I_{on}/I_{off} characteristics (Figure 2.8a). In the particular case the electrical variation is caused by variations in printed shape as shown in Figure 2.8b. The variations in the printed shape are, in turn, a result of increased sensitivity of the lithographic process to process variations. Parametric failures are expected to be one of the dominating yield detractors at and below the 45-nm node as process variabilities increasingly affect chip performance.

In addition to increased costs for the design and an increased probability of initial design failures, the implications of design failures have also become more severe. As product cycles have dropped from about two years to 12 months, delays in product delivery and the resulting loss in revenue constitute a serious issue.

In this chapter we provide a brief overview of the economic factors that affect manufacturing and design at the advanced nodes. On the one hand, we see rising manufacturing costs as exemplified by the rising capital investment necessary for wafer fabrication lines, skyrocketing process development costs, and increasing mask costs. On the other hand, companies that lack fabrication facilities are being squeezed increasingly by the rapid rise in design cost and the rising number of design failures, combined with decreasing tolerance to product delays due to shorter product cycles. Thus, there is a strong interest in exploring any opportunities through which these issues can be alleviated. Before addressing solutions to these challenges, we provide a more detailed overview as to how lithography contributes to manufacturing problems.

2.2 LITHOGRAPHIC TOOLS AND TECHNIQUES FOR ADVANCED TECHNOLOGY NODES

2.2.1 Lithographic Approaches to Sub-90-nm Lithography

The intent of this section is to introduce general patterning concepts relevant for the 65- and 45-nm nodes, with a glimpse of the 32-nm node. The discussion is restricted to those solutions most likely to be relevant for mainstream semi-conductor manufacturing [i.e., either memories (DRAM and flash) or logic products such as microprocessors, graphics chips, and FPGAs]. We therefore omit some techniques, such as optical maskless lithography [3], e-beam [4,5], imprint lithography [6], and extreme ultraviolet [7]. A recent comparison of various techniques is available in a paper by Liu [8].

2.2.2 Lithographic Infrastructure

The process used for manufacturing integrated circuits includes a sequence of microlithographic steps in which patterns are formed by projection printing. This printing process uses elements such as masks, exposure tools, resists, and resolution enhancement techniques.

Masks are the first physical incarnation of the abstract design information. Layer by layer they provide templates for the patterns to be placed on the wafer. The simplest mask option is the chrome-on-glass (COG) reticle, or binary imaging mask. These are chrome-covered quartz substrates with the desired patterns etched into the quartz to form transparent (chrome removed) and opaque regions. Those masks are used primarily for noncritical layers or in any other process where cost is of primary concern.

Attenuated phase shift masks are a more advanced, albeit more costly mask option. Those masks offer improved image capabilities over COG reticles. The dark region of attenuated phase-shift masks use a partially trans-mitting film (typically, a few percent) rather than the essentially nontranspar-ent chrome layer. In addition, these regions have a thickness and refractive index such that the light that passes through the absorber is phase-shifted relative to the light that passes through the clear regions. Due to the phase shift, light passing through the absorber region interferes destructively with light passing through the clear region, thus creating improved image contrast. Attenuated phase-shift masks are the most widely used mask type for critical layers.

The most advanced masks are alternating phase-shift masks. Their distinguishing feature relative to COG reticles is that in portions of the clear regions the quartz substrate is etched to a specific etch depth. The etch depth is chosen such that light passing through the etched regions exhibits a 180° phase shift. Contrary to attenuated masks, however, the transmission of these regions is identical to that in the clear regions (i.e., close to 100%). In its typical application a dark feature is bordered by two clear regions, one with and one without the phase shift. Due to the opposite phase of the two adjacent bright regions, negative interference occurs, creating a dark region with very high contrast in between.

Critical masks for sub-100-nm technologies are fabricated in a process that is not unlike a wafer fabrication process. Mask blanks, consisting of a quartz substrate about $\frac{1}{2}$ in. thick and typically 6 in. square in size, covered by a multilayer stack are the equivalent of the wafer. The multilayer stack depends on the type of mask to be fabricated. In the case of attenuated phase-shift masks, the stack typically consists of a layer of molybdenum silicide covered by chrome. The optical properties and thickness of this material have been tuned to provide the correct transmission and phase change (180°) at the exposure wavelength (typically, 193 nm). The chrome layer serves as an etch hardmask during mask manufacturing.

The most commonly used exposure tool for mask writing is an e-beam writer where electrons rather than photons are used to create the image. Contrary to an optical system, however, where a large amount of information in the form of spatial patterns can be transferred simultaneously onto the wafer, e-beam tools transfer the spatial information serially onto the mask. All advanced masks are written using a vector-shaped beam writer. For this tool type the smallest unit that can be imaged onto the wafer is essentially a rectangle of varying aspect ratio and size. There are some limitations on both characteristics: for example, there are limits on the smallest square that can be printed. The exposure field is imaged by exposing the desired patterns serially as a sequence of rectangles. The time required to print a rectangle is determined by the current density delivered by the e-beam, the resist sensitivity, and a shot settling time. Similar to step-and-scan optical systems there is a maximum exposure field size that can be patterned in this fashion. As a result, a mechanical stage movement is required to center the next exposure field under the e-beam. The exposure of the entire mask blank therefore consists of a sequence of image field exposures where the mask blank is stationary followed by the movement of the stage to the next image field, during which the write process is interrupted. The imaging process cannot start immediately after the stage movement, as the vibrations due to the stage movement need to subside. One of the simplest models is given by [9]

$$T = N_S \left(\frac{S}{J} + T_s \right) + T_{OH} \tag{2.1}$$

where T is the total write time, N_S the number of shots, S the resist sensitivity, J the current density, T_S the shot settling time, and T_{OH} the overhead time, which includes the time required for mechanical stage movements. In the simplest approximation the overhead time is roughly constant, assuming a fixed number of exposure fields on the reticle. The preeminent factor determining mask writing times is the number of shots.

Once the reticle has been built, it is sent to a wafer fabrication line, where it serves as a template for transferring the pattern onto the wafer. The main steps in the pattern transfer process are coating the wafer with a thin layer of a photosensitive polymer, transferring the pattern on the reticle as a light pattern into the photoresist using an exposure tool, converting the light distribution into a structural pattern in the resist through a series of chemical reactions, and transferring the pattern in resist onto the substrate through a reactive ion etch.

The purpose of the wafer exposure tool is to take the pattern template (i.e., the mask) and project its image onto the wafer, more precisely onto a thin layer of a photosensitive polymer with which the wafer has been coated. Key metrics for this operation are the smallest feature sizes that can be imaged, the precision with which the pattern can be aligned relative to the underlying pattern, and the throughput of the system, measured in wafers per hour exposed. Exposure tools comprise a variety of components, an optical system to create the image on the wafer, mechanical components to hold and transfer wafers and reticles, a variety of sensors to monitor wafer movement and accurately determine wafer position, and an enclosure to control temperature and humidity.

The imaging system consists of a light source, typically an excimer laser, optics to steer the laser beam into the exposure tool, an illuminator that projects light uniformly and from well-defined directions onto the reticle, and the projection optics, which forms an image of the reticle on the wafer. The exposure tools for the most critical layers in advanced nodes all use excimer lasers operating at a wavelength of 193 nm. Excimer laser are gas lasers. The active components of the gas are argon and fluorine; thus, the name *ArF lithography* is used for lithography using light at a 193-nm wavelength. Excimer lasers using a different gas combination—krypton and fluoride— result in light at a 248-nm wavelength. Illuminators play an important role in the lithographic optimization process. Their most important feature is the ability to control the angular distribution of light that illuminates the reticle. Light passing through the reticle is imaged onto the wafer through the projection optics. Up to the 65-nm node, the projection optics in almost all tools used exclusively transparent lenses. For a 193-nm primary lens, the materials are fused silica and calcium fluoride. The projection optics used in advanced exposure tools are quite complex, with more than 30 individual lens elements.

Mechanical stages allow precise wafer movement in the x, y, and z directions. They are typically equipped with optical interferometers that allow precise determination of stage positions. For the smoothest possible movement, the wafer holder glides on a highly planar surface using air bearings and propelled by contactless magnetic linear motors. In a step-and-scan system, not only is the wafer being moved but also the reticle. Accurate placement of the pattern not only requires accurate knowledge of the position of the wafer stage in the x and y directions but also requires accurate knowledge of the distance of the wafer front surface to the projection lens (focus) and accurate knowledge of the location of the previously formed pattern on the wafer relative to the x–y position of the wafer holder (alignment). One of the primary approaches to measuring the distance between the wafer front surface and the projection lens is to use an optical light beam at a glancing incidence. In this configuration the up-and-down movement of the wafer is transferred into a change in position of the reflected beam. This provides a relative measure of the wafer height. The best imaging distance is determined experimentally by evaluating the image quality in resist at a series of focus settings. Determination of the location of previously patterned layers on the wafer is also done through optical sensors. Those sensors are able to detect the precise position of previously patterned reference marks (commonly referred to as scanner alignment marks) on the wafer. In conjunction with the x–y positions available from the wafer stage interferometers, a precise map of the wafer location can be established.

Exposure systems also include mechanical components for transferring wafers and reticles. By means of a robot, wafers are first transferred from the track to a station for coarse wafer alignment. From there they are transferred to a metrology stage, where the wafer height (i.e., the focus) map and the map of the wafer position are generated. They are then passed onto the wafer exposure stage until they are finally returned to the track for further resist processing. Finally, the entire system is enclosed for environmental control. Within this enclosure, humidity and temperature are maintained so as to prevent changes in lens characteristics that affect the image. Also, particles and potential contaminations that may decompose under exposure to ultraviolet light are removed from this enclosure.

The exposure process starts with transfer of the wafer into the exposure tool. Most modern systems feature dual- or twin-stage systems, which have two independently operating stages for x–y movement of the wafer. One stage is used for wafer metrology (overlay and focus mapping), the other for wafer exposure. The primary advantage of these systems is higher throughput as the metrology and exposure tasks are executed in parallel. While one wafer is being exposed, the subsequent wafer is undergoing metrology.

All modern exposure tools are step-and-scan systems. In these systems the entire reticle field is not exposed at once and without reticle movement (similar to a slide projector). Rather, at any given point in time, only a slitlike portion of the reticle is exposed. The image of the entire reticle field is created by a synchronized movement of reticle and wafer. Once the exposure of the entire reticle field is completed, a shutter is closed, preventing light from reaching the wafer. Then the wafer stage moves the wafer into position ready for the exposure start of the next chip.

The main photoresist processing steps prior to exposure are resist coating and a post-apply baking. A post-exposure bake, and a development step follow after the exposure has been completed. Photoresists are essentially polymers dissolved in a suitable solvent. During the coating step a small amount of resist is dispensed on the wafer. The wafer is then spun at fairly high speed (on the order of 1000 to 3000 rpm). During this step the effects of solvent evaporation and the rotation-driven fluid flow combine in a fashion that results in a highly uniform layer of photoresist covering the wafer. During the subsequent post-apply baking, most of the remaining solvent is removed. Exposure to short-wavelength light results in the release of an acid from functional side groups [referred to as photo-acid generators (PAGs)] of the polymer chain. The photo-generated acid concentration is proportional to the light intensity, thus transferring the optical image into a "chemical image." During post-exposure baking, the photo-generated acid diffuses within the polymer matrix. The acid catalyzes another chemical reaction, whereby other functional groups within the polymer matrix are converted from a water-insoluble to a water-soluble form. At the end of the reaction the acid is released and available to participate in further reactions. Therefore, a single photon, through multiple reactions of the photo-acid, is able to convert several molecules from water-insoluble to water-soluble form, a process referred to as *chemical amplification*. The amplification process terminates at the end of the post-exposure baking as the photo-acid diffusion is frozen. In other cases the acid may recombine with photo-acid quenchers in a reaction that binds the photo-acid permanently and prevents further reactions. Those quenchers may have been added deliberately as part of the resist formulation, or they may have entered the polymer unintentionally as contaminants. After the post-exposure baking, an aqueous alkaline developer solution is dispensed on the resist surface. In regions where a large enough number of polymer side groups have been converted to a water-soluble form, the polymer dissolves in the developer solution. Those regions that have not been exposed do not have a high enough solubility, and resist remains in those locations. Thus, a resist pattern is formed on the wafer that mimics the original light distribution. Due to the chemical amplification process, the dissolution speed of most resists as a function of light intensity very much resembles a step curve: Within a very narrow range of light intensity the polymer film transitions from undissolved to completely dissolved. Thus, even gradually changing light intensities can be converted to sharp resist images. The photo-resist processing tool, called a track, consists of a large number of units, such as bake plates, developer and coating bowls, and chill plates. Wafers are transferred between those stations by a robot arm that also transfers wafers to and from the exposure tool. In advanced nodes the complexities of photoresist processing have increased as additional layers (e.g., organic antireflective coatings, trilayer resist systems, and top antireflecting layers) are used. Bottom antireflective coatings (BARCs) are coated before the photoresist is applied. Their primary purpose is to suppress the effect of substrate reflectivity variations due to the underlying pattern. Some advanced processes use multilayer resist systems that combine a thin photosensitive layer on top of a relatively thick nonsensitive underlayer. In addition, it has become quite common to use top antireflective coatings (TARCs). These layers are added after the resist coating. Their primary role is to reduce reflectivity from the top resist surface. In addition, they provide some protection against environmental contamination. They also prevent the leaching of resist components into water, a feature that has become significant with the introduction of immersion tools. The TARC layers are removed during the development process, whereas BARC layers must be removed after the lithographic process is completed, typically using reactive ion etching.

An increasingly significant role within the lithographic process is now played by techniques that convert the design shapes to shapes on the mask. Ever since lithographic processes are imaging at dimensions smaller than the wavelength of light, the polygons of the design are no longer simply transferred onto the reticle. Rather, fairly sophisticated software algorithms have been developed to counteract the image distortions inherent in subwavelength imaging by predistorting the shapes, on the reticle. In addition to modifying existing shapes, other resolution enhancement techniques may add additional shapes to the design in an effort to optimize the imaging process. These techniques are discussed in more detail in the following sections.

2.2.3 Immersion Exposure Tools

To meet the demands of the semiconductor roadmap, for decades lithography has followed an approach that consisted primarily of reducing the exposure wavelength and increasing the numerical aperture. Exposure wavelengths have been reduced from 450 nm to 193 nm, and numerical apertures have been increased from 0.2 up to about 0.95. However, lithography relies increasingly on resolution enhancement techniques such as optical proximity correction, subresolution assist features, and alternating phase shift masks to push further into the subwavelength domain. For the 45-nm node a unique approach is taken: Immersion lithography, characterized by the addition of a liquid between the last lens element and the wafer, is expected to become the preeminent exposure tool option. Immersion lithography adds a new element to the lithographers' repertoire of resolution enhancement techniques. Immersion technology is an extension of the 193-nm exposure tools that have entered mainstream manufacturing at about the 90-nm node, and thus this exposure wavelength will continue to be in use throughout the 45-nm node and most likely into the 32-nm node.

The concept itself is relatively old. In 1678 it was proposed to add water for the purpose of enhancing optical microscopy. In 1840 the first such lenses were manufactured. Application of this technique to lithography had been suggested in the 1980s; however, not until 2003, when the exposure tool manufacturers showed their roadmaps for immersion tools, was the approach considered seriously. Fortuitous circumstances such as the challenges and delays faced by 157-nm lithography and the fact that immersion appeared to have the fewest issues among the various solutions proposed for the 45-nm node resulted in quite rapid development and deployment of the tools to manufacturing. The first commercial immersion tool, a modified version of an existing 193-nm dry tool, was delivered in 2004. Since then, a large number of semiconductor manufacturers have announced their plans to use 193-nm immersion lithography for the 45-nm node.

Immersion lithography offers two advantages: improved depth of focus for a given resolution, and the ability to achieve numerical apertures larger than 1. As a first step to understanding the depth of focus advantage offered by immersion lithography, consider *Snell's law*. Snell's law describes an optical invariant for light that passes through a multitude of optical layers with refractive indices n_{lens}, n_{water}, n_{resist}, and n_{air} (see Figure 2.9):

$$\begin{aligned} NA &= \sin\theta_{air} \\ &= n_{lens}\sin\theta_{lens} = n_{water}\sin\theta_{water} = n_{resist}\theta_{resist} \end{aligned} \tag{2.2}$$

It states that in such a system the quantity $n\sin\theta$ remains an invariant, where n is the refractive index of the material and θ is the angle of incidence. Simplified lens systems with magnification m can be treated in a similar fashion; in this case the relationship reads

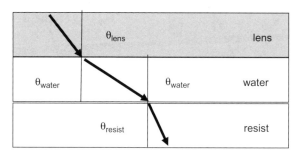

Figure 2.9   Snell's law in a multilayer stack.

$$m = \frac{n_{object} \sin \theta_{object}}{n_{image} \sin \theta_{image}} \qquad (2.3)$$

where the subscript "object" is used for the object side of the lens, the subscript "image" for the image side of the lens, and m is the magnification of the system. For fixed magnification and a fixed diffraction pattern that defines the object-side diffraction angles, the two previous equations may be rewritten as

$$n_{resist} \sin \theta_{resist} = n_{water} \sin \theta_{water} = \frac{1}{m} n_{object} \sin \theta_{object} \qquad (2.4)$$

Thus, replacing the air gap between the lens and the resist with water does not change the angle of incidence in resist. What does change is the angle of incidence at the resist–water interface. In the case of water, with a refractive index of 1.44 at a wavelength of 193 nm, the angle of incidence is smaller than the corresponding angle for the case of a resist–air interface. A sketch to provide insight into the resulting DOF improvement is shown in Figure 2.10. The right-hand side of the figure applies for an air gap, the left side depicts the case with an immersion medium of refractive index larger than 1. In this illustration we use arguments from geometric optics to compare shifts in the best focus position in response to a shift in the distance between wafer and lens. We assume equal resolution: identical angles between the diffraction orders in resist. Under those conditions the angle of incidence at the air–resist interface is much larger than for the water–resist interface, due to the larger refractive index of water. The initial scenarios are shown as solid lines. The best focus position is marked by the intersection of rays. Next, we assume a downward movement of the resist–air or resist–water interface and consider the corresponding shifts in the best focus position. In the air case the larger angle of

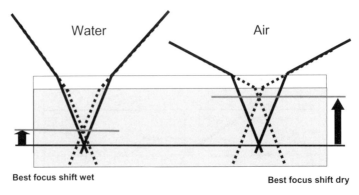

Figure 2.10   Shift of best focus position in immersion and dry systems.

incidence at the air–resist interface results in a larger shift of the best focus position within the resist. Thus, the image position within the resist is much more sensitive to changes in the relative distance between lens and wafer, resulting in a significantly reduced depth of focus. A more detailed calculation [10] provides

$$\frac{\text{DOF}_{\text{immersion}}}{\text{DOF}_{\text{air}}} = \frac{1-\sqrt{1-(\lambda/p)}}{n_{\text{water}} - \sqrt{n_{\text{water}}^2 - (\lambda/p)^2}} \tag{2.5}$$

The expected gain in depth of focus relative to the dry case is shown in Figure 2.11 as a function of λ/p. For small values of λ/p the DOF gain is given by the refractive index of water at 193 nm (i.e., a factor of 1.44). For smaller patterns the improvement in depth of focus increases substantially and reaches nearly a factor of 2 for ratios of λ/p close to 0.95.

Immersion tools not only provide the benefit of improved depth of focus but also allow the design of lenses with numerical apertures larger than 1. For a qualitative explanation we return to the optical invariant given in equation (2.2):

$$\text{NA} = n_{\text{resist}} \sin\theta_{\text{resist}} = n_{\text{immersion}} \sin\theta_{\text{immersion}} = \frac{1}{m} n_{\text{object}} \sin\theta_{\text{object}} \tag{2.6}$$

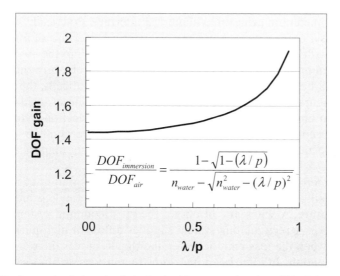

Figure 2.11   Immersion-induced gain in depth of focus as a function of λ/p, λ being the wavelength of light and p the pitch of the pattern. The refractive index of water at 193 nm is 1.44.

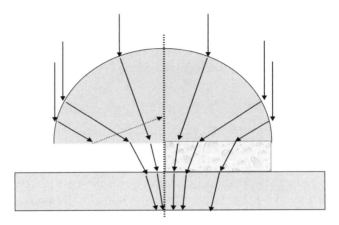

Figure 2.12   Total internal reflection at the glass–air interface.

A small modification has been made: We have added the numerical aperture and inserted the more general concept of the immersion medium, which may be water, air, or any other suitable medium. Considering that NA is a constant within the system, we note that the maximum achievable value of NA within this system is determined by the maximum possible value of $\sin\theta$, which is 1 and the smaller of the refractive indices for resist or the immersion medium. On the object side, the typical demagnification factor of lithographic systems of $\frac{1}{4}$ avoids the fact that the refractive index of air becomes the limiting factor. Refractive indices of resists are on the order of 1.7, and the refractive index of water is 1.44. Under these circumstances we see that the immersion medium defines the maximum achievable numerical aperture. Practically, water-based immersion systems achieve a maximum numerical aperture of about 1.35. If one attempts to impose a numerical aperture larger than the "weakest link" in the optical path, light passage through this layer will be prevented. Rather, the light will be reflected back at the interface. Schematically, the situation is shown in Figure 2.12, where light with the highest angle of incidence is prevented from entering the air gap by total internal reflection. When water is added between the lens element shown in the figure and the wafer, total internal reflection occurs at higher angles, and thus higher numerical apertures can be achieved.

All commercial immersion systems use an implementation of the concept where a small puddle of water is maintained between the lens and the wafer during exposure. A schematic of one of the most common systems is shown in Figure 2.13. A recirculating system supplies a flow of high-purity water to the gap between the lens and the wafer and then removes the water from the edge of the puddle. In some but not all implementations, an airstream cushions the water and helps maintain the water puddle underneath the lens even under the high-speed relative movements of wafer and lens used in step-and-scan systems. A key requirement for proper confinement of the water puddle

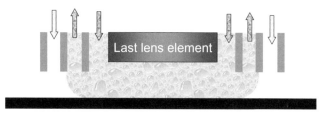

Figure 2.13 Immersion implementation.

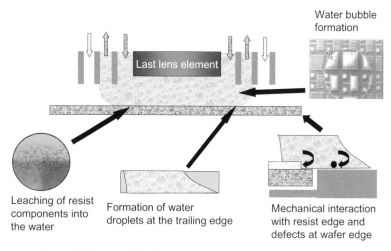

Water bubble formation

Leaching of resist components into the water

Formation of water droplets at the trailing edge

Mechanical interaction with resist edge and defects at wafer edge

Figure 2.14 Key defectivity mechanisms in immersion lithography.

underneath the lens is the proper hydrophobicity of the resist surface, which determines the contact angle at the resist–water–air interface. This is one reason for the widespread use of TARCs in immersion processes. Improper surface properties of the resist may lead to the formation of water droplets and concomitant defectivity issues.

The introduction of water as an integral part of the exposure system is not without challenges, which are related primarily to thermal control and defectivity. The main defect mechanisms contributing to immersion defectivity are the formation of air bubbles in the immersion liquid, mechanical interaction between the moving immersion liquid and the wafer edge, instabilities of the water–wafer meniscus, and leaching of resist components into the water (Figure 2.14). Air bubbles are formed as a result of turbulent liquid flow and/ or a high concentration of dissolved gases in the liquid. If the water has a high concentration of dissolved gases, bubbles are formed as a result of the pressure variations encountered in turbulent liquid flows. Another source of air bubbles is an unstable leading-edge meniscus, particularly for hydrophobic resist surfaces. With the introduction of degassing systems in the recirculating water system, the inclusion of an air cushion in the immersion showerhead, and the use of TARCs to control the surface properties of the resist water interface, water bubbles have been eliminated fairly efficiently as a major source of defectivity in immersion systems. Resist leaching is an effect whereby certain components of the resist are dissolved in water. The primary method to control resist leaching is the use of TARCs, which provide a diffusion barrier and the practice of prerinsing wafers with water before exposure. Lens contamination as a result of resist leaching has been greatly reduced through the use of effective filtration systems in the water flow that remove dissolved hydrocarbons. One of the primary challenges of immersion technology for step-and-scan systems are the fairly high speeds (500 mm/s) with which the wafer moves relative to the lens. Under static conditions at the water–substrate interface, a water droplet maintains a contact angle that is characteristic of the surface properties of the substrate. Hydrophilic surfaces are characterized by contact angles smaller than 90°; hydrophobic surfaces have contact angles larger than 90°. For a moving puddle the interface angles change, the contact angle at the trailing edge of the puddle is decreased, and the angle at the leading edge of the puddle is increased. The height of the air gap, the velocity, the viscosity, and the surface tension of water, as well as the hydrophobicity of the surface, determine the shape of the meniscus. Too high an advancing contact angle leads to an unstable leading-edge meniscus. This may result in the entrapment of air between the resist surface and the water puddles, creating air bubbles. Too shallow a receding contact angle leads to water droplets being left on the wafer. As those droplets are no longer part of the recirculating water systems, the dissolved organics remain as stains on the wafer once the water has evaporated. These remaining organics may be removed in a rinse step following the exposure. However, if these stains blocked subsequent exposure, the resulting patterning defects will appear in the final developed image. These defects could be a significant contributor to immersion lithography processes. One of the more serious issues in immersion lithography is the defectivity created through mechanical interactions. The mechanical movement of the water may result in the creation of new particles due to delamination of films as well as the transport of preexisting particles from the wafer bevel onto the wafer. Examples of such defects are residues due to delamination of the organic films, resist, and ARCs at the edge of the wafer. Preferred schemes regarding the relative placement of the various films at the edge of the wafer have been developed and have helped reduce the occurrence of delaminations. Defects from the wafer bevel pose a larger issue. Through the combined action of the air cushion and the forced water flow in the puddle, defects from the wafer bevel or even the back side of the wafer are transported onto the front surface of the wafer and contribute to wafer edge defectivity. Furthermore, such defects may be transported onto the stage, where they act as a defect reservoir contaminating subsequent wafers. In general, with well-controlled processes, immersion defectivity has reached levels comparable, but not equal to, those of dry systems. The exposure tool itself, contrary to dry systems, remains a source of defects for immersion systems. Furthermore, due to the addition of new mechanisms for defect generation requiring additional process controls preventing excursions in immersion systems is challenging.

The other main issue in immersion lithography is to maintain precise control over the wafer temperature. Failure to maintain tight temperature control results in degradations of overlay and difficulties in maintaining focus. For example, a 0.1° temperature change increases the diameter of a wafer by about 80 nm through thermal expansion of silicon, a significant number relative to the overlay requirements of the 45-nm node. The difficulties in maintaining precise temperature control are a result of evaporative cooling of the wafer and the wafer stage. Several mechanisms contribute to this cooling: a strong airflow in close proximity to the water puddle, the presence of a thin water film absorbed on the wafer after the puddle has passed, the presence of water droplets on the wafer, and the potential presence of a water–air mix in the return path of the air cushion. Water droplets on the front or back side of the wafer lead to localized temperature fluctuations, which contribute to localized overlay errors. Although the overlay performance of the initial immersion tools lagged behind their dry counterparts, their performance is now comparable, due to the incorporation of active temperature control on the wafer stages and evolutionary improvements in stage design. Similar to the defectivity case, however, the potential for excursions has increased dramatically, and good overlay control requires much more diligent process control.

Water-based immersion systems are expected to be one of the main lithographic approaches for the 45-nm node. However, for true 32-nm half-pitch imaging, a numerical aperture (NA) of the projection optics of approximately 1.35 is not sufficient from a resolution perspective. Even with an index of refraction of 1.44, the gap between lens and wafer remains the weakest link in the optical train. Thus, a search for liquids with refractive indices higher than that of water has begun that would enable NAs of about 1.65, sufficient for 32-nm half-pitch imaging. The search for these materials is not trivial, due to a variety of requirements which fortuitously are met by water but not necessarily by the new high-index fluids: for example, compatibility with the optical elements and the resist surfaces, the small absorption coefficient at 193 nm, a sufficiently small dependency of the refractive index on temperature, and the ability to form puddles that can be moved at sufficiently high speeds. These hurdles have delayed the introduction of immersion tools with NAs in the 1.65 regime, and it is generally believed that these tools will not be available in time for true 32-nm half-pitch imaging as required in memory applications.

2.2.4 Overlay

The patterning process needs to meet two independent requirements: the formation of an image at the required dimensions and accurate placement of the image relative to an already existing pattern. Accurate pattern placement relies on two key components of the exposure system, a highly accurate means of measuring distances in x and y as well as a system to create a highly localized detector signal from the wafer. Measurements of the wafer position are made with a laser interferometer. The operating principle of the interferometer is essentially that of a Michelson interferometer, where the path length of a mirror placed on the moving stage is compared to that of a reference beam reflected off a fixed reference mirror. Relative changes in the mirror positions by a distance of half a wavelength correspond to a total path difference of one wavelength or the transition from one maximum intensity to another. Typical wavelengths of the laser interferometer are in the optical range, approximately 600 nm; that is, a movement of 300 nm corresponds to one transition between fringes. Finer resolution can be accomplished by digitizing the intensity levels in between bright and dark levels. Some of the challenges in controlling the positional accuracy are related to maintaining a stable stage and wafer temperature despite the heat generated by the linear magnetic motors driving the stages. Also, as we have seen, temperature control in the presence of an immersion medium poses significant challenges. Other challenges are the control of temperature and humidity in the optical path of the interferometer, and avoiding turbulent airflow. Vibrational control of stage movement and avoiding vibrations of the projection optics relative to the wafer stage are also important.

Sensing the wafer position entails two components: a set of reference marks placed by the previous patterning step on the wafer and a detector system that creates a sharp localized detector response as a function of the relative position between the reference mark and the detector. One implementation for accurate sensing of reference marks is shown in Figure 2.15. A grating placed on the wafer is illuminated with a laser beam. A lens system is used to project

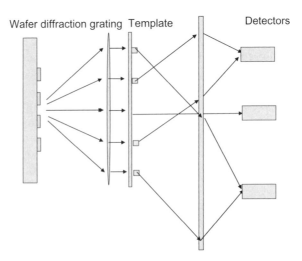

Figure 2.15   Schematic of an alignment sensor.

an image of the diffraction pattern onto a grid with the same spatial frequency. A detector placed behind these gratings will indicate maximum light intensity if the peak intensities of the wafer pattern coincide with the transparent parts of the filter. Through custom optics, individual diffraction orders can be projected onto separate gratings, sharpening the x–y response of the sensor. The scanner places alignment marks underneath the sensor based on expected mark positions and scans over a small range to determine their exact position. The wafer stage on the scanner and the interferometers connected to it allow accurate positional readings when the marks trigger a signal on the position sensitive detector. This procedure is repeated for a small set of marks. Typically, approximately 10 locations on the wafer in both x and y directions are measured. Based on these measurements, the scanner calculates the location of the center of each chip that is to be placed on the wafer. The model is determined primarily by the x and y stepping distances for each chip; however, it also allows for small deviations from the expected center chip positions. Those corrections are linear in the wafer coordinates; they account for some of the simple geometric distortions on a grid: shift x, shift y, magnification x and y rotation, and skew. Chips are then imaged on the wafer; in general, the same model is applied to all wafers from a particular lot. After the lot has finished, a small sample (two wafers) is selected and measured on an overlay metrology tool to measure the actual misalignment. The data for a lot are analyzed using a model similar to the model used on the scanner, and the data measured are uploaded to an automated process control system. This system provides feedback to future lots exposed on the same exposure tool and supplies initial corrections to the overlay model (see Figure 2.16).

A common metric for measuring overlay performance is the "tool to itself" overlay performance, whereby the overlay performance is measured between two exposures carried out on the same tool under fairly ideal situations (no degradation of mark quality). The published overlay performances for this best-case scenario are summarized in Figure 2.17 for typical tools used in the 90-, 65-, and 45-nm generation.

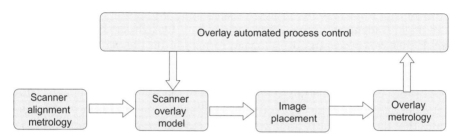

Figure 2.16 Schematic diagram of overlay process control system. The overlay metrology data of several lots may be averaged to generate new settings for the scanner overlay model.

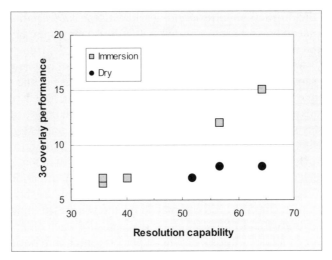

Figure 2.17 Trend of overlay performance specifications for dry and immersion tools.

2.2.5 Cooptimization of the Mask, the Illuminator, and Apodization

The most advanced concepts in resolution enhancement are approaches where two, sometimes three, elements of the lithography process are optimized simultaneously to obtain the best possible lithographic solution. The elements of this optimization process are the angular distribution of light in the illuminator, the shapes placed on the reticle, and in same cases spatial filters placed in the projection optics, a process called *apodization* (for rigorous mathematical details the reader is referred to articles by Rosenbluth et al. [11] and Fukuda et al. [12]). Figure 2.18 shows the basic imaging concepts: Light from the illuminator is projected onto the mask. Patterns on the mask diffract the light. The light transmitted from the mask is imaged onto the wafer using a projection optic. Due to the finite size of lenses, only light up to a certain angle of incidence on the wafer contributes to the image. The sine of this maximum angle of incidence is the numerical aperture (NA) of the lens. In the simple situation shown, where the projection optic does not have a demagnification, this angle is identical to the upper angular limit for light emanating from the mask that can be captured for image formation on the wafer. In a similar fashion, one can define a numerical aperture for the illumination system which describes the maximum angle of incidence for light projected onto the mask, NA_i. More commonly, however, the ratio between NA_i and NA is used. This ratio is referred to as the *sigma* (σ) of the illumination system. A σ of 1 indicates that the illuminator completely fills the entrance pupil of the projection lens. To explain the imaging fundamentals in a simple fashion, we discuss the following basic principles: illumination of the mask, diffraction of light at the mask, and recombination of a portion of the diffracted light at the projection lens to form an image on the wafer.

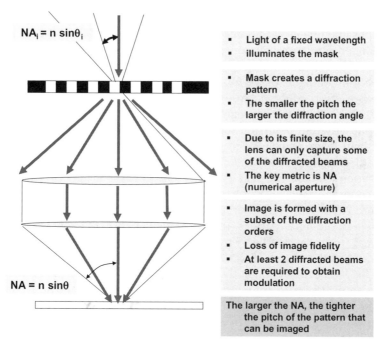

$NA_i = n \sin\theta_i$

$NA = n \sin\theta$

- Light of a fixed wavelength
- illuminates the mask

- Mask creates a diffraction pattern
- The smaller the pitch the larger the diffraction angle

- Due to its finite size, the lens can only capture some of the diffracted beams
- The key metric is NA (numerical aperture)

- Image is formed with a subset of the diffraction orders
- Loss of image fidelity
- At least 2 diffracted beams are required to obtain modulation

The larger the NA, the tighter the pitch of the pattern that can be imaged

Figure 2.18 Basic imaging principles: Illumination of the mask, light diffraction at the mask, removal of high order diffracted light through the projection lens, and recombination of the light passing through the lens to form an image on the wafer.

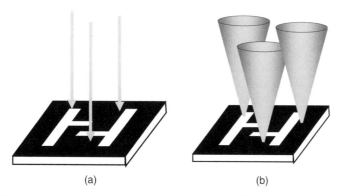

(a)   (b)

Figure 2.19 Illuminators provide not only the same integrated light intensity for each point on the mask but also the same distribution of angles. (a) Coherent imaging: Each point on the mask is illuminated with light from a single direction. (b) Partially coherent imaging: Each point on the mask is illuminated with light from a multitude of directions.

Illuminators provide uniform light intensity across the reticle. Not only does the illuminator provide the same integrated light intensity on every point on the mask, but every point on the reticle also receives light with the same angular distribution (Figure 2.19). Coherent illumination is the limiting case, where only light from a single direction reaches the mask. In partially coherent imaging, the most commonly used mode in lithography, light from multiple directions illuminates each point on the mask. The illuminator uses light from an extended effective light source. One characteristic of this extended light source is that light emanating from different points of the source is incoherent. As the common light source for advanced exposure tools is a laser, the light source is a priori coherent: Additional optics (e.g., a diffuser) is required to convert the coherent laser source into an effective distributed light source where the different points on the source are incoherent. Through additional lenses, light from a single point of the effective light source is converted to light with a single angle of incidence and direction covering the reticle (Köhler illumination). As a result, light with different angles of incidence and direction illuminating the reticle is incoherent.

Figure 2.20a depicts a coherent imaging case: light from a single direction illuminating the mask pattern. In this particular case, the angle of incidence is normal to the mask. This is represented in a simple diagram on the right with a dot in the center of a circle. The pattern within the circle is used to depict the light distribution at the entrance pupil of the lens. The radius of the circle is equivalent to the numerical aperture of the system. The radial distance from the center of the circle corresponds to the sin of the angle of incidence. As the mask is illuminated with light of normal incidence, the dot is in the center of the circle. In our simple example the entrance pupil is the surface represented by the first of the two lenses shown in Figure 2.20.

The next important aspect is the diffraction pattern created by the reticle. A pattern of lines consisting of dark and transparent regions on the mask, with equal width of line and space and pitch p, is known to create a diffraction pattern where the diffraction angles are defined by

$$p\sin\theta = n\lambda \tag{2.7}$$

where θ is the diffraction angle, n is an integer number $(0, \pm1, \pm2, \ldots)$, and λ is the wavelength of light. This equation may be derived from the condition that neighboring mask openings interfere constructively (i.e., the optical path difference between adjacent openings needs to be an integer multiple of the wavelength). Note that for pitches equal to the wavelength of light, the first diffraction order occurs at an angle of $90°$. There is no solution for this equation other than $n = 0$ for pitches smaller than the wavelength. For illustrative purposes we slightly rewrite equation (2.7) and express it in the shape of k-vectors, defined by $k = 2\pi/\lambda$. The vector points in the direction of light propagation:

$$\frac{2\pi}{\lambda}\sin\theta = n\frac{2\pi}{p} \qquad k_\perp = nk_p \tag{2.8}$$

In this form the equation states that the normal component of the k-vector for diffracted light is an integer multiple of the k-vector of the grating, which

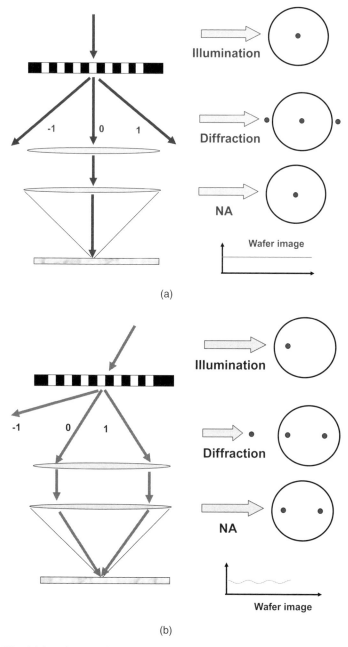

Figure 2.20 (a) Imaging condition where no image is formed as only one diffraction order passes through the lens. (b) Using off-axis illumination two diffraction orders pass through the lens and the resulting image now has spatial modulation.

in turn is equivalent to the Fourier transform of the grating. Returning to the light distribution on the entrance pupil we note that the diffraction pattern is represented by a series of points, spaced at a distance of $\sin\theta$ to the left and right of the zero order or undiffracted light. According to equations (2.8), the distance of the diffracted orders to the zero order will increase with decreasing pitch. In fact, in the example shown, the diffraction angle is so large that only the zero order passes through the lens; the first diffraction orders lie outside the circle of radius NA. The lens refocuses the diffraction orders such that they interfere in the vicinity of the focal point. In our case, only the zero-order light passes through the lens. In the absence of interference between various diffraction orders, only a uniform background with no spatial modulation appears on the wafer (i.e., no pattern is visible on the wafer).

Figure 2.20b shows the same imaging scenario as Figure 2.20a, with only a small modification: Instead of illuminating the mask only at normal incidence, we chose to illuminate with light at nonnormal incidence. The implications for imaging are dramatic, however. In this scenario the illumination is again represented by a single point in the entrance pupil but now shifted relative to the center. Next, we have to consider the diffraction pattern for light at nonnormal incidence. In this case, equation (2.2) is modified to take account of the non-normal incidence:

$$p\sin\theta_{out} - p\sin\theta_{in} = n\lambda \qquad (2.9)$$

Equation (2.9) may be paraphrased as stating that the diffraction pattern under oblique illumination is obtained simply by shifting the pattern at normal incidence (see Figure 2.20b). Note the effect of illuminating the mask at an angle θ_{in} larger than zero. Solutions for this equation now exist for pitches smaller than the wavelength of light if the sign of θ_{out} is opposite to the sign of θ_{out}. For $\theta_{in} = 90°$ this equation has solutions for pitches as small as half the wavelength. The key difference to the previous imaging example is that under these circumstances two diffraction orders pass through the lens: the zero order and one of the first orders. Due to the interference of two diffraction orders at the wafer, the intensity pattern now shows spatial modulation. The pitch of the pattern is determined by the difference in diffraction angles. As none of the higher diffraction orders is transmitted, the image does not have the sharp transitions between the transparent and opaque regions present on the mask. Rather, there is a gradual transition between the bright and dark regions in the image. The thresholdlike dependency of photoresist dissolution on light intensity still serves to convert the gradual modulations to sharp resist images. The principal effect of the finite slope of the image intensities is a dose sensitivity of the line width: The shallower the slope, the larger the CD versus dose variation. It is important to note that it is not necessarily the absolute modulation of the image that is important for characterizing image stability; rather, the slope of the image at the resist threshold is of primary concern. Several factors influence that slope, one being the number of diffraction orders that can be brought into interference: Both the fact that only one of the first-

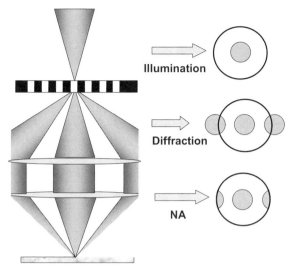

Figure 2.21 Partially coherent imaging.

order diffracted beams and the fact that none of the higher-order diffracted beams contribute to the image leads to a degradation in the image slope. The relative intensity (more specifically, the amplitude of the electric field) is another factor. Finally, higher diffraction orders are advantageous for another reason, as they contribute higher spatial frequencies and thus larger slopes at the image threshold.

Figure 2.21 extends the principles outlined so far to the case of partially coherent imaging: that is, where the mask is illuminated with light from a multitude of directions. The example shown is a relatively simple one in which the light distribution resembles a cone. This particular type of illumination is referred to as *conventional illumination*. Its characteristic number is the maximum angle of incidence for light on the reticle, or the maximum σ value. It is represented by a circle centered on the entrance pupil. The mask diffraction in this case creates additional circles spaced equidistant and with the spacing determined by the pitch of the pattern according to equation (2.7).

Figure 2.22 demonstrates the link between the imaging framework developed so far and the design. The design largely determines the patterns present on the reticle, which in turn determine the diffraction pattern. This pattern, combined with a specific illumination pattern, leads to a specific distribution of light on the entrance pupil, as shown in the figure for the case of conventional illumination, and a mask pattern that is a simple line space pattern. Specific characteristics of the pattern on the entrance pupil determine the image quality, in particular the sensitivity of the image to dose and focus. One quality metric of the image as mentioned above is the image slope at the resist

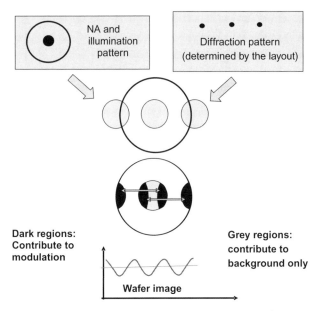

Figure 2.22 Interaction among layout, illumination, and the image formed on the wafer.

threshold. For the incoherent case the final image intensity is determined by adding the coherent contributions from the various points of the light source. As we can see, the coherent contributions fall in two categories: one for which there is no modulation, as only one diffraction order passes through the lens, and one where two diffraction orders contribute to the image. Thus, the effective image modulation is determined by the ratio between those portions of the illuminator that have first-order diffracted light passing through the lens versus those where only one of the diffraction orders passes through the lens. These portions are indicated schematically in Figure 2.22 as dark regions together with their corresponding first-order diffracted equivalents. With decreasing pitch, these regions get smaller and thus the modulation in the image decreases. In the extreme case where the dark regions have disappeared completely, no spatial modulation is present in the image and the resolution limit of the system is reached. The other portions of the illuminator, shown in grey, do not contribute to image contrast. Worse yet, they add to the background intensity and thus actually reduce the contrast. So it is desirable to minimize those portions of the illumination pattern that do not contribute to diffraction. There are other specific characteristics of the pattern on the entrance pupil that determine the sensitivity to focus. A desirable quality of the image is related to the decrease in contrast as a function of focus: The slower the decrease, the more stable the image is with respect to fluctuations in underlying substrate topography, for example. These imaging properties can be expressed with respect to certain properties of the light distribution created on the entrance pupil. The main principle for improving depth of focus is to keep to as narrow a ring as possible the light that is contributing to the imaging process. Fundamentally, this may be explained by the fact that defocus is equivalent to a phase shift between diffracted orders, where the magnitude of the phase shift depends on the distance from the center of the circle.

As a next step we discuss these conditions and criteria in a more rigorous fashion. The Debye approximation is the foundation for computation of images formed by projection optics with a high numerical aperture [13]. This method starts by expressing the electric field at the exit plane of the mask created by each source point as the sum of plane and evanescent waves:

$$E(\alpha,\beta)e^{i(2\pi/\lambda)(\alpha x+\beta y+\gamma z)} \tag{2.10}$$

$$\gamma^2 = n^2 - \alpha^2 - \beta^2 \tag{2.11}$$

The vectors $E(\alpha,\beta)$ are obtained by setting z equal to its value at the mask exit plane, and matching boundary conditions with equation (2.10). Physically, they represent the diffraction spectrum created by the mask. Mathematically, they are equivalent to the Fourier transform of the fields in equation (2.10). In equations (2.10) and (2.11), λ is the wavelength in vacuum of light forming the image, and for $\alpha^2 + \beta^2 \leq n^2$, α, β, and γ are direction cosines of plane waves propagating with respect to the x, y, and z axes multiplied by the refractive index of the medium in which they propagate. According to the Debye approximation, only those plane waves whose propagation directions fall within the cone defined by the entrance pupil of the projection optics contribute to the image. This condition is expressed mathematically by stating that only plane waves for which $\alpha^2 + \beta^2 \leq NA^2$ form the image, where NA is the numerical aperture in object space. To complete the calculation it is necessary to transform this truncated plane-wave spectrum into image space, but the details of this are not important for the topic in this section. The essential elements are that the image is also expressed as the sum of plane waves, and the transformation of α and β into the corresponding direction cosines in image space is linear. This permits computation of the image from the Fourier transform of the electric field at the exit plane of the mask, truncated according to the expression $\alpha^2 + \beta^2 \leq NA^2$.

Decomposition of the image into sums of plane waves of the form in equations (2.10) and (2.11) provides immediate insight into contrast and depth of focus. The intensity of the image due to the source point defined by ξ and η is proportional to the mean square of the electric field:

$$I_{\xi\eta}(x,y,z) \propto \sum_{\alpha^2+\beta^2\leq NA^2} E(\alpha,\beta)e^{i(2\pi/\lambda)(\alpha x+\beta y+\gamma z)} \sum_{\alpha'^2+\beta'^2\leq NA^2} E^*(\alpha',\beta')e^{-i(2\pi/\lambda)(\alpha'x+\beta'y+\gamma'z)} \tag{2.12}$$

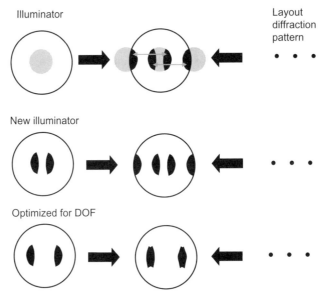

Figure 2.23 Illumination optimization.

High contrast means that I changes rapidly as a function of x and y, which from equation (2.12) is equivalent to the magnitude of E being large at large values of α and β. Depth of focus is large if the magnitude of E is significant only for values of α and β such that $\alpha^2 + \beta^2 \approx$ constant. Under these conditions, γ is also approximately constant according to equation (2.11), so the exponential terms involving it may be taken outside the parentheses in equation (2.12), causing the z dependence to vanish. The condition for high contrast is therefore that the image receives large contributions from plane waves with $\gamma \approx (n^2 - NA^2)^{1/2}$, and a large depth of focus is obtained when the image is formed primarily of plane waves with $\gamma \approx$ constant. It is intuitively evident that these conditions are most easily met for highly regular patterns, as they have fairly simple and well-defined diffraction patterns.

For a simple graphical explanation of the illumination optimization process, we turn to Figure 2.23. We start with the same layout and illumination pattern as was assumed for Figure 2.22. As a first step, the grey portions of the illumination pattern are removed. Those portions are characterized by the fact that they do not result in a first-order diffracted beam that lies within the entrance pupil. This reduces unnecessary background light and helps improve contrast. The illuminator at this point consists essentially of two slivers. As a next step we attempt to optimize the expected depth of focus, which can be accomplished by pushing the two poles of the illuminator farther apart, up to the point where the first diffracted order of one of the poles coincides with the other pole of the illuminator. Under this condition we have constrained all portions that contribute to image formation within a narrow donut around the center. This corresponds to the condition γ that is approximately constant as mentioned above. More details and applications of this methodology may be found in Section 2.4.7.2.

Having gone through this optimization problem, we can now assess the difficulties in printing a multitude of patterns. The simplest is one where we assume that we have to print simultaneously two simple line space patterns with two different pitches located on a different portion of the mask. In the simple case that we have selected, the diffraction pattern is qualitatively the same; it differs only by the spacing between the diffraction orders. As we can choose only one illuminator we have to find the best solution for both diffraction patterns. The last step in our previous example of illuminator optimization, however, was to place the two poles of the illuminator at a very specific distance from the center. The spacing between the poles was determined by the condition that the first diffraction order from one pole coincides with the zero order of the other pole, and vice versa. Thus, the optimum illumination pattern is determined by the pitches present in the layout, two different pitches lead to two differently shaped illuminators. In the case of a layout that is characterized by more than one pitch, only a less optimal solution can be provided. One may, for example, choose to select an illuminator that is a combination of the two optimized patterns. This illustrates the general principle whereby the lithographic optimization process becomes less and less efficient the more complex the Fourier transform of the pattern under investigation. Complex Fourier patterns may be the result of nonrepeating patterns or a large multitude of repeating patterns, with different pitches in different locations of the reticle.

A more complete and refined optimization process may consist of optimizations in the illuminator shape, the mask type, the type of patterns placed on the mask, and lens apodization. Apodization is a process whereby light passing through a lens is attenuated depending on the diffraction angle. In our simple example, a filter is placed between the two lens elements drawn in Figure 2.18. Apodization has been discussed in the literature. However, it is not generally used, due primarily to the difficulties of reliably inserting exchangeable optical elements into the projection lens.

Other elements of the optimization process are the type of mask being used as well as the shapes placed on the reticle. Even though one may assume that the layout uniquely determines the diffraction pattern, this is, in fact, not the case. For example, instead of representing the layout pattern as transparent and nontransparent regions, one may choose to add additional shapes on the reticle. For example, phase shapes can be added to the transparent regions of the mask, creating a mask type that is referred to as an alternating phase-shift mask. Figure 2.24 illustrates the working principle of an alternating phase-shift mask within the framework laid out above. We use simple illumination: low-sigma conventional illumination. The wafer pattern desired consists of equidistant lines and spaces. In our first implementation this pattern is imaged simply by using transparent and opaque regions on the reticle. The pitch is such that the first diffraction orders do not pass through the lens, and thus no image is formed. In the bottom row of Figure 2.24, the mask pattern has been

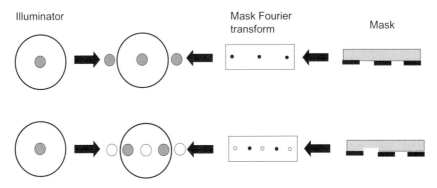

Figure 2.24   Alternating phase-shift mask.

modified. The original mask pattern consisted of chrome features on a mask blank, with clear openings in between. In the lower part of Figure 2.24, the mask has been altered by etching trenches into the quartz in every second opening on the reticle. The etch depth is chosen such that

$$(n-1)d = \lambda \qquad (2.13)$$

where n is the refractive index of the the mask substrate, d the etch depth, and λ the wavelength of light. Under these conditions the light passing through these openings has a $180°$ phase shift relative to the unetched opening. This modification has a profound impact on the resulting Fourier transform of the pattern. Introducing phase openings for every second space changes the basic periodicity of the pattern to twice the pitch, and thus the Fourier transform has additional contributions at $1/2p$. In addition to introducing additional frequency components, there is another unique characteristic to this pattern: Not all of the diffraction orders at $1/2p$ actually exist. Considering the zero order, one realizes that it consists of equal amounts of light at 0 and $180°$ phase shifts. These contributions interfere destructively and thus eliminate the zero order. A more rigorous analysis shows that, in fact, all even orders are eliminated by destructive interference. Thus, the resulting diffraction pattern is one with spacing between the diffraction order corresponding to $2(1/2p)$. In addition, the zero order is not present and thus two diffraction orders pass through the lens, and the diffraction angle between the two orders is now equivalent to $1/p$. This diffraction patterns has quite desirable properties: First, the introduction of alternating phase-shift masks allows imaging of the pitch p which in the same optical setup was not possible before. In addition, the diffraction orders that do pass through are symmetric to the center, resulting in good depth of focus. Finally, the fact that the diffraction orders are symmetric relative to zero is independent of the pitch of the pattern, contrary to the example in Figure 2.23, where the symmetric arrangement of diffraction orders can be achieved only for one particular pitch. Alternating phase-shift masks are an

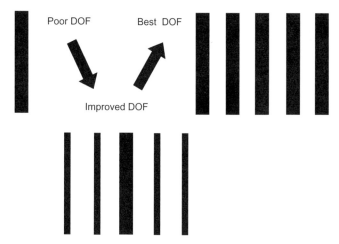

Figure 2.25 Use of subresolution assist features.

example of the possibilities that exist in optimizing imaging of a particular layout by adding additional, nonprinting features to the mask.

Subresolution assist features are another common approach whereby the diffraction patterns are changed through additional features on the mask. Off-axis illumination is quite efficient in improving the depth of focus of nested dense lines, in particular if the illumination conditions are tuned to the particular pitch that is being used. However, the depth of focus for isolated lines is quite poor. This is related to the fact that the Fourier transform of an isolated line does not have distinctive maxima as the nested lines do; rather, the spectrum has a wide frequency distribution with a strong zero-order contribution. Under those circumstances, off-axis illumination does not improve the depth of focus. The wide frequency distribution precludes achieving a narrow range within which the key contributions to the image are located in the entrance pupil. One solution is to make the isolated line appear more dense. This can be accomplished by adding additional lines next to the feature at a pitch for which the illuminator has been optimized. Use of more than one feature per edge is advantageous, due to a finite coherence length. Figure 2.25 gives an overview of the transition from poor depth of focus for an isolated line to the best DOF for a nested line. From the perspective of the diffraction pattern, the assist features accomplish two things: They reduce the strong zero-order component of an isolated line, and they shift some of the weight of the Fourier spectrum toward frequencies for which the illuminator has been optimized, thus contributing to improved DOF. Subresolution assist features demonstrate yet another principle important for the combined optimization process: The shape of the image is important only in the vicinity of the resist threshold, where the edges of the final pattern are defined. Outside of that domain, the details of the intensity distribution are not important as long as

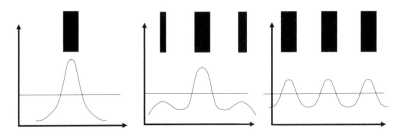

Figure 2.26 Improvement in image properties at the edges of the desired patterns by modulating image properties in regions outside the desired patterns.

they do not create unwanted patterns. Thus, this portion of the image can be modified (Figure 2.26) for improved image qualities at the pattern edges. Subresolution assist features improve the depth of focus of the pattern by modifying those portions of the image that are not relevant for the patterning process. To avoid printing the assist features, their feature sizes are kept small.

The last element of the optimization process is to modify the as-designed patterns. The layout should be altered to allow more efficient optimization of the overall imaging process. In general, this amounts to simplifying the diffraction pattern. Such layout restrictions are discussed in more detail later in this chapter.

In this chapter we have seen how the individual components of the lithographic process—illuminator, mask, and layout—are increasingly being optimized in a joint fashion to address the challenges of a low-k_1 regime. The basic principles underlying this optimization process have been laid out, and many of the common resolution enhancement techniques have been positioned within this framework. One of the key messages of this chapter is the close relationship between layout style and the efficiency of this optimization process: The simpler the layout style as expressed by the complexity of the spatial Fourier transform of the pattern, the more efficient the optimization process can be made. Most layout restrictions have their roots in this relationship: radical layout restrictions reducing the number of allowed pitches to one or two. Arranging patterns with a single orientations, avoiding highly two-dimensional layouts with many turns and corners all contribute to simpler diffraction patterns and thus allow for better, more variation tolerant lithographic processes.

2.2.6 Optical Proximity Correction

Optical proximity correction has become an integral part of the overall patterning process essentially since the 0.18-µm generation. The introduction of optical proximity corrections is the result of the loss of image fidelity inherent in low-k_1 imaging. Typical wafer-level effects that occur as a result of a loss of image fidelity are shown in Figure 2.27a, which presents examples of isonested

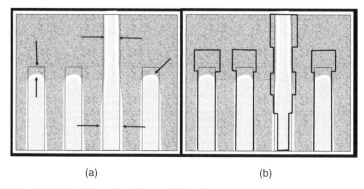

(a)   (b)

Figure 2.27 (a) Typical distortion characteristics for showing line-end foreshortening, corner rounding, and iso-nested biases; (b) typical mask distortions used to counteract deviations of the image from the nominal design shapes.

biases, line-end foreshortening, and corner rounding. The primary objective of OPC is to counteract those effects. Figure 2.27b presents the typical type of corrections applied during OPC to counteract the deviations between the as-designed shape and the printed image. Line-end foreshortening is compensated by using hammerheads and extending the mask shape beyond the original design, iso-nested biases are compensated by modifying the line width of the mask shape, and corner rounding is alleviated by using inner and outer serifs.

In its practical implementation, OPC is a collection of fairly sophisticated algorithms. The main elements of OPC are a hierarchy manager that breaks the entire design into smaller pieces consistent with the optical radius, an algorithm to introduce additional points to the design polygons (as can be seen in Figure 2.27b), a simulation engine that efficiently simulates expected wafer contours based on the mask shape, and an optimization routine to minimize deviations between predicted contours and the target. These elements are depicted schematically in Figure 2.28.

The first step in OPC flow is to convert the layout hierarchy into an optical hierarchy. The difference between the two is best demonstrated using the case of a simple layout, a small cell such as an SRAM cell, laid out in a regular array. An example of such an arrangement is shown on the left side of Figure 2.28. Consider the two cells marked with circles, one at the edge of the array, one within the array. From a design perspective, the cells are equivalent. From a lithographic perspective, however, the cells are not equivalent. This is due to the fact that the optical image at a given point depends on the arrangement of all shapes within the optical radius. This radius is on the order of several wavelengths and in our example is significantly larger than the cell itself. Thus, for the purpose of optical simulation, the relevant layout is not only the unit SRAM cell but includes all the layout cells within the optical radius. Even in the simple layout case of a regular SRAM array that uses a single design cell, a multitude of "optical cells" differentiated by differences

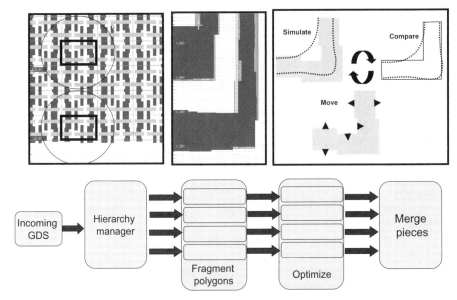

Figure 2.28   Schematic overview of OPC flow.

in the arrangement of unit cells within the optical neighborhood is required to perform proper OPC corrections. The challenge of the hierarchy manager is to identify, for each cell used in the design, all the different surroundings given an optical interaction radius of several wavelengths. For each variation, the cell, together with its surrounding, has to be made available for subsequent simulations. The hierarchy manager is called into action once again at the end of the OPC algorithm, when the individual optical cells have to be recombined to form the post OPC layout. In our simple memory case the array no longer consists of a regular layout of identical cells; rather, it is composed of a multitude of cells, each characterized by its unique environment and the resulting unique corrections.

To achieve the necessary turnaround times for OPC, computations are typically done with clusters comprised of a large number of CPUs. Use of 1000+ CPU clusters have been reported for OPC applications. The individual cells of the optical hierarchy are distributed onto the nodes of the cluster for parallel processing, indicated by the division of flow in Figure 2.28.

Once the optical cells have been loaded in the memory of one of the individual nodes, this node will execute the entire OPC optimization loop for this cell. As a first step, the layout polygons are prepared for the optimization loop by adding additional points to the polygon, a process called *segmentation*. Initially, those points are simply added onto existing segments without actually changing the shape itself. The way in which the polygons are segmented has a significant impact not only on the quality of the final solution but also on OPC run times and mask cost. In general, the quality of the OPC solution, measured as the average deviation of the final wafer contour from the target, decreases with decreasing segmentation length. At the same time, mask cost and OPC run time increase. The simplest segmentation approach is to divide each side of the polygon into segments with lengths approximately equal to some fixed length. The smallest meaningful segmentation length is comparable to the Nyquist frequency of the optical system. Creating segments at distances significantly smaller than the Nyquist frequency no longer improves the quality of the solution. To find the best compromise between run time and OPC quality, segmentation rules are usually significantly more complex than simply dividing polygon edges into equidistant pieces. Commonly, these rules start by placing *seedpoints*, for example, at the corner of a design polygon. Seedpoints are also created by projection from the corner of adjacent polygon. Starting from these seedpoints additional segmentation points may be generated based on rules that specify the number and distance of these points away from the seedpoint.

Once the polygons have been segmented, the next step is to optimize the segment positions in an iterative loop. Each iteration loop consists of three parts (see Figure 2.28): The simulation of the wafer contour based on the current mask shapes (simulate), a comparison between the simulated contours and the target shape (compare), and a segment movement step where segment positions are changed in an attempt to minimize deviations between predicted contours and target (move). For all commercial simulators, calculation of the predicted wafer contour is primarily a problem of calculating the optical image from a given mask pattern. Nonoptical processes such as acid generation, acid diffusion in the resist during the post-develop baking and resist dissolution are handled primarily as a perturbation on the optical image. In one of the more common modeling approaches, deviations between the optical image predicted and the actual resist image are modeled as a function of a variety of optical image parameters. When OPC was first used, this approach was quite valid, as the significance of nonoptical effects was small relative to the optical effect. However, several processes have challenged this assumption, such as the large etch biases encountered in gate etching for flash memories and the resist reflow processes used in some contact printing processes.

The deviation of the simulated contour from the target is quantitatively measured using a cost function. The key quantity in this cost function is the difference in location between the simulated and desired contours a quantity referred to as *edge placement error*. The cost function is the square of edge placement errors summed over a number of evaluation sites. The goal of the optimization is to minimize the cost function by moving the polygon segments that have been created in the initial step. In general, this is a multidimensional optimization problem. One possible simplification is to use localized optimizations. In this case the edge placement error in a particular location is corrected by a movement of the local segment only. A more complex optimization calculates segment movement taking into account the effect of all other segments simultaneously. This approach tends to result in a significantly smaller number of iterations to reach convergence than are used in other algorithms. At the same time, the computational effort per iteration increases. The optimization process is terminated once the cost function has dropped below an acceptable maximum or the maximum number of iterations has been exceeded. Once the optimizations for all optical cells have been completed, the entire layout is reconstructed using corrected optical cells.

It is important to bear in mind some of the limitations of optical proximity corrections. At least in principle, the different print biases for long, isolated and long, nested lines can be corrected perfectly—with one caveat. The correction is exactly correct only for a fixed process condition. As process conditions such as dose and focus vary, the iso-dense bias changes. Similarly, line-end foreshortening can be corrected perfectly, but again only for a specific process condition. Thus, the quality of the final OPC solution is a strong function of the ability to maintain the correct process conditions. Other errors that contribute to OPC inaccuracies are model errors; line-end corrections primarily suffer from such inaccuracies. This is due, in part, to the fact that resist contributions to line-end foreshortening tend to be quite significant, contrary to the basic resist modeling assumptions used in most OPC models.

2.2.7   Double Patterning

One of the challenges of the 32-nm node, particularly for memory applications, is that the only exposure tools capable of providing the necessary resolution in a single exposure are either immersion tools using high-index fluids and a numerical aperture of 1.55, or EUV tools with a wavelength of 13.5 nm. The k_1 factors in these cases are 0.26 and 0.55, respectively. Neither of these options is expected to be available for early manufacturing of this technology node by about 2009 [14]. As a result, double-patterning techniques have received renewed interest.

There have been a variety of double-exposure techniques, one of the most common ones being the use of alternating phase-shift masks. This approach is characterized by a first exposure using a phase mask with 0 and 180° phase openings, followed by a second exposure with a trim mask, typically a COG mask (also called a binary imaging mask) to remove unwanted patterns at phase boundaries, for example. This technique is characterized by two masks (a phase and a trim mask), one resist coating step, two exposures, two development steps, and one etching step (Figure 2.29). Furthermore, the two exposures are typically done without the wafer leaving the exposure stage after the first exposure has been completed. After the reticles have been exchanged and the illumination settings have been switched, the wafer is exposed for a second time, and only then leaves the exposure tool. As a result, the intensities of the two exposures are added in a linear fashion in the resist (the light intensities of the first and second exposures are stored as photo-released acid in the resist). With this approach one can not exceed the fundamental resolution limit of an optical system of $k_1 = 0.25$.

Figure 2.29 Comparison of double-patterning and double-exposure techniques: (a) process flow for a double exposure; (b) process flow for double patterning.

In contrast, double-patterning techniques make it possible to go below the resolution limit of $k_1 = 0.25$, essentially by combining the effect of the two exposures in a highly nonlinear fashion (as opposed to the linear addition of intensities inside a single resist). Double-patterning techniques are processes that generally consist of a first lithographic step, complete with a resist coat, exposure, and development, a first etching and resist strip, followed by a second exposure, again complete with resist coat, exposure, and development and a second, final etching step to form the patterns on a single mask level (LELE: litho–etch–litho–etch). The two process flows are compared schematically in Figure 2.29.

To cover back-end as well as front-end processes, a negative as well as a positive version of this approach is required. A schematic comparison of the negative and positive double-patterning process is shown in Figure 2.30. In the positive tone approach, distributing the pattern onto two reticles is relatively straightforward. The features themselves are being assigned to one of the two reticles, a process similar to the coloring in alternating phase-shift mask processing. Connections between features of different colors are achieved through partial overlap. For the negative tone process, splitting the design into two portions is less intuitive. In this case the spaces between features are colored, and based on their color they are assigned to one of the two reticles. One of the big issues in this approach is that the features on the wafer are formed with the two edges defined by the two different masks: one edge by one mask, the other edge by the other mask. As a result, the line width control in this approach becomes a function of overlay, clearly not a very desirable situation.

Figure 2.31 shows in more detail one possible flow for a positive tone, double-patterning process for a poly layer. The materials stack consists of a poly layer on top of which two hardmasks are deposited. We assume the pattern to be printed is a dense line space pattern. In the first exposure step, resist lines are printed for every second line. Those are subsequently etched into the first hardmask. The resist coating for the second exposure now occurs over a nonplanar stack caused by the lines present in the first hardmask layer. The second mask step patterns those lines in between the previous ones. At this point in the process flow the entire pattern is present; however, half of the lines are present as lines in resist, the other half, as lines present in the first

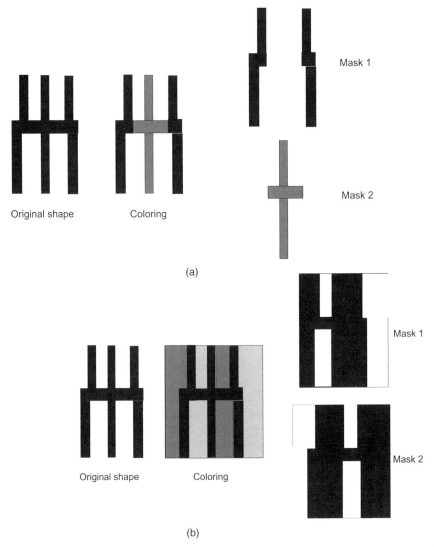

Figure 2.30   (a) Positive tone design splitting for a double-patterning process; (b) negative tone design splitting.

hardmask. This combined pattern is transferred via reactive ion etch into the second hardmask. The final polysilicon lines are created by etching with the second hardmask in place.

An alternative double-patterning process is the self-aligned process, which borrows from the well-known spacer process used in gate manufacturing. The process flow is shown in Figure 2.32. Similar to the previous scheme, the process starts with a first patterning step on a double hardmask stack, and the pattern is etched into the first hardmask. However, the subsequent process

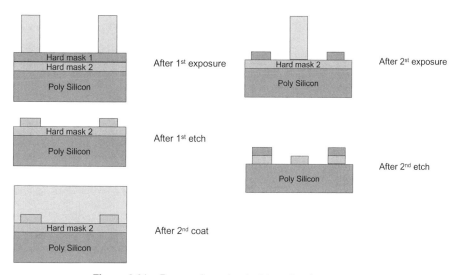

Figure 2.31   Process flow of a double-patterning process.

sequence is dramatically different, as it uses a spacer process, in which a conformal nitride film is deposited over the patterned hardmask. At the edges of the hardmask step, the nitride film is significantly thicker; essentially, the nitride thickness just beyond the hardmask step is twice the nitride thickness plus the height of the first hardmask. A uniform nitride etch leaves a pattern on the wafer similar to the one shown in Figure 2.32. Then the remaining material from the first hardmask step is removed selectively. A further etch results in a pattern formed in the second hardmask, as shown in the figure. We note that as each line has two spacers, one on each side, the original line pattern is effectively doubled. This processing approach saves one lithographic step—however, at the expense of a spacer etch process, which adds an additional deposition and etching step.

There are several challenges inherent in the double-patterning process. The first and most obvious challenge is cost. Not only are the number of masks doubled but the processing costs are significantly higher, as every patterning step requires two lithographic, two etching, and two resist strip steps, and two wet cleans, as well as twice the number of hardmask depositions. Another significant contributor to the cost of this process are reduced yields. Compared to a single patterning process, the defectivity is twice as high, as twice the number of lithographic, etching, and strip processes are required. Furthermore, a closer look at the processing flow reveals the process control challenges. There is an inherent asymmetry between the two sets of lines patterned in the two lithography steps. The lines formed in the first lithographic step are subject to a second blanket etching step, which contributes additional CD biases and CD variability not encountered by the second set of lines. In addition, for the second set of lines, resist coating occurs over the topography created during

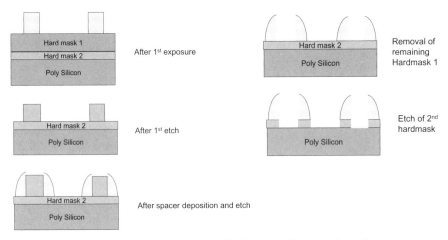

Figure 2.32 Process flow for the self-aligned double-patterning scheme.

the first hardmask etching step. This is particularly difficult if the double patterning is used in a random logic circuit where the effective resist coating thicknesses are changing as a result of varying pattern densities. The changing resist and ARC thicknesses, in turn, contribute to additional CD variations. Thus, double-patterning technologies face significant cost, yield, and CD control issues.

Other challenges of double patterning are related to the fact that the design needs to be separated into two pieces, which will be printed with two separate reticles. In the case of a flash memory pattern, with essentially only long lines and spaces, the design splitting that needs to occur is relatively straightforward. Splitting the layout of a random logic pattern is another matter, however. As we have seen in Figure 2.30, the layout split process is reminiscent of the coloring problem encountered for alternating phase-shift masks, and the requirements are similar to those for alternating masks: No two neighboring minimum features should have the same color, as they would create a pitch beyond the resolution capability of the exposure tool. An exemplary pattern exhibiting a coloring issue is shown in Figure 2.33. The layout is reminiscent of those encountered for phase coloring conflicts with alternating phase shift masks. The line and space patterns are assumed to be on minimum pitch and therefore need to be placed on separate reticles, different colors corresponding to the dark and light gray areas. The shape with the hatched pattern cannot be assigned uniquely to either one of the masks. Fortunately, the conflict resolution is significantly simpler, as the patterns can always be cut and stitched. Conflict resolution for the pattern of Figure 2.33 is shown in Figure 2.34. The main drawback of the cut and stitch operations is that they make the design susceptible to variations in relative overlay of the two patterns. In extreme

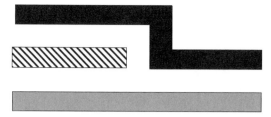

Figure 2.33 Layout scenario with coloring conflict. A unique color cannot be assigned to the hatched shape.

Figure 2.34 Layout scenario, colored for double exposure with cut and stitch.

cases this could lead to a catastrophic failure of the stitch region. It is therefore advisable to provide a certain overlap of the cut mask beyond the actual cut region for enhanced stability, as detailed in Figure 2.35. Even so, stitch regions remain problematic, even if they have been designed with the appropriate overlaps: Line-width control in these regions is dependent on overlay. The sensitive direction is the direction perpendicular to the stitch cut, as the shapes on two separate masks determine the line edges in this region.

2.2.8 Lithographic Roadmap

Table 2.1 provides an overview of current and expected patterning solutions for 65-, 45-, and 32-nm nodes. The primary process solution for the 65-nm node is use of 193-nm dry exposure tools at fairly high NAs (0.85 to 0.93). In some instances, primarily for gate layer, double-exposure techniques such as alternating phase shift exposures and true double-patterning techniques such as LELE solutions are used. Immersion tools have been used primarily to demonstrate the feasibility of the technology, and their use appears to be mostly a research effort for the 65-nm node.

The 45-nm node is a good example of how rapidly technologies can change. Up until the beginning of 2004, Sematech surveys showed that the semiconductor industry viewed extreme ultraviolet (EUV) and e-beam projection

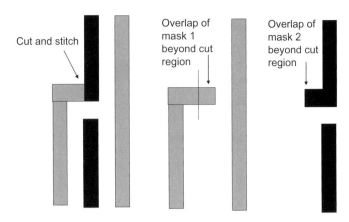

Figure 2.35 Overlap in the stitch region.

Table 2.1 Patterning solutions for the 65-, 45-, and 32-nm nodes

Technology Node	Primary Solution	First Alternative	Second Alternative
65-nm	193-nm dry, single exposure	193-nm, double exposure	
45-nm	193-nm immersion	193-nm dry, double patterning	
32-nm	193-nm immersion, double patterning	193-nm immersion, single exposure, high-index fluids	EUV

lithography as the leading candidates for exposure tools at the 50-nm node. In 2004, the general opinion suddenly changed. Immersion lithography entered the stage and became the solution of choice for the 50-nm node.

Several leading semiconductor manufacturers have announced their use of immersion lithography at 45 nm. A few manufacturers, driven either by the early issues of immersion lithography, such as high defectivity and poor overlay, or the financial burden of a new, fairly expensive exposure approach, have or are planning to extend the use of 193-nm dry exposure tools with double-patterning exposures and aggressive resolution enhancement techniques.

For the 32-nm node, the most conservative approach appears to be double patterning on immersion tools using water, since all the key technology elements have been tested in previous generations. Alternative approaches such as wet exposures using high-index fluids or EUV introduce new technology elements. Obviously, the technology challenges of EUV are drastically more daunting than those of introducing high-index immersion fluids. The primary reason for choosing double patterning over an immersion solution using high-index fluids is one of pure timing. Suitable high-index immersion fluids and high-index lens materials do not appear to be available in time for the development of capable 32-nm manufacturing tools. While the 32 nm memory manufacturing is likely to use double patterning techniques, the bulk of 32-nm logic manufacturing will probably be done using single exposure due to the cost advantage over double-patterning exposures. The technology challenges and the enormous cost issues of EUV still appear too daunting to make adoption of this technology for 32 nm likely.

2.3 LITHOGRAPHY LIMITED YIELD

The preeminent role of lithography in DFM is a result of its strong impact on the economics of the semiconductor manufacturing process. The increased cost of lithographic equipment and a dramatic rise in mask costs are some of the preeminent contributors to high manufacturing costs. Longer turnaround times for mask sets due to longer OPC run times, delays in product delivery due to catastrophic patterning failures, and severe yield limitations due to increased variabilities that affect parametric yields also have a negative impact on revenues from a semiconductor device. The economic aspects of DFM were discussed earlier. In this section we provide more detail on the technical aspects that drive lithography limited yield. There are three factors that we discuss in more detail: increasingly severe deviations of the printed image from the design shapes despite the use of OPC, the stronger impact of process variations on lithographic printing, and catastrophic failures due to suboptimal use of resolution enhancements.

2.3.1 Deviations of Printed Shape from Drawn Polygon

As the lithography process is pushed to lower and lower k_1 values, there is a growing disparity between the rectangular shapes used in the physical layout versus what these patterns will look like on a wafer (Figure 2.36). The impact of image distortions such as line-end foreshortening, iso-nested bias, corner rounding, and image ringing is becoming increasingly severe (Figure 2.37). Even though OPC attempts to counteract these image distortions, the wafer results may still show significant deviations. For example, it is not possible to eliminate corner rounding, as rounding is ultimately related to the limited bandwidth of the optical system. Also, even for cases where at least in principle a perfect correction is possible, de facto poor model quality and increasingly large and thus error-prone corrections limit the quality of the final solution. Thus, despite the presence of OPC, increasingly severe differences are observed between the perfectly square layout and contours as they appear on the wafer. Figure 2.38 gives an example of how realistic contours of a

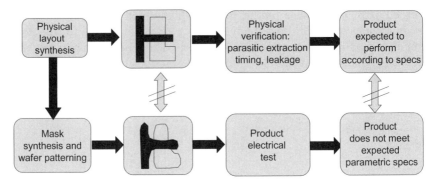

Figure 2.36 Discrepancy between shape representation on the physical layout side, including physical verification and actual product patterns.

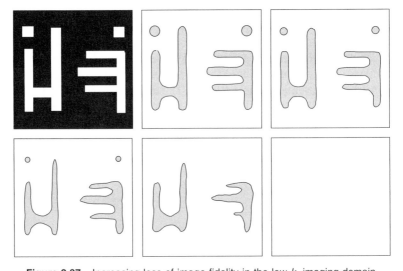

Figure 2.37 Increasing loss of image fidelity in the low-k_1 imaging domain.

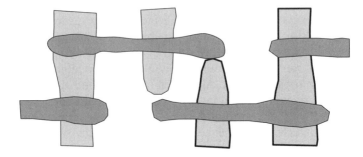

Figure 2.38 Actual 65-nm SRAM contours.

65-nm SRAM may look. Focusing on the intersection between gate and active, one notices significant deviations between the ideal contours, which would be represented by rectangles and the actual contours. For example, toward the end of the polyline, significant line-width variations can be observed, due to an incorrectly sized hammerhead. Further examples are shown in Figures 6.16 and 6.22 to 6.24, and the reader is referred to the corresponding discussion in Chapter 6.

2.3.2 Increased Variabilities

Variability has become a major challenge for designs in the 65- and 45-nm nodes and is considered one of the primary limiters to continued technology scaling. It is affecting device and as well as interconnect parameters. In integrated-circuit designs, sufficient tolerance to delay variations among various signal paths is required to achieve the desired chip operating frequencies. It has been difficult to provide a detailed assessment of the variabilities seen in manufacturing. In particular, accurately capturing correlated as opposed to uncorrelated changes in the delays in two signal paths has been a challenging modeling, problem. This leads to fairly conservative corner modeling, which will overestimate the real circuit delays and make it difficult to achieve the desired design goals in a straightforward and thus time-efficient manner. Also, timing closure variabilities exacerbate the already critical power dissipation problems. Even though a chip may reach its intended speed performance, it may do so only at an unacceptable level of power consumption.

A wide range of effects lead to device variability, and lithography is certainly not the only contributor. Effects unrelated to the process, such as the local operating environment of the transistor on the chip, play a significant role as well. For example, the effective supply voltage of a transistor may vary across the chip due to changes in the voltage drop along the power grid. The local operating temperature of a transistor is affected by local variations in power dissipation. There are spatial as well as temporal effects. Negative bias instabilities and hot carrier effects modify device thresholds over time. From a process perspective, almost all sectors, etches, thin-film deposition, hot processes, and even wafer clean processes, influence device parameters and thus contribute to varibilities. They affect structural parameters such as gate length, gate width, and gate oxide and spacer thickness. They also modify dopant activation and electron mobility. Some of the more prominent process effects are mentioned in the following section. The level of dopant activation during rapid thermal anneals may vary due to temperature control problems in the equipment used or as a result of layout-dependent thermal coupling between the heat source and the patterned wafer. The extensive use of mobility enhancement technologies has added a source of variability not present in prior technologies: local strain may vary due to variations in stress layer thickness or as a result of layout-dependent stress variations. There are pattern density–driven variations in etch rates affecting L_{poly}. Across-wafer temperature variations during gate oxide formation affect gate oxide thicknesses. Etch- and deposition–driven variations in spacer thicknesses affect the implant locations. The primary topic of this section, however, is lithography (or more generally, patterning)-driven variability. We first provide some guidelines on the extent to which critical dimensions are expected to change, based on the International Technology Roadmap for Semiconductors. After describing the targets, we describe how variability can be classified further from a process perspective. This ultimately allows us to convert the variability target to specifications for process equipment and process control specifications. Finally, we discuss the impact of lithographic budgets on such properties as SRAM sizes and address strategies to reduce variability.

2.3.2.1 International Technology Roadmap for Semiconductors
There are two major categories of lithography-driven variabilities: variations in the critical dimensions, and variations in the relative placement of features on subsequent patterning steps (alignment). The International Technology Roadmap for Semiconductors [15] provides some guidelines on the overall targets of the patterning process. Table 2.2 lists expected variabilities for CD of general patterning steps, the required gate CD control, and targeted overlay performance for several years and several technologies into the future. All numbers are given as 3σ values. The CD control numbers include all sources of variability of the patterning process. The severity of the issue regarding variability is highlighted by the fact that according to the 2006 version of the roadmap, there are no known solutions to meeting CDs from 2007 and overlay control requirements from 2008 onward.

2.3.2.2 Classification of Process Variabilities
There are two common schemes for characterizing variabilities of physical dimensions: phenomenological and mechanistic. The phenomenological description divides variability in terms of:

- Lot-to-lot variability
- Wafer-to-wafer variability

Table 2.2   Excerpt from the 2006 ITRS[15] regarding variability

	Year of Manufacture					
	2007	2008	2009	2010	2011	2012
Technology node	65	57	50	45	40	36
CD control (3σ) (nm)	6.6	5.9	5.3	4.7	4.2	3.7
CD control (3σ) (MPU)	2.6	2.3	2.1	1.9	1.7	1.5
Overlay control (3σ) (nm)	11	10	9	8	7.1	6.4

- Chip-to-chip (or within-wafer) variability
- Within-field variability
- Layout-driven variability
- Microscopic variability

Although most of these categories are self-explanatory, layout and microscopic variations warrant a few comments. Microscopic variations are those which affect device variability even for devices with absolutely identical layout and in close proximity to each other: for examples, the two pull-up, two pull-down, or two pass gates of an SRAM cell within an SRAM array. One well-known mechanism driving microscopic fluctuations in threshold voltage is random doping fluctuations. Similarly, line-edge roughness is a patterning-driven microscopic mechanism affecting device width. Common layout-dependent mechanisms are variations in critical dimension as a function of gate orientation, pattern density, or gate-to gate-spacing.

The mechanistic description attempts to describe variability based on a number of known process contributors:

- Dose and focus variations
- Imperfections of the optics used to image a pattern, such as lens aberrations and flare
- Thickness variations of optically transparent layers below the resist coating as well as topography variations
- Variations in mask parameters
- Temperature dependencies of the resist processing, primarily from the post-exposure bake
- Overlay-driven variations
- Atomistic mechanisms

This set of categories considers primarily lithographic mechanisms, although contributions from etching are equally important. Examples of etching mechanisms influencing variability are temperature variations of the substrate temperature and the extent to which the chamber surfaces have been coated with polymer from previous etch processes. The total budget is typically derived assuming that the various mechanisms are independent of each other, and thus the square of the total 3σ variability is given as the sum of squares of the individual components. This approach is useful in obtaining estimates for total variability, although there are some complications. For one, the components in both model descriptions contain random and systematic contributions. In addition, the probability distributions created by many of these mechanisms show significant deviations from normal distributions. The example in Figure 2.39 shows the response of the critical dimensions to a particular process variation. The response curve is essentially quadratic, typical for the impact of focus on the critical dimensions. Given the shape of the response curve, the resulting

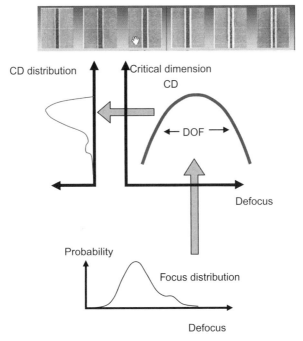

Figure 2.39   Example of strongly non-Gaussian CD distributions: focus-driven CD variations. The CD response curve to focus is roughly quadratic, resulting in strongly asymmetric CD distribution, even in the case that the underlying focus distribution is Gaussian.

probability distribution is strongly skewed, even if the underlying process variation has a Gaussian distribution. Due to the presence of systematic and random components with sometimes strongly non-Gaussian distributions, a generally much better description of the CD variability can be obtained using probability distributions.

Both descriptions are closely intertwined, as a particular mechanism may affect several components of a phenomenological description. For example, focus variations may have a lot-to-lot component, due to changes in the thickness of an underlying optically transparent layer. Those variations are known to have a detrimental impact on the commonly used optical focus sensors in exposure tools. The thickness variations may, in turn, be the result of variations in the as-deposited film thickness or the lot-to-lot variabilities of CMP processes. Wafer-to-wafer focus variations may be caused by lens heating. As the lens warms up due to the exposure light passing through it, the refractive index of the lens material changes, due to the temperature change and thus the focal length of the projection optics. Even though the exposure tools have built-in correction algorithms for this effect, the advanced illumination schemes quite frequently challenge the capabilities of these models. Chip-to-chip variations in focus are quite commonly the result of CMP-induced nonuniformities of wafer thickness, primarily at the edge of the wafer. In immersion exposure tools, temperature differences between the wafer and the chuck holding the wafer, combined with the differences in thermal expansion coefficients of chuck and wafer, may cause a nonplanar wafer surface in an effect similar to the bimetallic bending of temperature sensors. A common cause of across-field focus variations is due to the focal plane deviations of the projection lens. This example shows how one mechanism, such as focus variation, may contribute to many phenomenological variations in critical dimensions. There are other cases where a mechanisms is much more specific in that it affects only one particular category. For example, mask variations and lens imperfections affect across-field variations exclusively. Atomistic mechanisms only contribute to microscopic variations. Such direct relationships greatly facilitate the diagnosis of variability issues from a process perspective.

2.3.2.3   *Variability Mechanisms*

Variations in critical dimensions are the main component that the lithographic process contributes to the overall variability seen in chip performance. Two factors generally determine the magnitude of the variation. The first factor is the range over which a process parameter such as dose or focus varies in normal manufacturing operations. The second is a proportionality factor that describes the relationship between the change in process parameter and the resulting CD change. This factor can be modified by choosing different lithographic processes. For example, modifying the illumination or changing the mask type may have a significant impact on the proportionality factor. In general, the smaller the k_1 factor of the lithographic process, the larger the sensitivity factor. Finally, the proportionality factor also has a layout dependency. Common layout parameters with strong influence on the proportionality factor are the minimum critical dimension and the pitch of the pattern. Line ends and corners typically show higher sensitivities to the same process variations than do straight lines. To illustrate these concepts, we discuss briefly the impact of variations in mask CDs on the wafer CDs. The wafer-level CD change, ΔCD_{wafer}, which results from the variations in mask CD, ΔCD_{mask}, is described by

$$\Delta CD_{wafer} = MEEF \times \Delta CD_{mask} \qquad (2.14)$$

The sensitivity factor in this case is referred to as mask error enhancement factor. In this example the process variation in question is the deviation of the critical dimension of the mask features from their nominal dimensions. The MEEF is the proportionality factor. Mask error factors larger than 1 are the result of the nonlinearities of the imaging process. MEEF tends to increase with smaller dimensions. For a given minimum dimension, MEEF increases with decreasing pitch of the pattern. Generally, the tightest pitched contacts or the tightest pitch lines have the largest mask error factors.

During normal manufacturing the process experiences a range of variations that change effective dose and focus in a noncontrollable fashion. Similar to the case of the mask error factor, there is a sensitivity factor that translates the observed dose and focus variations into fluctuations of the critical dimensions. In the case of dose, the relationship is linear. The linear relationship is sometimes expressed in terms of the percentage change in dose, and the proportionality factor is then referred to as exposure latitude. For focus the relationship is nonlinear. To a good approximation the change in CDs is a quadratic function of defocus. The curvature of that relationship is related to the depth of focus in the sense that the larger the depth of focus, the smaller the focus-induced CD variation. Similar to the other cases, depth of focus and exposure latitude tend to get worse as the k_1 factor of the imaging process decreases. Possible sources of focus variations are topography variations on a local as well as a wafer scale, as may be caused by CMP nonuniformities. Even though the sensitivity of CMP to local pattern densities can be reduced through the introduction of dummy fill, significant topography variations still exist that require a large enough depth of focus. Another source of focus variation is the exposure tool itself. For example, deficiencies of the scanner metrology used to determine wafer topography, thermal effects related to heating of the lens elements during exposure, and lens imperfections (focal plane deviations) also contribute to the focus budget [16]. Effective dose variations are due primarily to local reflectivity variations from the underlying substrate geometries. Even though antireflective coatings are commonly used, those coatings do not provide 100% decoupling from variations in reflectivity from the underlying patterns. Another common mechanism that contributes to effective dose variations is due to resist swing curve effects. Depending on the thickness of the resist, light reflected off the top and bottom resist surfaces will interfere either constructively or destructively. Thus, the effective light intensity within the resist becomes a function of resist thickness, which in turn may vary due to underlying topography.

These examples have focused primarily on macroscopic effects that contribute to lithographic variabilities. Similar to the random doping effect, microscopic variabilities are having an increasing impact on lithographic performance. The manifestation of microscopic variabilities in lithography is line-edge roughness. Lithography controls line width through light intensities and the subsequent chemical reactions that occur inside the photoresist. At the dimensions involved, the underlying stochastic nature of these processes is playing an increasingly significant role. It is to be expected that these effects will ultimately limit the achievable dimensional control in lithographic processes. Photon shot noise is one of the contributions to the statistical variation in the lithographic process. Using typical lithographic conditions at 193-nm wavelength, the number of photons per square nanometer is about 100. Therefore, the shot-noise-induced variation is about 10%. Further statistical effects are due to the random distribution of photo-acid generators within the resist and the random nature of the chemical amplification process. Including all these contributions, the 1σ variation in the photo-generated acid concentration in a 10 × 10 nm square reaches about 10%.

Overlay variations are the result of process or exposure tool issues. The magnitude of these variations is not typically dependent on the layout. An example of process-related sources of overlay degradation are deformed wafers. Wafer deformations may occur as a result of nonuniform wafer cooling after the spike anneals used for junction activation or due to the presence of strained layers on the front as well as the back side of the wafer. Other process-related sources of overlay degradation are asymmetric erosion of overlay metrology marks due to CMP and the metrology errors they introduce. Examples of sources of overlay error on the exposure tool side are warped wafer holders, nonflatness of the mirrors used on stage interferometers, and thermal drift due to uncontrolled heat sources within the exposure tool. Some of these issues can be controlled only to a limited extent, and thus design rule manuals provide numbers for expected overlay performances.

The phenomenological approach to variability generally leads to a fairly simple breakdown of the budget. Each of the six components receives equal weight; that is, a total CD control of 2.6 nm for the 65-nm node requires that each of the components not exceed 1 nm, a requirement quite close to atomic layer distances. Quite often, such tight budgets cannot be met. In particular, critical dimensions as a function of pitch, determined primarily by the model accuracy of the OPC model, fail to meet those requirements. Some insight into what can be accomplished may be gained from an article by Maharowala [17], who quantifies three contributions to the layout-driven device variability: gate orientation, topography effects, and pitch effects. In addition, across-field and across-wafer components are quantified. Quantitative information was collected using test structures that span one of the layout parameters while keeping the remaining parameters constant. For topography effects an essentially negligible contribution was achieved after lithography; the 3σ variability of this component after etching was approximately 2 nm. For the pitch-dependent component, after optimization of the OPC model, the systematic contribution to CD variation is essentially negligible; the random contribution is on the order of 2.5 nm (3σ). The low systematic component related to OPC is due primarily to the fact that critical gates were present only at two distinctive pitches. The value would be significantly higher if all pitches were considered. Similarly, small systematic contributions of less than 0.1 nm were reported for the horizontal/vertical offset. The across-field contribution to line-width variability was about 2.6 nm. Finally, the origin of the across-field component was analyzed further with respect to certain layout attributes, such as the x and y positions in the field and layout densities on different length scales. In addition, the pattern density effects resulting from the reticle and the wafer were separated. Reticle pattern densities can be assessed by detailed measurements of critical dimensions on the reticle. A summary of all the quantitative results from this study is given in Table 2.3. This detailed phenomenological analysis of CD variability blurs the boundary to the mechanistic analysis of CD

Table 2.3 Quantitative estimates of various phenomenological CD variability mechanisms

Mechanism	Contribution	Subcontributions	Comment
Orientation	0.2 nm systematic		
Topography	2 nm (3σ)	—	Post-etching
Pitch	0 nm systematic	—	Restricted to two pitches
Across field	2.6 nm (3σ)	26% x-position in reticle field (across-slit contribution) 4% y-position in field (along scan) 37% wafer pattern density 100 µm to 1 mm 25% reticle pattern density 100 µm to 1 mm 18% reticle pattern density < 100 µm 13% wafer pattern density > 1 mm	
Chip to chip	2.6 nm (3σ)		

Source: Based on ref. 17.

variability. The layout parameters are so specific that they narrow greatly the mechanisms that may cause these variabilities. Through this analysis, very specific process actions can be initiated.

An increasingly large effort within the lithography community is dedicated to characterizing the variabilities seen in the process. Efforts come from the exposure tool suppliers [18,19] as well as the process community [20–22]. Data on the CD budget of a 32-nm double-patterning process have been published by Dusa et al. [19] and provide some insight into the relative significance of the various mechanistic contributions to the CD budget. In the double-patterning process it is important to distinguish between features where both edges of the feature are defined by a single-patterning process as opposed to a process where one of the edges is defined by the first patterning step, and the second by the second patterning step (Figure 2.40). In this figure we first discuss the line-patterning case, the case where both edges are defined in a single patterning step. The breakdown of contributions is given in Table 2.4 based on a model of CD variability. The total variability of 4.8 nm exceeds the allotted budget. Key contributors are mask CD variations, estimated at about 3.6 nm, and etch/track contributions of about 2.8 nm. For comparison, raw experimental data are between 4.9 and 6.3 nm. The difference may be explained by an etch-driven offset between the first and second exposure steps. Even though this difference may be compensated by a corresponding adjustment

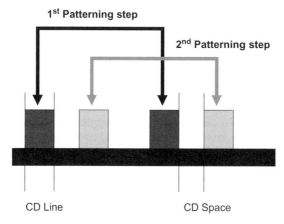

1st Patterning step

2nd Patterning step

CD Line   CD Space

Figure 2.40 Critical dimensions in a double-patterning process. The dark features are patterned in the first step; the light gray features are patterned in the second step. Note that we also assume an overlay shift between the first and second exposures. In this case critical dimension control for the line is determined by only one patterning step. Critical line-width control for the space is, however, determined by two patterning processes (one for each edge) and thus is sensitive to misalignment between the two exposures.

Table 2.4 Predicted CD variability

Component	Magnitude (3σ), CDU Model[a]
Total	4.8 nm
Metrology	0.5 nm
Etch/track process	2.8 nm
Mask CD	3.6 nm
Mask registration	NA
Overlay	NA
Dose	0.7 nm
Focus	0.6 nm

Source: Ref. 19.

[a]NA, not applicable.

in the second exposure, it is evident that an additional variability mechanism for double exposure processes is variability in the etching bias between the two patterning steps. The CD budget for the space features includes the overlay tolerances in a 1:1 fashion. It is worth noting that variations in the space between features are quite relevant, even at the gate level. For example, in SRAM memories a degradation in the space width control on the order of an overlay tolerance is significant with respect to contact to gate shorts and thus ultimately, the size of the SRAM cell. Dusa et al. [19] quote a figure of approximately 8 nm for the overlay contribution to CD control of the gate space.

2.3.2.4 Impact of Variabilities

Changes in critical dimensions affect such electrical parameters as line and contact resistivities, device leakage and drive currents, and thus related electrical characteristics such as signal delays. In this section we make an attempt to provide a link between the lithographic description of line-width variabilities and timing analysis. In a greatly simplified view, we assume that the delays are determined only by the L_{poly} dimensions, and those in turn are determined only by the lithographic effects. We use a simplified diagram that considers only two devices, devices 1 and 2 in Figure 2.41. In this chart we plot L_{poly} dimensions for devices 1 and 2 simultaneously. We are interested primarily in how the various variability mechanisms discussed above influence device tracking: that is, a correlated change in the two devices in Figure 2.41 as opposed to the uncorrelated behavior of the two devices. Charts such as Figure 2.41 may be obtained empirically through the corresponding measurements. The points shown in the charts indicate the results of such an experiment. The probability distribution for one of the devices is indicated schematically below the x-axis. As mentioned above, this distribution need not be Gaussian. To highlight this point, a strongly skewed distribution generally seen for distributions where focus errors are the main source of variability is plotted in Figure 2.41. For a particular device, this distribution is determined by the combination of all relevant mechanisms: lot-to-lot, wafer-to-wafer,

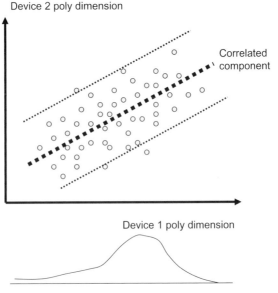

Figure 2.41 Hypothetical L_{poly} correlation chart for two devices.

within-wafer, across-field, and microscopic variations determine the shape of the distribution. Device 1 is located in a specific layout configuration. In our first case we assume that device 2 has the absolutely identical layout configuration from a lithographic perspective, including orientation. We note that identical in an optical sense means identical within the optical radius of several times the exposure wavelength. Also, as we will see below in the discussion on overlay (see also Section 2.2.4), it is important that the layout be identical not only at the gate level but also in the underlying active regions. As the two devices shown in the chart belong to the same chip, lot-to-lot, wafer-to-wafer, and chip-to-chip variabilities affect the two devices identically. Once again it is important to note that we neglect nonlithographic effects. This correlated behavior is indicated in the figure by a thick dotted line. If only those variability mechanisms were present, the two devices would vary in an entirely correlated manor. As a result, only the slow–slow and fast–fast corners would have to be considered from a timing perspective. Mechanisms that contribute to across-field variabilities as well as microscopic mechanisms such as line-edge roughness result in a spread of the L_{poly} dimensions around the correlation line in Figure 2.41. In addition in this picture, long-range effects such as optical flare or fogging on the reticle also contribute to the 3σ spread. The effect of all these mechanisms is to create a random distribution around the correlation line. This is indicated by two thin dotted lines, which are meant to represent the 3σ spread around the correlated behavior. The spread is a function of the distance between the two devices on the chip. The spread is smallest for devices in close proximity and increases up to distances of about 100 μm. At the smallest distances this correlation is limited by microscopic effects; in the lithography, primarily line-edge roughness. Line-edge roughness affects narrow devices of a width comparable to the spatial correlation length of line-edge roughness. Thus, the 3σ spread in L_{poly} dimensions of two devices will be smallest if they are wide compared to the high-frequency spatial frequencies of line-edge roughness, have an identical layout within the optical radius, and are in close proximity. The closeness of two devices is expressed by a correlation length that is determined by the mechanisms that control across-field variations, their relative magnitude, and their length scales. Almost all exposure tool contributions to the across-field variations change on relatively large length scales of several 100 μm. Thus the spread increases for distances up to about 100 μm and then remains roughly constant. There are, however, some mask and optical flare-related effects that vary on shorter length scales.

Next we discuss the case where the two devices use different layouts: for example, different pitches. Under these circumstances the two devices no longer have the same sensitivity factor for any of the CD variability mechanisms listed in Section 2.3.2.2. This, in conjunction with a situation where there is more than one mechanism contributing to the lot-to-lot, wafer-to-wafer, and chip-to-chip mechanisms, reduces the correlation and increases the seemingly

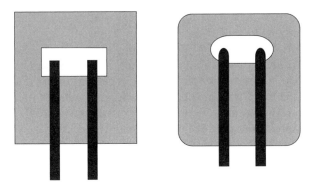

Figure 2.42 Layout scenario with increased overlay sensitivity.

random component in charts such as Figure 2.41. In addition, OPC errors or orientation-dependent variability mechanisms reduce device correlation, even for cases where they are placed in close proximity. Providing adequate information on the extent to which the devices are correlated requires significantly more detailed knowledge of the variability mechanisms. Simply determining the spread caused by such components as across-wafer variation is no longer sufficient.

Variations in electrical properties may not only be driven by variations in feature sizes. Quite often they are also determined by the intersection between features on two levels. One of the most prominent examples is that of transistors, where the channel is defined by the shape of the intersection between the active region and the poly. Another example is the lateral distance between contact and gate, affecting such properties as capacitive coupling, or if a stress nitride layer is used, the effective strain of the device. In the latter case it is difficult to avoid overlay-driven variations by means other than simply increasing the distance. In the first case the shape of the contours in the vicinity of the intersection plays a significant role in determining sensitivities. Changes in device lengths and/or widths occur only if the contours of the active and/or gate area are curved in the vicinity of the intersection. An example illustrating the changes in device characteristics due to overlay is shown in Figure 2.42. In this layout configuration two gate line ends have been terminated inside an active cutout. There is severe corner rounding at the inner corner of the active region. Misalignment of the gates in the x-direction will lead to increased device leakage as the gate end approaches the left or right edge of the cutout region. Cutout regions are also problematic for other reasons. Cutouts generally force the lithographer process engineer or OPC team to choose among three evils: There is generally little margin for error on the gate line ends, too much correction, and the line ends will touch the other end of the active area in the presence of misalignment. Too little line-end correction and there is little margin for overlay error, as significant gate leakage will set in once the

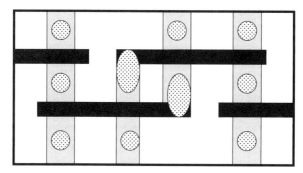

Figure 2.43 Schematic layout of an SRAM showing active region (gray), gate (black), and contacts (dotted).

rounded gate line end touches the border of the active area. The third alternative, creating a "bulbed" line end, leads to varying effective gate widths in response to a misalignment in the y-direction.

Not surprisingly, CD and overlay budgets play an important role in determining the layout of circuit components, most notably memory cells, where area efficiency is critical. A discussion of the lithographic sensitivity of SRAMs has been presented by Nackaerts et al. [23]. Figure 2.43 shows a schematic layout of an SRAM with active region, gate, and contacts. The height of the cell is determined by the minimum gate pitch for which a contact can be inserted (contacted pitch). The lower limit on the contact-to-gate space is determined by the onset of functional failures: namely, contact-to-gate shorts. CD tolerances for gate line width and contact size as well as the contact-to-gate overlay budget are the main culprits in limiting this distance. Besides these macroscopic contributions, microscopic fluctuations such as line-edge roughness on both the contact and gate need to be considered in order to determine the failure rate for contact to gate shorts. One of the challenges of connecting lithographic budgets with functional yields lies in accurately determining the failure rate at the part per million (ppm) level. Lithographic budgets are typically determined with data volumes of up to several thousand points. Thus, the probability distributions derived from such measurements are accurate down to probabilities of about 1 in 1000. As pointed out above, even at these levels the distributions typically do not follow a Gaussian distribution. Furthermore, for a detailed and reasonably accurate analysis of SRAM functional yield, the probability distributions need to be known much more accurately, as the yield relevant failure rates are on the order of 1 in a million. Only electrical metrology methods hold the promise of collecting a sufficiently large volume of data. So far, no such analysis is available in the literature, and a complete bottom-up approach to determining acceptable SRAM cell sizes from a lithographic budget perspective seems to be within reach, but the gap has not yet been closed.

2.3.2.5 *Strategies to Reduce Variabilities*

There are several strategies to reduce variabilities. One of the more obvious ones is to reduce the spread of a process parameter encountered in manufacturing. For example, some of the exposure tool–driven focus variations may be reduced by more frequent tool calibrations. The wafer contribution to the focus budget can also be reduced by putting more stringent control requirements on the CMP process and the deployment of more sophisticated dummy fill mechanisms. There is a tendency for such measures to increase the manufacturing costs. For example, more frequent calibrations and exposure tool adjustments increase the nonproductive time of the tool.

Another strategy is to reduce the sensitivity factor. This may be accomplished by avoiding layout scenarios with known high sensitivities. For example, the forbidden pitch rules (Figure 6.12) eliminate pitches with high sensitivity to focus variations. Also, lines at those pitches generally require large, and thus error-prone, OPC corrections. Another approach is to use more aggressive resolution enhancement techniques. Implementation of assist features is one example of reducing primarily focus-driven sensitivities. Optimizations of the lithographic process as outlined in Section 2.2.5 constitute another avenue for minimizing sensitivities.

One of the most efficient strategies for reducing variability is to eliminate systematic contributions to CD variabilities. There has been a strong effort from the exposure tool side to address systematic variabilities. Recent exposure tools offer the ability to correct systematic across-wafer and across-field variations. In both approaches the variations in CD are corrected for by an adjustment in the exposure dose. Figure 2.44 provides a schematic overview of the compensation scheme. In the uncorrected process the same exposure dose is applied to every field on the wafer. Nonuniformities of the patterning process, which include wafer exposure, resist processing on the track, and reactive ion etch, lead to a nonuniform distribution of chip average CDs. A hypothetical example with a common wafer center-to-edge signature is shown in the top half of the chart. These across-wafer distributions are measured and fed into a software tool that calculates an exposure dose correction for each exposure field. Besides the across-wafer CD maps, the dose sensitivity factor (i.e., the CD versus dose slope) of the feature to be corrected is required. The optimized process uses an exposure field–dependent dose correction to counteract the across-wafer variations in CDs created by subsequent processing steps. With the modified exposures, significantly improved across-wafer distributions are accomplished. In a similar fashion, some exposure tools also offer correction capabilities for across-field variations, primarily in the form of corrections parallel and perpendicular to the exposure tool scan direction. Among the practical complications of these methodology are that these corrections are product dependent. For example, etch-driven across-wafer variations depend on the global wafer loading and thus may vary from one product to another. Across-field signatures, if dominated by mask effects, may even vary

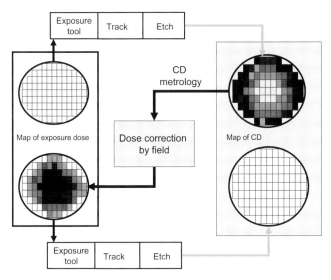

Figure 2.44 Dose compensation scheme to counteract systematic across-wafer variations. Non uniform across-wafer CD variations created by the patterning process (exposure tool, track, and etch processes) may be measured. A software tool determines exposure dose corrections for each field to counteract CD variations.

from one reticle to another. Under these conditions, each reticle requires its own setup and thus quite extensive metrology work. As the dose sensitivity factor depends on the layout (e.g., the pitch), not all features can be optimized simultaneously; rather, a particular configuration is chosen for the optimization process.

To address lot-to-lot variations, advanced feedforward control systems have been deployed successfully. They are used primarily for gate patterning, where the etch process has a resist trim component. Resist trim processes have an isotropic etch component that allows to reduce the resist CD. The basic principle is to adjust the trim time, based on measurements of the lot average CD after the lithographic step. Increased trim times result in a larger reduction of the resist CD and thus may be used to compensate for an above-target CD after the lithographic step. A key component of this feedforward loop is a CD metrology technique such as scatterometry that is not sensitive to resist line-edge roughness.

The physical layout is a key factor in modulating the variability of the circuit. We will discuss a few recommended layout practices that help reduce variabilities. Primary among those is to keep the layout as close as possible to a line-space pattern. One dimensional features in general have better process stability than line ends and corners. Further improvements can be accomplished by decreasing the variability in the layout: for example, by limiting the number of pitches used (discussed in Section 2.4 and Chapter 6). Strategies to minimize the impact of overlay on circuit performance are to avoid curved contours in the vicinity of gate–active intersections or at least to keep them as far away from the intersection as possible. Thus, corners as well as line ends should be kept as far away as possible from a gate–active intersection. Examples may be found in Chapter 6.

In summary, we have seen that various lithographic parameters contribute to variabilities in electrical characteristics. The main macroscopic parameters are overlay, mask CD variations, dose, and focus. The impact of those process variations is typically described by a sensitivity factor, which increases as the imaging process is pushed toward lower k_1 factors. In addition to macroscopic process variations, the microscopic variations of the lithographic process are becoming increasingly more significant. Line-edge roughness is primarily the result of such microscopic factors.

2.3.3 Catastrophic Failures

More extensive use of resolution enhancement techniques not only leads to improved process windows but carries with it an increased risk of catastrophic failure. In some cases, certain layout configurations prohibit the use of resolution enhancement techniques. The best known case of interactions between layout and resolution enhancement techniques are phase conflicts that occur when using an alternating phase-shift mask. Such a phase coloring conflict is shown in Figure 2.45. Coloring is a process whereby a unique phase, either 0 or 180°, is assigned to each opening on a mask. The phase coloring process is a sequential process. In our example we assume that it starts from the left-hand

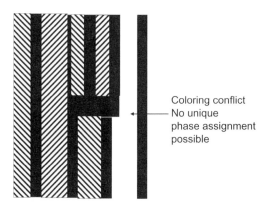

Coloring conflict
No unique
phase assignment
possible

Figure 2.45 Example of a layout with phase conflicts.

side, assigning one unique phase shape to the left side of the first line. It then proceeds to the right; the space between the first and second lines is filled with one unique phase of the opposite color. The coloring path then splits; one path proceeds along the top half of the structure, the other path proceeds along the bottom half of the structure. Key to the creation of the coloring conflict is splitting, followed by remergence of the coloring path and a difference in the number of 0/180 transitions in the two segments of the path. In the particular example there are an odd number of lines and thus an odd number of 0/180 transitions along the top path, whereas there are an even number of lines and thus an even number of 0/180 degree transitions on the bottom path, resulting in a phase conflict when the two paths merge. The inability to assign a unique phase combination to the rightmost line would result in catastrophic failures, as the dramatic reduction in process stability for a minimum line that cannot be colored is too large. Fortunately, those types of failures are rarely observed on wafers; most coloring programs have built-in checks to verify that proper phase assignment for all lines has been accomplished. However, resolving phase conflicts in real a design can be quite a tedious procedure. A correct-by-construction approach to phase-compliant layouts has been described in a U.S. patent [24]. This process requires a revision of the design process. Rather than drawing individual lines, the layout process uses an aggregate shape that consists of the actual line plus two satellite shapes that represent the two opposite phases. Phase-compliant designs are accomplished by enforcing relatively simple placement rules. For example, satellite shapes of the same phase are allowed to overlap, whereas satellite shapes of opposite phases are not allowed to overlap. Satellite shapes are no longer required if the line width exceeds a certain minimum width. Those basic rules are shown in Figure 2.46.

Alternating phase shift masks are not the only cause for catastrophic failures. Similarly, failure to place an assist feature may result in undesirable layout instabilities. Figure 2.47 shows a layout consisting of a horizontal line together with a set of vertical lines. The rules-based algorithm attempts to place assist features along the horizontal line. Due to the rapidly varying environment, which alternates between regions where either one or two assist features are being placed, a sequence of very short assists is generated. Additional rules that govern the minimum length of an assist feature in this case eliminate the entire assist coverage of the horizontal line, leaving it susceptible to strong line-width variations as a function of focus and potentially catastrophic failures.

As a result of the complexities of the algorithms involved, issues in properly executing OPC remain a source of catastrophic failures observed on wafers. Figure 2.48 provides an overview of some of the issues. In the early days of model-based OPC, occasionally the post-OPC layout contained portions where the original layout shapes had been lost. An example of such a failure is shown in Figure 2.48a. Most of these failures can be traced to issues with the hierarchy manager. Fortunately, such errors are now quite rare. Other

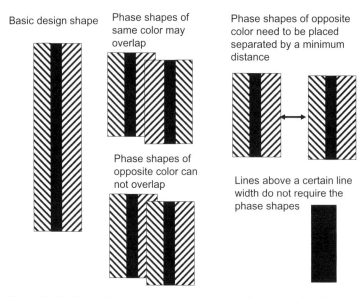

Figure 2.46   Correct by construction approach to phase-compliant layouts.

Figure 2.47   Elimination of assist feature placement.

possible failures (see Figure 2.48b) are related to improper segmentation. In this case the segment length of the hammerhead is too short. To compensate for line-end foreshortening, this short segment had to be moved quite far in the direction perpendicular to the main feature. The resulting mask shape is too sensitive to focus variation and has a strong tendency for pinch-off. Situations

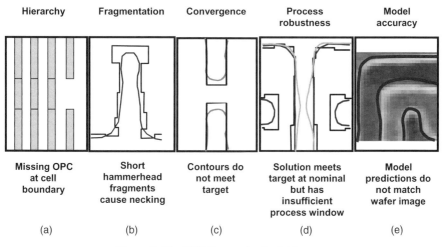

| Hierarchy | Fragmentation | Convergence | Process robustness | Model accuracy |

| Missing OPC at cell boundary | Short hammerhead fragments cause necking | Contours do not meet target | Solution meets target at nominal but has insufficient process window | Model predictions do not match wafer image |
| (a) | (b) | (c) | (d) | (e) |

Figure 2.48 OPC flow and potential issues.

where the contours do not match the expected target shape are another class of OPC errors. A possible cause is a premature termination of the optimization loop either by setting the maximum number of iterations too small or by constraining segment movement (see Figure 2.48c). Segment movement is typically constrained to ensure that the mask is mask rule compliant and, for example, does not violate minimum segment lengths or minimum feature-to-feature spacings. Such errors are not uncommon for line ends, where the large line-end foreshortening requires large segment movements. One of the more subtle OPC errors is shown in Figure 2.48d. In this case the target contours are met under nominal conditions, but the solution has a very poor process window that severely limits the yield on a given layout. Finally, one of the most difficult OPC errors to assess is a modeling error. These occur when the model predictions do not match the actual silicon results. Due to the large resist contributions to line-end foreshortening, line ends commonly exhibit these type of issues.

2.4 LITHOGRAPHY-DRIVEN DFM SOLUTIONS

2.4.1 Practical Boundary Conditions for DFM

For the economic reasons outlined in Section 2.1, there is generally no disagreement on the need to implement DFM. There are, however, practical aspects of DFM that need to be addressed. These issues are related to the following questions:

- What information should be transferred?
- What are the motivations and impediments of such an information transfer?
- Who is the recipient of this information?
- How is this information used in the design flow?
- What is the expected benefit of implementing a particular DFM solution expressed in as quantitative a manner as possible?

Wafer contours, representing the features as they appear on the wafer and their expected variability, are of primary interest for the upstream transfer from manufacturing to design. In the subwavelength domain the as-drawn rectangles are no longer a good representation of the printed shapes on the wafer. Furthermore, faced with increasing variabilities, only providing information on the nominal shape is no longer sufficient; rather, detailed insight into the range of variations that will be encountered during normal manufacturing is desirable.

Within the existing design flow there are several potential recipients of such information. Examples are lithography-based design checks identifying potentially troublesome two-dimensional geometries that cannot be captured with traditional tools used for design rule checks (DRCs). Those applications are a natural extension of the current DRC sign-off, replacing the more cumbersome attempts with which lithographers have attempted to incorporate lithographic boundary conditions into a design rule manual. Therefore, there is the desire to augment conventional DRCs with an additional design check for compatibility with the lithographic process. Naturally, this idea has been received with skepticism in the design community, due primarily to the fact that the burden (and delays) associated with correcting the additional failures is carried by the design teams. Fortunately, as we will see in the following chapters, the capabilities of these tools have been extended beyond those required for pure checking tools. Software tools for correcting failures without the intervention of a designer have been introduced. Such tools will greatly facilitate the acceptance of the new lithography checks, as they relieve the designer of the necessity to become a "lithography expert" in order to meet tapeout deadlines. Physical analysis tools such as parasitic extraction are another potential target. More realistic wafer shapes provide more realistic simulations, albeit at a significantly higher computational cost. Realistic representation of the expected printing of gates allows more detailed device simulations, which in turn enable more refined leakage and delay analysis. Knowledge of the variabilities expected also enable an estimate of the delay variations expected for ASIC cells. Optimization of logic IP for manufacturability, delay, and power consumption based on detailed wafer contour information and variability is one of the already existing real-world applications of lithography information.

There are also benefits for the downstream flow. OPC and mask manufacturing are primarily shape-handling operations. The most stringent requirements determine the effort spent on treating all shapes, including those that clearly do not warrant such an effort. This leads to excessive time spent on data preparation, unnecessarily large post-OPC data volumes and unnecessarily expensive masks. Primarily, what is required to improve the current situation is to convey, for each shape, a measure of its criticality, that is, the tolerances that need to be met to ensure chip functionality.

The expected benefit of implementing DFM is to provide a manufacturing advantage to the resulting design. Examples are reduced variability in chip performance as a result of reduced lithographic variabilities, reduced mask cost, improved yield, and fewer design failures that increase time to market.

As to the impediments of implementing a DFM flow that conveys manufacturing information, there is the general concern about loss of intellectual property and competitive advantage. In particular, foundries consider the details of their manufacturing, lithographic, and OPC approaches part of their competitive advantage. As a result, they are reluctant to provide information that would reveal anything about these processes. The files required to perform lithographic simulations contain a significant amount of detail regarding the lithographic processes and OPC. Primarily, these issues are being addressed through the use of decryption techniques to protect sensitive information. One of the more significant impediments to DFM is that the benefits expected are not expressed as a real dollar value. The standard DRC is at the transition between design and manufacturing. It, in fact, establishes a contract. The traditional viewpoint is that as long as the design team submits a DRC clean design, the foundry will fabricate it, at a certain cost per wafer, within a certain time and in a certain yield. In other words, a quite real and significant dollar value is attached to the DRC. Furthermore the responsibilities remain nicely separated between design and manufacturing. One of the issues of DFM is that with very few exceptions, it has not been possible to estimate (and commit) the expected monetary benefits upfront (e.g., as increased yield). Thus, it has not been possible to quantify the value of DFM. On the other hand, there is a real cost associated with DFM; in most cases the chip sizes increase. That, in turn, can be converted quite easily into lost revenue. As a result, the best strategies for DFM are to avoid affecting chip area or to establish, in a quantitative fashion, its impact on yield.

A common concern for DFM tools in general, but in particular for lithography-based tools, is that they are point tools. They are quite often poorly integrated into existing design flows and in some cases require quite a bit of lithographic knowledge on the part of the designer to interpret the results or take corrective action. Fortunately, there are efforts to remedy the situation. All the main vendors of design automation tools are integrating DFM tools into their flows and systems, like automatic hotspot fixing to reduce the lithographic knowledge burden on the design. Finally, lithography-based DFM tools face a challenge unique to them. Even though lithography can be one of the main contributors to yield degradations and variabilities of electrical performances, it is by far not the only contributor. For example, drive current variations of transistors are affected by variations in the physical gate length, and there are a variety of other sources. They include random doping variations, variations in gate oxide and spacer thickness, variations in implant activation due to layout-induced changes in thermal coupling, and temperature inhomogenities due to spatially varying power dissipation. Thus, focusing on improving lithographic performance for a chip design may not be sufficient to gain the desired improvements in yield. It has been difficult to bring together people with expertise in all of those areas to provide comprehensive and unbiased solutions.

2.4.2 Classical Approach

The classical approach has been to use design rules and the corresponding design automation tools for DRCs to establish the manufacturability of a design. There are generally two sets of rules: interlayer and intralayer. We discuss the latter first and return to intralayer rules later.

Overlay errors drive a variety of design rules. For example, the minimum contact-to-gate distance is controlled primarily by misalignment tolerances. Rules governing the extent of gate past-active extension and the size of the landing pad on a metal layer required to ensure a reliable connection between a via and metal are further examples of design rules that are a consequence of the overlay capabilities of the process.

Intralayer rules are written with relatively simple guidelines describing lengths, widths, and minimum areas of shapes. Some of the complexities of subwavelength imaging have entered the design rule manual as forbidden pitch rules, rules governing the minimum distance of line ends to neighboring shapes, and rules that attempt to prevent sidelobe printing issues. These are quite often complicated rules when expressed in the language of "geometrical" DRC: The DRC decks consider shapes and potential distances of shapes to the next neighbor. Unfortunately, this approach is, from a lithographical perspective, fundamentally insufficient, and increasingly so as we move deeper into the low-k_1 regime. Optical interactions are not local effects; rather, they take into account all features within a certain optical radius. The printability impact of a shape drops as a function of the distance from the center, a good estimate for the optical interaction radius is about 1 μm. Effectively, this means that the specific arrangement of shapes within a 1-µm radius determines if the shape in the center is printed adequately. A large variety of geometries exist within a 1-µm radius in advanced nodes. Some of those layout configurations will cause issues; others won't, and separating the two categories from each other is certainly beyond the capabilities of conventional DRC tools. Anticipating all possible layout scenarios, their interaction with resolution enhancement techniques and converting those to one-dimensional layout rules is simply not possible. As a result, failures may be observed even for layouts that have passed all the DRCs. Furthermore, the design rule manuals that result from such a procedure are not necessarily the most area-efficient ones. Lithographic process engineers, faced with the realities of current DRC decks, generally write very conservative rules that ensure design printability under the most adverse layout scenarios. Quite often, tighter design rules could be used if more details of the layout scenario were known—a fact that most SRAM designs use to achieve the desired cell sizes. Even though the SRAM cell is once again small compared to the optical radius, the surrounding of the cell is quite well defined by virtue of the regularity of the memory array. This principle is even enforced at the edge of the array. The dummy cells at the edges of memory arrays are a result of this mutual agreement whereby tighter geometries are implemented given a limited number of possible layout scenarios within the optical radius.

Another commonly misunderstood aspect of the design manual is that the rules are not digital. The layout will not transition from one that can be manufactured to one that will fail depending on whether the design passes or fails the design rules. Rather, the failure rate per instance is a continuous function of the critical dimensions. To convert this into the digital language of the design manual, assumptions need to be made regarding the frequency with which such features occur in the chip. The cutoff to this function is again set quite conservatively, as no a priori knowledge on the frequency of a particular layout scenario in a given design is available. Intelligent trade-offs between tighter geometries for a small number of instances cannot be made. In a design manual and the DRC decks that are derived from it, all those details are hidden from the end user. In general, conservative decisions have been made at the manufacturing site, due primarily to the limitations of current DRC tools, which can only consider nearest-neighbor relationships.

In an effort to introduce DFM within the current framework, design manuals quite often provide recommended rules. Those driven by lithographic considerations tend to list increased line or space widths where the process team believes better line-width control is possible or risk of failure is less likely. Use of these rules is at the designer's discretion Their primary benefit is to achieve more printable designs where there is no area trade-off. An example is: drawing isolated lines at a line width larger than the required minimum.

Another attempt at implementing a DFM methodology within the framework of current DRC decks are the *radical* (restricted) *design rules* (RDRs).

These provide an approach to enforce the regularity of the design from the bottom up: for example, by enforcing single orientation of patterns and enforcing a regular grid by limiting the number of pitches allowed [25]. An alternative of this approach is to enforce designs on a regular grid [26] The advantages demonstrated for these approaches are improved ACLV performance (up to threefold improvement has been reported) and significantly easier and more efficient implementation of RET approaches, in particular for alternating phase shift masks.

2.4.3 Printability Checkers

Lithographic simulation tools capable of computing the resist contours of small layout sections have been commercially available for quite some time. They arrived even before the advent of model-based optical proximity correction. Their primary purpose was to explore the impact of process modifications before committing to expensive experiments in the wafer fabrication line. These types of simulators are not discussed here; rather, we focus on simulation tools capable of providing simulations for an entire design. To gain the computational speed required some trade-offs, primarily on the accuracy of the optical models and the complexities of the models used to describe the nonoptical contributions to the resist images. Such nonhoptical contributions are due to the photoresist development process, photo-acid diffusion, and the chemical amplification process that characterizes modern resists. We return to these points in some more detail later in the chapter.

The development of full-chip optical simulation capabilities was a necessity for the implementation of model-based optical proximity correction capabilities. Not only did OPC require the ability to simulate wafer contours on a full-chip scale, early OPC systems were also prone to a large number of issues. For example, issues due to the design hierarchy management, the segmentation of polygons, or segment movement were not uncommon. Quite often those issues were only found when they caused catastrophic failures at wafer level. Therefore, the initial application of a full-chip simulation tool was simply to ensure that the complexity of the operations involved in creating a post-OPC GDS did not result in any catastrophic failures, as such printability checkers have greatly facilitated the development of robust model-based OPC software applications.

The core optical simulation engine at the heart of the initial full-field simulation engines were essentially identical to the ones used for OPC. There are two separate approaches for calculating images: dense and sparse simulation approaches [27] In dense simulations the image intensities are calculated on a regular grid, as shown in Figure 2.49; the smallest required size of this grid is determined by the optical bandwidth of the system. Dense simulators calculate the image in the frequency domain and then obtain the spatial image through an inverse Fourier transform.

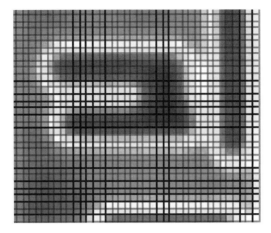

Figure 2.49   Example of image intensities calculated on a regular grid.

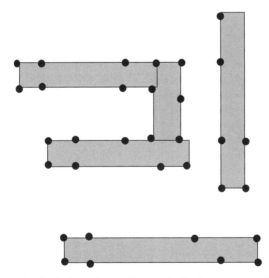

Figure 2.50   Example distribution of sampling points for image calculations in the sparse mode.

The sparse simulation approach is characterized by selecting a few evaluation sites on the existing design polygons. Image intensities are calculated at these sites only. This approach was instrumental in enabling the original full chip calculations as the number of evaluation points, and the number of geometrical shapes was relatively limited. The original sparse OPC approaches were used in the 0.18- and 0.13-μm generation. A typical distribution of sampling points adequate for those nodes is shown in Figure 2.50. However, as the technology nodes have shrunk, more stringent line-width control requirements have forced a denser distribution of sampling points, increasing the number of evaluation points faster than expected based on the geometrical shrink factor. A second factor also contributed to diminishing the original advantages of sparse simulation approaches. In the optical simulations the image is calculated by summing up the contributions from the mask shapes within the optical radius. Within this approach, further improvements in speed could be made by cutting the geometrical mask shapes into geometrical primitives such as trapezoids. In the original OPC implementations due to limited fragmentation, the number of trapezoids to be considered was fairly small. In advanced nodes, however, not only has the number of geometrical shapes within the optical simulation radius increased but also polygon fragmentation due to OPC. As a result, the number of geometrical primitives that need to be taken into account for calculation of the image intensities at a particular point has risen. Thus, the computational advantages of sparse simulations have diminished. The increase in computational effort required for dense simulations scales approximately with the density of the pattern, as it is only determined by the number of shapes within the optical diameter. The effort for sparse simulations, however, increases proportional to density^n, where n lies somewhere between 2 and 3. For example, the transition from the 65-nm node to the 45-nm node leads to a 50% increase in computational effort for dense simulations compared to a 200 to 400% increase for sparse simulations [28]. There is yet another advantage of dense simulations. Contrary to sparse simulations, they allow the detection of printing issues that occur in locations where no evaluation points have been placed; for example, they allow the detection of sidelobe printing, most commonly observed for contact masks.

Dense simulations have further benefited from the development of hardware-accelerated platforms which have led to dramatic improvements in computational speed [29]. Traditionally, optical simulations have been performed on general-purpose CPUs arranged in increasingly large clusters. In these systems the task of computing the full-chip contours is broken up into separate tasks, each node computing the results for a small domain, one domain at a time, using the general-purpose computing capabilities of the node CPU. In hardware-accelerated systems the computationally intensive tasks are handled by customized field-programmable gate arrays (FPGAs). A leaf, consisting of two CPUs and four FPGAs, offers a 10-fold improvement in computational speed over a conventional node with two CPUs. The pixel-based approach, representing both the mask and the resulting image on a regular grid, greatly facilitates the hardware acceleration.

As mentioned above, primary application of the lithographic verification tools has been to ensure the quality of the OPC output in avoiding catastrophic failures. For that purpose the image contours are further analyzed and categorized. Typical failure criteria shown in Figure 2.51 are CD errors,

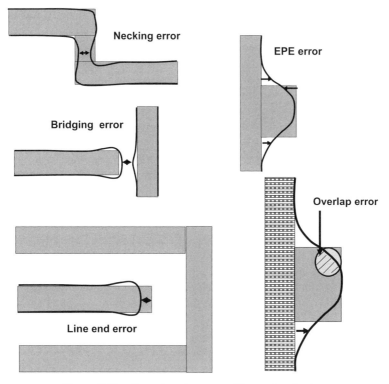

Figure 2.51   Examples of common lithography checks.

edge placement errors, line-end pullback, overlap errors, or missing/extra pattern detectors. CD errors involve determination of the distance between two adjacent image contours. Distances that are below a predetermined threshold are categorized either as necking (line width too small, leading potentially to electrical opens) or bridging errors (space between two adjacent features too small, leading potentially to shorts). Edge placement errors are those where the image contours deviate excessively from the desired target location. These errors do not necessarily cause a problem in themselves, as this error may not lead to an open or short. Primary concerns for those errors are either significant deviations from the desired electrical properties or the risk of shorts to features on the layer above or below. One such error is a line pullback where the line end, despite OPC corrections, does not reach the desired target, thus potentially risking contact opens by insufficient coverage of the via above or below. Quite a few systems allow special treatments for line ends, typically with more relaxed requirements relative to the case of straight lines. Overlap errors are primarily area checks, the enclosed area of the metal shape is intersected with the via, and the size of the remaining area is compared with the area of the fully covered via.

At the advanced nodes it has become apparent that simulations of the design under nominal conditions are no longer sufficient to guarantee adequate stability of the design under varying process condition. Simulations are now performed at a multitude of process conditions that generally represent the corners of the lithographic process. As outlined in Section 2.3.2.2, the semiconductor manufacturing process does not run under fixed lithographical conditions; rather, various parameters, such as dose and focus, are changed within certain ranges for a particular manufacturing process. The typical ranges for these parameters are captured in process budgets [30]. They provide estimated upper limits for the process variation seen in normal manufacturing. Those corners are determined by process engineers and are provided as input for the corresponding simulations. An example of printability simulations considering a variety of process conditions is shown in Figure 2.52. In this case, not only dose and focus variations but also mask CD variations and overlay errors are considered. The combined effect of these varia-

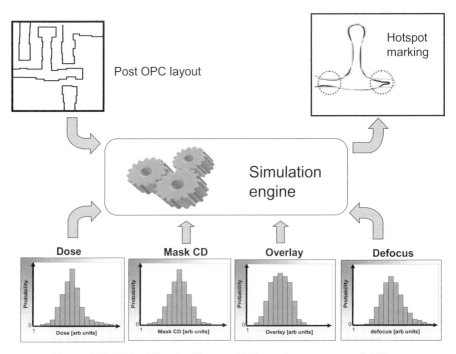

Figure 2.52 Printability checking considering various process variabilities.

tions is evaluated, providing a complete assessment of the printability of a design.

Despite OPC and RET technologies, printing in advanced technology nodes is characterized by significant deviations between the shapes as the appear in silicon relative to the rectangular shapes of the design. Realistic contours offer significantly enhanced accuracy for various extraction and analysis tasks. An application example is shown in Figure 2.53, where the impact of various OPC versions on SRAM stability is analyzed. The figure shows a specific SRAM layout for which two different OPC recipes have been evaluated. The dark contours represent the wafer contours of the active region. Notice the strong curvature of the active region. The gate contours reproduced by the two different OPC versions are shown as horizontally and vertically hatched regions. One of the solutions results in gate contours that remain fairly straight across the active region. Even more important, the gate contours remain

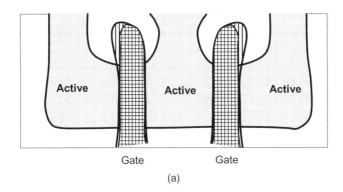

(a)

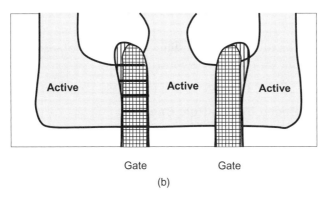

(b)

Figure 2.53   (a) Comparison of two gate OPC solutions for an SRAM. The active regions are shown in gray, the gate contours as a result of two different OPC solutions are shown as hatched regions. Note the differences in the line end shape for the two solutions. (b) Same SRAM cell including misalignment. The differences in gate contours under misalignment lead to drastically different device characteristics.

perpendicular to the active region for some distance away from the gate to active intersection. In the other case, OPC has been optimized to provide the largest possible gate past active extension while simultaneously avoiding the contact that is located between the ends of the two gate lines. The resulting excessive and one sided hammerhead leads to a widening of the gate lines just beyond the active region. The impact of the curved surfaces on the electrical characteristics becomes most prominent in the misaligned situation depicted in Figure 2.53b. In this figure the gate level is shifted primarily in the *y*-direction relative to the active region. For the purpose of estimating the electrical impact of this misalignment, the gate is divided into small sections. To a first approximation the performance of the entire gate is approximately equal to the combined performance of a set of transistors with smaller widths working in parallel. The effective width of these transistors is marked by lines in Figure 2.53b. Combining this information with a similar analysis for the pass gates, the change in the beta ratio of this cell can be evaluated under different lithographic process conditions. These lithographic corner conditions may include different dose, focus, and misalignment conditions.

One important aspect of these simulations is that under the different lithographic conditions, the variations of the six transistors in the SRAM cell are not independent from each other. A particular overlay shift will impose correlated changes on the transistors in the cell. With the proper layout, these correlations may be used to minimize beta ratio variations relative to the case where the individual transistor variations are considered independently. In the example the contours together with maximum misalignment numbers can be used to gauge the stability of the beta ratio for the two OPC versions: The OPC solution with the straighter lines also has a smaller range of beta ratio fluctuations. A more extensive discussion of these approaches is given by Balasinski [31], who has used the predicted wafer contours, with the channel region discretized as outlined above to estimate the *I–V* characteristics and their variation on process conditions for the pull-up, pull-down, and pass gates in an SRAM cell. These device characteristics are then combined with circuit simulations of the SRAM cell to estimate the signal-to-noise margin for the SRAM cell in the various process corners.

2.4.4 Model-Based Design Rule Checks

So far we have discussed applications based on lithographic simulations. One of the implicit assumptions has been that the post-OPC layout, including all the RET features, that has been added is available. Predicting wafer contours before the layout has actually undergone the full OPC run (i.e., in the design phase) is an entirely different challenge, one that is, however, significantly more applicable for true design for manufacturing solutions: Three categories of solutions have been developed.

One solution is for device manufacturers to provide a file to the design team capable of executing the entire OPC and RET flow. The confidentiality of the foundry process is maintained, as the files are typically encrypted. When simulations of the wafer contours are required, either to check for the occurrence of specific hotspots or to minimize the expected variabilities of a particularly sensitive circuit, the entire flow is mimicked. The system, invisible to the user, will execute the OPC and RET script on the particular layout. Once completed, it will perform the image simulations on the resulting mask image. In some cases the simulations are performed under different process conditions to provide insight into the lithography-driven variabilities. This process ensures the highest accuracy possible, as the process models have been calibrated at the foundry and the OPC and RET recipe fully matches the process used at the manufacturing site. This process is, however, computationally quite intensive. In advanced nodes, even using several hundred CPUs, full-chip run times for OPC recipes are on the order of days.

An entirely different approach was proposed by Gennari et al. [32]. This solution deploys a fast pattern-matching algorithm to identify problematic layout scenarios rather than use detailed simulations of the entire post-OPC output. Thus, it can be used to operate directly on the layout. The problem remains, however, how to obtain the problematic patterns that are used as templates for the matching algorithm. In its original implementation the pattern-matching algorithm was used to search for layout scenarios susceptible to lens aberrations of the exposure tool. For this application the problematic templates required can be derived from first principles. Fortunately, focus can be described as an aberration, and therefore a first-principles template for focus-sensitive patterns can be implemented. Other critical layout scenarios may have to be provided by the wafer manufacturer based on their experiences [33], thus effectively implementing a two-dimensional DRC tool. The tool has the essential characteristics of a DRC engine. The input is a design and a set of problematic layouts provided by the manufacturing team. The software will search through the layout. Locations within the layout where the matching score exceeds a certain threshold are flagged for further review.

One of the appealing aspects of this methodology is that for a variety of nonoptical effects, templates can be derived. Examples of such nonoptical processes are laser-assisted thermal processing and reflective notching. The matching algorithm itself borrows from the similarities of the problem to the optical simulation: The geometries present in the layout are divided into geometrical primitives, and the contributions from the various geometries are added together. Similar to the approach used for optical simulations, a significant amount of the computational effort can be done upfront by precalculating the convolution of the bitmap pattern with the geometric primitives.

A third approach has received significant attention. Similar to the pattern-matching approach, the solution starts from the layout. Using a black box model, generally described as a *nonlinear optical model*, the wafer contours can be calculated directly from the original design. The model has been foundry certified and is offered to clients by several foundries as part of a DFM package. According to Perry et al. [34], the software satisfies accuracy requirements and provides fast full-chip simulation capabilities. Unfortunately, none of the details of the model are publicly available.

The ability to check a design based on lithographic printing conditions has become an integral part of the DFM offering provided by most foundries or, for IDMs, has become part of the internal design flow. Commonly used names for these checkers are LRC (lithography rules checker), LCC (lithography compliance checker), and LFD (lithography-friendly design). Even though the checks may not be mandatory, they are useful by providing flags on the areas of primary concern for lithographic printability, which may require design modifications. These checks generally provide the designer with a database that marks areas of printability concern. Process simulations are typically executed under a variety of process conditions. Thus, the designer obtains contours not only under nominal conditions but also the extent to which the contours will vary based on the typical process condition variations encountered in manufacturing.

Removal of the hotspots encountered during an LCC inspection of a design is another matter, however. In general, the correction of lithographic hotspots is a time-consuming, quite often iterative process that requires highly skilled designers. If the design is synthesized using standard libraries and automatic place and route routines, the necessary layout resources may not be available at all. Recently, automated hotspot fixing systems, which attempt to resolve hotspots without human intervention, have been described [35–37]. The automatic hotspot fixing procedure starts with a simulation of the printability of the design, potentially under different process conditions. Similar to the case of printability checkers, hotspot locations are marked in the design and sorted into various categories: for example, necking or pinching defects. They are ranked based on their severity. The hotspot correction uses modification rules. The rules are specific for specific hotspot categories. Figure 2.54 shows a horizontal line sandwiched between two vertical line ends. We assume that a hotspot has been located on the horizontal line right between the two line ends. We use this example to illustrate some of the applicable layout modification rules. The three possible options to resolve the pinching defect marked with a circle are (1) to increase the line width L of the centerline and decrease the space width S simultaneously, (2) to increase the line width L and maintain a constant spacing S, or (3) to maintain the line width and decrease the space by pulling in the opposing line ends. Hotspot checkers are aware of design rules and will attempt to find a DRC clean solution to the problem. We assume that the distance from the centerline to the adjacent line end is already at the minimum. Thus, options (1) and (3) are not applicable in this case. Therefore, the only feasible solution is option 2. In this solution the line width of the centerline is increased and the line ends are pushed away by an equal amount.

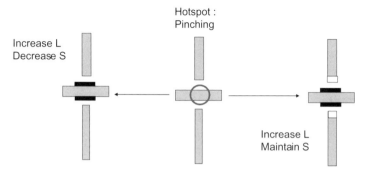

Figure 2.54 Sample modification rules for a hotspot fixing system.

Assuming that the vias located at those line ends can also be moved, a clean design rule solution is obtained and the algorithm stops. If the vias cannot be moved, no workable option exists. In this case the hotspot fixer is able to select the solution with the least detrimental impact based on a ranking of various design rules. For example, resolution of the pinching defect may be more of a concern than a slightly subminimum coverage of the via at the line end. The efficiency of the hotspot checkers developed and deployed by several companies is already quite impressive, with successful fix rates in excess of 70%. In some cases, success rates of 99.9% have been reported with quite acceptable run times.

Lithography rule checking, as an addition to DRC, provides highly valuable information with respect to the manufacturing capabilities. It is important to realize that this information can be used in more than one way. LRC should not only be viewed as a tool that provides restrictions in addition to the DRC manual, one should also realize that there are opportunities where designs may be allowed to follow tighter design rules based on LRC feedback. As mentioned earlier, the DRC design rules tend to be conservative in an effort to capture worst-case scenarios in a large space of potential configurations. The opportunities that lie in LRC checkers and work for the benefit of design teams are that in certain scenarios tighter designs may be possible. It is unlikely that this will affect minimum line or space dimensions. However, line-end extensions (e.g., for gate past active) depend heavily on what restrictions the layout environment poses on hammerhead formation. In these situations a LRC check may lead to more aggressive designs than would otherwise be allowed in the design manual. Other opportunities exist in providing information for gates that will exhibit tighter L_{poly} control than specified for the most general device in the design manual. Identifying opportunities for the design team in addition to flagging scenarios where there are lithographic concerns is likely to go a long way toward finding widespread acceptance for such checking tools.

2.4.5 ASIC Cell Optimizations

The availability of printability checkers and detailed lithographic information as it is primarily available within an IDM offer the opportunity to provide lithography-optimized ASIC libraries. The conceptual flow for this approach is shown in Figure 2.55 (see Kobayashi et al. [36]. Based on the design rule manual and the functional requirements, an initial physical layout is created for the library elements. This initial layout is then evaluated using the LCC, which executes data-preparation and OPC operations, performs a printability simulation on the post-OPC layout, and then searches for hotspots. If hotspots are found, they are supplied to the automatic hotspot fixer, which creates a modified layout. The modified layout is again sent through the LCC check to verify successful correction of the hotspots. Even though the particular flow shown uses a layout simulation approach whereby the actual OPC and RET are executed, this is not a necessity. Hotspot checking can be done with any of the tools mentioned in Section 2.4.4. One of the implications of the relatively large optical radii is that it is not sufficient to perform hotspot detection on a single library element in isolation. Rather, the cell in question needs to be embedded within a sufficiently large layout piece surrounded by other elements of the library. Such layout samples may be quite large, as ideally, they account for all the possible arrangements of the individual library elements.

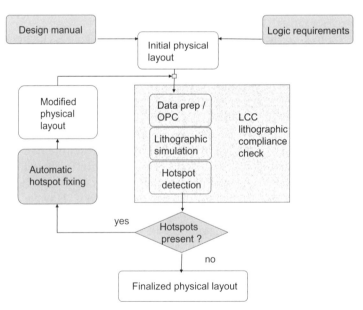

Figure 2.55 Conceptual flow incorporating lithography compliance check, hotspot detection, and automatic hotspot fixing in the library development (adapted from S. Kobayashi et al., *Proc. SPIE*, vol. 6521, paper 65210B, 2007.)

The creation of library elements is an example of quantitative information being available on the economic impacts of DFM optimized libraries [38]. To estimate the economic impacts, yield models are used. These models consider the effects of random defectivity as well as systematic, layout-driven yield effects. Printability effects and layout-environment-dependent via failure rates are only one of the few examples of systematic yield detractors. Exemplary lithography simulations are shown in Figure 2.56, where the contours at multiple process conditions are being evaluated. Both the lithographic and general yield models are calibrated for the actual manufacturing process using the results from test chips. The resulting yield models are quite accurate. Figure 2.57 provides a comparison of predicted and actual yields of several products on several different processes. The differences are on the order of a few percent. Such accurate models enable a detailed comparison of ASIC libraries with respect to area/yield trade-offs. An example of such yield characterization, in the form of generalized hotspots which include lithographic issues as well as other process failure mechanisms, is shown in Figure 2.58. Three cell variants are shown, ranging from a high-density variant with a significant number of hotspots, to an intermediate cell version, and finally, to a high-yield version with essentially no generalized hotspots. The latter carries a 50% area penalty, however. Thus, several yield variations of a specific logic function can be created, and the yield score of the cells can be used while creating the physical layout. In fact, the yield assessment can be made dynamic, reflecting the current or even predicted yield improvements as the manufacturing process matures to allow for the most effective trade-offs between the different cell variants. Thus, a yield-enabled ASIC library flow can be implemented based on the following elements:

- Creation of modified library elements for the same logic functions and drive strengths but different yield strengths

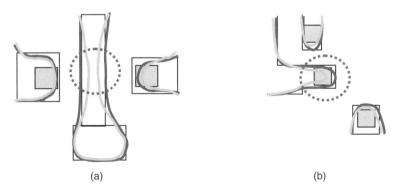

(a)   (b)

Figure 2.56 M1 lithographic simulations revealing (a) necking and (b) pull-back issues.

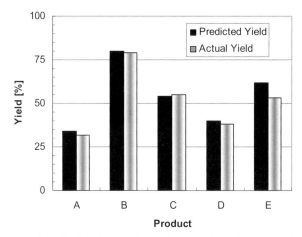

Figure 2.57 Application of a full-chip yield model: comparison of actual and predicted yields for five SoC designs.

- A data file that contains accurate information on yield attributes
- An extension of the physical synthesis tools that recognizes yield and incorporates it in the logic optimization cost function

The benefit of this methodology has been quantified, expressed as the percentage improvement of good die per wafer, a number that has a very direct relationship to cost. Example predictions are shown in Figure 2.59, where for five different designs the predicted improvement in good die per wafer is shown. Charts such as Figure 2.59 enable quantitative assessment of the financial trade-offs between improved yield and increased chip size.

The advantages of highly regular designs have been mentioned in Section 2.4.2 in the discussion of radically restricted design rules (see also Section 6.7.1.1). In this approach, regularity of the design is enforced on a local neighbor scale, allowing only certain spacings to occur in the design. This approach is amenable to the capabilities of current DRC tools. An alternative to the micro-regularity approach is to enforce design regularity from the top down, as is done in SRAMs and FPGAs. The top-down approach for regularity, effective over distances beyond the actual cell size, is accomplished by virtue of repetitive (i.e., array-like) placement of the unit elements. Thus, a regular pattern is created over distances comparable to the optical interaction length, in particular if the pitch of patterns inside the cell is an integer fraction of the cell dimensions. Design fabrics extend these two basic concepts, micro- and macro-regularity, into the ASIC design methodology. Thus, design fabrics combine the predictability and regularity of gate arrays with the customization

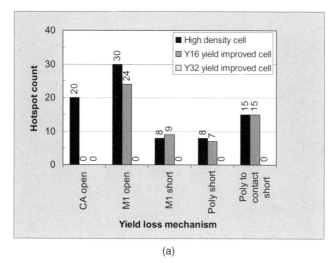

(a)

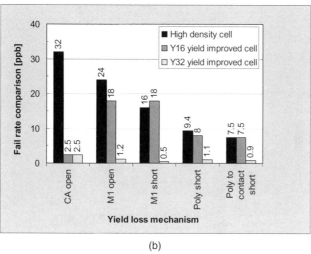

(b)

Figure 2.58   (a) Hotspot counts and (b) comparison of various cell variants regarding cell failure rates (in ppb) for various failure mechanisms.

of a standard cell library. An additional goal of design fabrics is to extend the concept of design regularity onto all mask levels. The elements of design fabrics pertaining to micro-regularity—for example, for poly, contact and first metal layer—are shown in Figure 2.60. All shapes are strictly implemented on a regular grid, poly lines oriented in one direction, first metal lines oriented in another, and contacts only at the intersection of two grid lines. A significant amount of attention has been put on evaluating the printing performance of

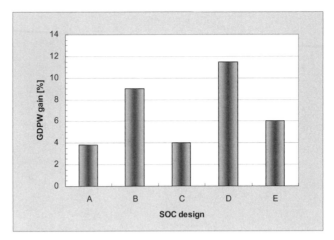

Figure 2.59 Measured percentage of good die per wafer (GDPW) improvement using a yield-aware methodology.

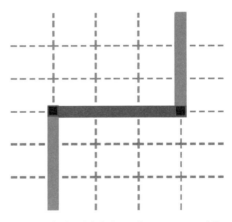

Figure 2.60 Regular design fabric for poly, contact, and first metal layout.

various line-end configurations, such as the ones shown in Figure 2.61. Three different line-end configurations, A, B, and C, have been evaluated for their printing performance using standard depth of focus versus exposure latitude plots, a chart that characterizes the sensitivity of the layout to process variations. Lithographic simulations and optimizations such as these ensure microregularity. Macro-regularity is accomplished through a novel design flow, consisting of brick discovery, physical brick implementation, and brick mapping, followed by placement and routing of the bricks. Bricks are the library elements that will be placed in a regular arrangement on a chip, ensuring macroregularity. The main difference to an SRAM is that not one but several bricks

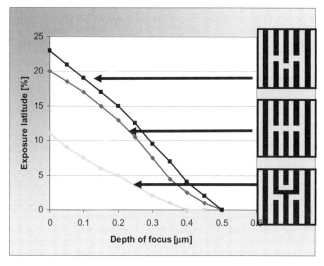

Figure 2.61   Line-end configurations and process window comparison.

are available, even though their number is much smaller than the number of library elements in a standard ASIC library. During brick discovery, commonly occurring combinations of logic functions are combined into a brick. Choosing the proper number of bricks that constitute the library is important in determining the right performance (e.g., timing) versus area trade-offs. Physical brick implementation is where the desired logic functions are converted into a physical layout using the principles of micro-regularity. Finally, the brick mapper converts the original netlist into a "bricklist." From a lithography perspective the regularity results in a dramatic reduction of the frequencies present in the Fourier spectrum. Even though the regularity of a memory array may not be accomplished, the Fourier spectrum of the brick pattern is dramatically simpler than that of a random ASIC library. This regularization in the Fourier domain enables more efficient use of resolution enhancement techniques. Design fabrics take advantage of the restricted layout configurations by applying more aggressive design rules than what is possible without the high degree of regularization. As a result, the area penalty for a ARM9 implementation using a 65-nm library is relatively modest, 7%. This should be compared, for example, with estimated ACLV benefits of about a factor of 1.8.

2.4.6   Lithography-Aware Routers

Lithography-aware routing is one of the more obvious steps in the ASIC design flow, where manufacturing knowledge, in particular printability information, can be incorporated in the design flow to generate correct-by-

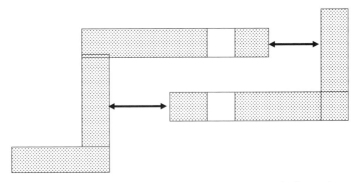

Figure 2.62   End-of-line spacing rules requiring increased distance of a line end to an adjacent feature.

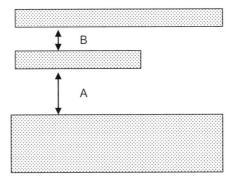

Figure 2.63   Line-width-dependent spacing rule.

construction designs. Routers have increasingly incorporated manufacturing knowledge by following certain rules that do provide for improved lithographic process windows.

Examples of such rules are end-of-line-rules, which require increased spacings of a line end to an adjacent line feature (Figure 2.62). These rules allow sufficient space for OPC to provide the necessary hammerheads and avoid regions of low image contrast, which are prone to bridging and thus shorts. Another implementation are line-width-dependent spacing rules, which require larger distances between features of larger width (Figure 2.63). These rules are also derived from lithographic considerations: The process window of a narrow space tends to decrease as a function of the line width of the adjacent features. A routing rule that eliminates a certain range of pitches above the minimum wiring rule is based on the forbidden pitch rules. The identification of problematic layout scenarios may be expanded using the pattern-matching approach described by Gennari et al. [32] to encompass more general geometries that cause lithographic problems.

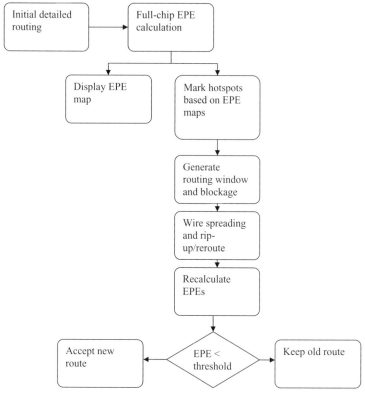

Figure 2.64   Radar flow.

The practical challenge of lithography-aware routing is, however, the fact that it requires a combination of two computationally intensive processes: the routing optimization as well as the execution of full-chip RET and OPC.

In one approach, entitled RADAR (RET-aware detailed routing) [39], the need for execution of RET and OPC on the routing layout is circumvented simply by using the initial edge placement error that occurs when the design is imaged as is, without OPC. As this is a noniterative process, the computational effort can be greatly reduced, making full-chip lithography-aware routing feasible. The EPE error is calculated using established techniques. Similar to the approach taken in OPC, a set of evaluation points is placed on the as-routed polygons for EPE calculations. For each evaluation point the contribution to the EPE error of each of the neighboring polygons can be quantified. The sorted list may be used to provide guidelines on how to modify the surrounding geometries. Details of the flow are shown in Figure 2.64.

In another case, a mixed rules- and model-based routing approach has been implemented [40]. The schematic routing flow is shown in Figure 2.65. In a first step, a rules-based engine is identifying potential weak spots in the routing

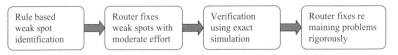

Figure 2.65   Conceptual flow of a lithography-aware router.

with relatively minor effort. Subsequently, those weak spots are fixed by rerouting with moderate effort. For the remaining weak spots that cannot be fixed in the previous step, a rigorous optical simulation is invoked, further reducing the number of weak spots that need to be addressed. The remaining weak spots are then resolved by the router with relatively extensive effort.

2.4.7   Advanced OPC Techniques for Improved Manufacturing

2.4.7.1   General Techniques

Most of the techniques described in this chapter are not strictly DFM techniques. Rather than transferring information from manufacturing to design, the techniques described in this section are used to remedy manufacturing issues primarily through a more extended use of design information. In the past, the patterning process has focused on reproducing all shapes in the layout with equal fidelity. From a design perspective, however, the control requirements for some shapes are more stringent than for others. For example, tight tolerances are necessary for widths of lines that form gates of transistors and ends of lines that must overlap contacts, but significant rounding of corners is often tolerable. OPC, at the gateway between the layout and the lithographic process, has significant control over the final pattern sensitivities (e.g., with respect to dose and focus variations) as well as mask manufacturing cost. In this section we introduce advanced OPC concepts related to the use of tolerances, modifications of the target patterns, and the use of process-variation-aware OPC corrections. We also cover the possible benefits for mask manufacturing. And discuss inverse lithography as a new approach to solving the OPC problem.

Within the OPC process there are several natural anchor points that allow more efficient use of design information. The first is the OPC target (i.e., the shape that OPC is ultimately attempting to accomplish). The second is information on how accurately the target needs to be reproduced (i.e., the tolerance with which the final shapes need to be reproduced). A related aspect is control over the segmentation; finer segmentation will result in a closer reproduction of the target shape on the wafer. Finally, means to reduce pattern variability have been incorporated into the OPC flow.

It is a little known fact within the design community that the data preparation flows in advanced nodes quite commonly include steps that actually change the desired dimensions of shapes on the wafer relative to the design. One of the simpler steps is the correction of targets in polygon corners. For

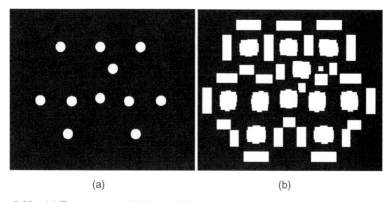

(a)   (b)

Figure 2.66   (a) Target pattern (b) inverse lithography solution for a set of contacts. The solution was constrained to reduce mask cost by simplifying polygons not corresponding to the target features.

convenience, layouts usually consist of polygons with right angles, but minimizing the radius of curvature at corners is usually less important than matching the desired contours elsewhere. The bandlimited nature of optical imaging does not impose strict limits on minimum corner radii. However, trying to force the solution to converge in these circumstances not only wastes computing cycles with no improvement in quality but generally also produces solutions that exhibit poor stability against dose and focus variations. One method to address this is to round corners of pattern layouts to create more suitable targets. Figure 2.66 shows an example of this practice, in which circular targets are derived from contacts drawn as squares. More severe modifications to the target are common in wiring layers. For example, isolated metal lines which generally suffer from pure lithographic process windows are widened (i.e., the OPC target is to print a design line that is wider than the line as designed). This correction will greatly improve the process window of these isolated lines; that is, it will reduce the risk of creating open metal lines. Common reasoning within the lithographic process community is that widening of an isolated line should not have any negative impact on the design. However, the size of these corrections may not be small. In the 45-nm node, with minimum metal line widths on the order of 70 nm, target adjustments on the order of 40 nm are not uncommon. That is, a line that was drawn at 70 nm will appear as a 110-nm-wide line on the wafer, with some impact on the overall resistivity of these lines and thus potentially signal delays. Other common target modifications are applied to dense features (i.e., those structures that are close to the minimum pitch). In these cases the targets are quite commonly modified to achieve an equal-line-space pattern on the wafer, as these patterns generally have the best lithographic process window. Target modifications have also been introduced in the context of interlayer-aware OPC corrections [41]. A possible goal of this correction is to improve contact coverage by extending the metal contours in the presence of a contact or via and thus providing better stability of the design to overlay variations. It is difficult for the designer to judge the extent to which these corrections have been incorporated in the OPC flow. One possible solution is to ask the OPC team to provide the target layout, usually an intermediate layer generated during OPC flow. Alternatively, one may ask for simulated wafer contours. Even so, the interpretation of this information will require expert advice, as other corrections may have been included in the target generation.

Minimizing the mean of absolute pattern errors at every point does not necessarily yield the best solution not only from a design perspective but even from a lithographic perspective. Tolerance is a measure of the extent to which the OPC algorithm will attempt to meet the target shape. Tolerance may be introduced into the cost function, or it may be used in the stop criteria, which terminate the execution of OPC iterations. As mentioned earlier, in OPC the deviations between the desired contours and the simulated wafer contour are defined for a set of evaluation points. Thus, at least in principle, tolerance can be defined on a point-by-point basis within the design. In practice, algorithms need to be defined that generate the tolerances. One such algorithm is to recognize regions such as ends of lines and corners and assign more relaxed tolerances to those. Taking this approach a step further with Boolean operations on multiple layers, we can identify transistor gates and overlaps with contacts for special treatment. This is one of the most common applications. Tighter tolerance requirements are imposed on the transistor relative to the poly wiring. Thus, the more critical CD control required for gates can be implemented without excessively affecting, for example, computational time. Further refinements of this approach are also feasible whereby gates located in the critical path may receive even tighter tolerances. The key is to be able to recognize such gates at OPC run time. Extensive analysis of the design at this point is not possible; rather, the information has to be transferred to OPC most easily through the use of marker layers. Other elements of design intent that would be useful in the optimization process are not easily inferred from the target patterns. For example, matching a pair of transistors in a sense amplifier may be more important than getting their absolute dimensions on target. Again, designers can create explicit marker layers to pass this type of information on to the OPC algorithm. Conventional OPC performs corrections to the layout only at nominal exposure conditions (i.e., best dose and focus). In reality, the patterns have to be stable through a range of these parameters given by the corresponding process budgets. Solutions with better process stability can be obtained by performing the simulations at multiple exposure conditions and checking each condition against a set of tolerance requirements [42]. If one of the conditions fails to meet the tolerance requirements, a conflict resolution algorithm is applied. In general, it will attempt to resolve the process window violation by relaxing the tolerance requirements under nominal conditions. Other approaches to include process window considerations in the OPC algorithm are to add a term that measures the amount of contour movement under varying process conditions in the cost function

[43]. It is again important to note that in most cases, process window improvement comes at the price of relaxing the tolerances for the nominal exposure conditions. Performing lithographic simulation at multiple exposure conditions is, however, computationally expensive. A solution has been described by Peny Yu et al. [42]. The reduction in computational cost is accomplished primarily by using an analytical method to evaluate the image at various exposure conditions rather than recomputing the image for each condition separately.

The ability to improve the process stability of a pattern by modifying existing shapes is limited. Significantly more efficient improvements can be made by adding additional shapes, such as subresolution assist features or phase shapes, to the reticle. Conventionally, this is done in a rule-based manner in a step that typically precedes execution of the OPC algorithm. Recently, however, there have been multiple attempts to generate these features in a model-based fashion [44–46]. All these approaches require a significant deviation from the conventional OPC approach. The mask is now considered in a pixelized version, and the intensities and phases of these pixels may be changed in an attempt to optimize new cost functions. Since one is primarily concerned with the creation of subresolution assist features, the pixelized approach is not necessarily detrimental. The complicated shapes generated through these algorithms may be converted to shapes that comply with mask manufacturing requirements. For example, in a technique called *interference mapping technique* [44] the optimization goal is to maximize the intensity at the center of a contact hole. This is roughly equivalent to mapping out mask locations where the light from these locations interferes constructively with the center contact hole. Locations that create destructive interference may be altered by changing the mask phase, thus converting destructive interference into constructive interference. A more obvious cost function proposed by Zach et al. [46] attempts to minimize contour variations in response to variations in the exposure conditions. The concept is general enough that it can be applied to a large variety of nonoptimal exposure conditions such as dose and focus, also lens aberrations and changes in resist diffusion length can be included. It is also shown that the mask patterns which suppress the impact of a certain process variation in the most efficient way will depend on the nature of the process variation. Inverse lithography, described in the next section, is the most comprehensive approach to these problems.

2.4.7.2 Inverse Lithography

The approach for optical proximity correction described so far has been tightly bound to the original polygon representation. In an effort to optimize and refine each element of the lithographic infrastructure as much as possible, drastically different solutions have been developed. Within the domain of optical proximity corrections and RET placement, inverse lithography has been developed.

Superficially, the inverse imaging problem in microlithography resembles phase retrieval or reconstruction of an object from its image [47,48]. In the reconstruction problem, we start with a two-dimensional representation of the entire image, such as would be obtained from a sensor array or a densitometer trace of a photographic emulsion. In the microlithography problem [49], we start with a desired set of contours of patterns in photoresist or an etched film, which contains far less information than does the entire image. Due to the high contrast of photoresist, many images can produce identical contours as long as they are dark enough and bright enough in the interior and exterior regions defined by the contours. The additional degrees of freedom in the bright and dark regions allow us to optimize more than just the desired contours. Contours obtained with different masks may be identical under nominal conditions but exhibit different sensitivity to errors in exposure time, focus, or sizes of mask patterns.

Range of Inverse Lithographic Solutions   Solution of the inverse lithographic problem is generally limited to optimizing the mask transmission function, or a combination of mask transmission function and illuminator. There are essentially three approaches to generating the mask transmission function. In a first class of solutions, mask features may be generated on a pixelized basis with pixels of equal size and below the resolution limit of the exposure tool. A second class of solutions to the inverse problem divides masks into larger regions with different transmission properties. Finally, the most restrictive class of inverse lithographic solutions consists of perturbations of target patterns. Edges of target polygons are divided into segments that are moved perpendicular to the original edges to minimize deviations of resist patterns from their targets. Solutions corresponding to the third class show the most resemblance to conventional OPC results. In particular, if the solution is restricted to prevent merging disjoint polygons, there is a one-to-one correspondence between polygons in the inverse solution and the target patterns, similar to conventional OPC results. The additional freedom inherent in the first two classes of solutions allows for the most unique post-OPC results. Strongly curved shapes and the ability to generate, in an integral fashion, subresolution assist features or phase shapes that did not exist in the layout prior to OPC are one of its differentiating features. Figures 2.67 to 2.71 provide an impression of the range of solutions that can be obtained. Sharp corners in the target patterns produce some characteristics of the solutions, which are particularly evident in Figures 2.68 and 2.69. It is impossible to obtain perfectly square corners because the numerical aperture of the projection optics limits the maximum spatial frequency of the image. However, it can be shown that the best approximation to a small, square target is the image of an ×-shaped object [50]. Contours in Figure 2.68 near exterior corners of the target are stretched out at 45° until they merge with those corresponding to adjacent target rectangles. Near interior corners, contours are pulled in at 45°, so that in some cases, regions corresponding to a single target polygon

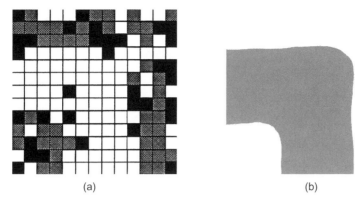

(a)   (b)

Figure 2.67 (a) Pixel mask and (b) calculated pattern contour for a simple elbow target. The mask uses three pixel classes: opaque, unit transmission, and unit transmission with the phase shifted 180° (From ref. 7.)

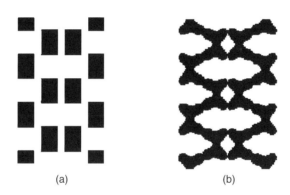

(a)   (b)

Figure 2.68 (a) Target pattern, (b) inverse lithography solution.

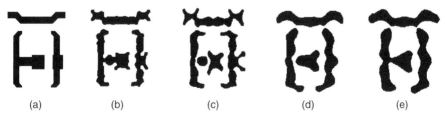

(a)   (b)   (c)   (d)   (e)

Figure 2.69 (a) Target pattern and (b–e) inverse lithography solutions scaled to different process critical dimensions. The patterns are scaled back to similar dimensions to make it easier to compare differences in shapes of the solutions.

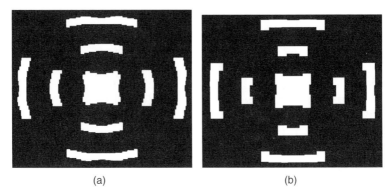

(a)   (b)

Figure 2.70 Inverse lithographic solutions for a simple square pattern corresponding to the features at the center. The other features improve depth of focus and contrast of the central feature, but do not print. Solution (b) was generated with larger rectilinear segments than solution (a), to reduce mask cost.

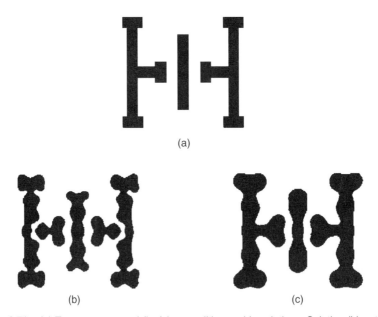

Figure 2.71 (a) Target pattern and (b,c) inverse lithographic solutions. Solution (b) optimizes pattern accuracy for nominal exposure conditions. Solution (c) exhibits less variation with small dose changes.

have been divided in two. In all cases, the resist patterns obtained using inverse lithography solutions approximate the target patterns better in their corner regions than do those corresponding to masks patterned to resemble the targets. Figure 2.70 shows inverse lithography solutions that differ from target patterns for other reasons. In this case, the target is a single small

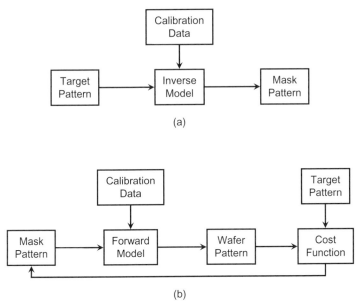

Figure 2.72 (a) Conceptual and (b) practical flows to solve the inverse lithographic problem.

square, but the objective of the inverse solution is to improve the contrast and depth of focus of the image rather than to sharpen corners. The regions surrounding the central polygon are responsible for this improvement, as described below.

Not surprisingly, one of the biggest hurdles for the practical implementation of such solutions are mask manufacturing constraints. Pushed to its limits to pattern subresolution pixels, the mask manufacturing process rounds corners and induces variation in the slopes of edges of pixels. Such manufacturing variations alter the optical properties of the mask and introduce variable errors. Cost and defects associated with mask manufacturing limit the number of available attenuations and phases to about two or three.

Methods to Solve the Inverse Lithographic Problem   If it were possible to construct an operator based on an inverse model that captures all the pertinent physics, we could solve the problem as outlined in Figure 2.72a, transforming target patterns directly to obtain inverse solutions. Although this serves to describe the concept, even the simplest inverse microlithographic problem is too complicated to submit to such a direct approach. Practical inverse solutions rely on iterative perturbation of the forward imaging problem, as shown in Figure 2.72b. Starting with an initial guess at a suitable mask pattern, we use a forward model to calculate the patterns it should produce in photoresist or an underlying etched film. We then compare the calculated patterns with their targets and determine how to perturb the mask to minimize a cost function that measures deviation from the targets. Contrary to the approach in standard OPC, the mask may be optimized in a pixelized fashion, with phase and transmission of the pixel being the variables used in the optimization process rather than the segment movement in more conventional OPC. We repeat this procedure using modified mask patterns until we have convergence. Obtaining desired results depends on many factors—the model, desired target, and cost function being three of the most important, similar to those of standard OPC approaches. Models and the topic of target generation were discussed earlier in the general context of advanced OPC approaches.

Capturing other benefits of inverse lithography requires a more complicated treatment of the cost function. If it is desired to improve contrast, depth of focus, or sensitivity to mask pattern errors, it is necessary to minimize the cost function and its derivatives with respect to an appropriate set of variables. A convenient way to do this is to take finite difference approximations by computing the cost function under nominal conditions, and with perturbations of dose, focus offset, and mask dimensions. Assigning weights to terms in the cost function that correspond to off-nominal process conditions permits control of the inverse solution. Figure 2.71 shows how a particular solution changes when the cost function contains off-nominal terms. It is also necessary that the cost function consider locations in the image away from desired pattern contours to ensure that no undesired patterns print. Inverse lithographic solutions such as those in Figure 2.70 are particularly susceptible to this problem. The outer polygons serve to improve contrast and depth of focus for the central target pattern, but will print if they are too large. To guard against this, the cost function must contain terms that cover each point of the image, checking that the intensity is above safe limits in bright regions and below them in dark ones.

Choice of a data structure to represent mask regions during iterative solution is another important consideration. A straightforward representation is a set of polygons defined by lists of vertex coordinates. Another representation is referred to as a level-set function [51]. These offer natural access to a richer set of solutions than the polygonal representation, because initial regions can split into multiples, and new regions can appear that do not correspond to the starting patterns. Figures 2.68 to 2.71 show some examples of these phenomena.

Due to the cylindrical symmetry of projection optics, unconstrained inverse lithographic solutions generally take the form of continuous curved contours, even if target patterns are rectilinear. Data formats and equipment used to generate patterns on masks are designed for rectilinear patterns. It is possible to fracture curvilinear patterns into many small rectangles to make masks, but manufacturing cost can be prohibitive. To deal with this issue, it is necessary to impose a rectilinear approximation on the inverse solutions. Typically, the coarser the rectilinear approximation to the naturally curvilinear solutions, the more the lithographic performance of the pattern is compromised. Figures 2.68, 2.69, and 2.71 use very fine rectangular grids to show details of inverse solutions, but these would be expensive to manufacture. In general, coarser rectilinear approximations require larger departures from ideal solutions, so it is important to be able to adjust this to obtain the desired compromise between lithographic performance and mask cost. The patterns in Figures 2.66 and 2.70 show inverse solutions appropriate to less expensive masks.

2.4.7.3 *Advanced OPC Methods for Improved Mask Costs*

In Section 2.2.2 we discussed the primary dependence of mask writing time on shot counts. Figure 2.73 provides insight into the relationship between modern resolution enhancement techniques such as model-based optical proximity correction and the shot count. In this simple example, design shapes, consisting of several rectangles and a polygon, are broken up into rectangles, the unit shape that the mask writer can expose in a single flash. The number of shots is only slightly larger than the number of design shapes, the difference in this example being driven primarily by the fact that the polygon is broken down into rectangles. With the introduction of OPC, primarily with the advent of model-based OPC, a drastic change has occurred. Due to the segmentation that occurs during OPC (see Figure 2.73b), the design shapes are converted to polygons with a large number of polygon edges (for more details, see Section 2.2.7). These segments are subsequently moved in an effort to reproduce an on-wafer contour that most closely resembles the original polygons. As a result, even a simple rectangle has now been broken up into a polygon with a large number of edges. To reproduce these jogs adequately, the mask writer converts the polygon into many smaller units, with a resulting dramatic increase in shot count (see Figure 2.73c). Further increases occur if breaks in the polygons are not aligned across a shape. This is illustrated by a comparison of the two polygons shown in Figure 2.73d. Both polygons have eight segments. In one case, the breakpoints are aligned across the shape (top of Figure 2.73d), and the reproduction of this polygon on the mask requires three shots. A drastic increase in shot count occurs if the breakpoints are not aligned across the shape. In this case the same number of segments leads to five shots. Also, shot counts are increased due to the fact that modern resolution enhancement such as subresolution assist features add shapes to the design outside the preexisting polygons. As we have seen, the increase in shot count is essentially driven by polygon segmentation during the OPC process, the alignment of segments across a shape, and the addition of features not originally present on the reticle.

There are two strategies to minimize shot counts:

1. Segment removal during the OPC process (in this case the process starts from a relatively aggressive segmentation)
2. More intelligent segment generation

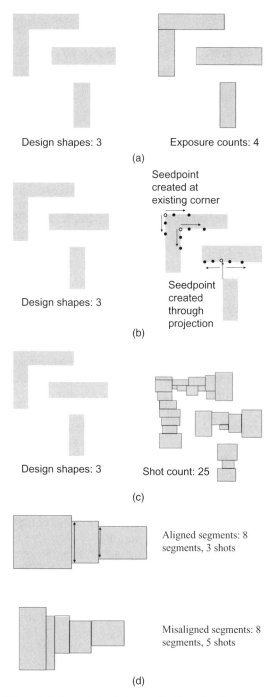

Figure 2.73 Evolution of shot count due to the application of optical proximity corrections: (a) reproduction of design shapes as shots (a modest increase in shot count is observed primarily as a result of the decomposition of polygons into rectangles); (b) polygon segmentation example using seedpoints and satellite points; (c) shot count increase due to polygon segmentation during OPC; (d) shot count comparison for aligned various misaligned segments

An example of the first strategy has been given by Yehia et al. [52]. Strategy 1 can be described loosely as a jog reduction approach. Initially, the segmentation is set to meet the most aggressive accuracy requirements. The initial OPC iterations use all available segments. Based on the segment movements during the first few iterations, segment priorities are assigned. A high priority is assigned to segments whose movement is significantly different from that of neighboring segments. The large difference is a good indication that removing this segment will result in a significant deterioration of OPC quality. Low-priority segments essentially move in unison with their neighboring segments (i.e., there are only small jog heights). At the intermediate iteration step at which the evaluation is made, small differences in segment position are not sufficient to indicate that the final solution will retain the small jog heights. Therefore, an additional requirement for low-priority segments is a small difference in edge placement error. Neighboring segments with small differences in their segment movement (i.e., small differences in jog height) and small differences in EPE error are most likely to retain a final OPC solution with only small differences in segment position. Thus, at some point the algorithm switches from moving all segments independently to a mode where low-priority segments are moved jointly based on their average EPE error. High-priority segments continue to be moved independently. In this approach the information necessary to eliminate segments is calculated during OPC execution, and thus no upfront segment reduction is possible. The fact that the merging is done during the OPC iterations make it possible to compensate for small OPC errors that are a result of the merging of adjacent segments. Studies of the efficiency of this approach show that a reduction in shot count of up to 20% can be achieved with relatively minor degradation of the overall EPE distribution (i.e., the overall OPC accuracy).

In an alternative strategy, segments are generated in a model-based fashion rather than the more standard rule-based approaches. The advantage of the model-based approach is that segments are generated on an as needed basis, and thus the minimum number of segments to reach a certain accuracy goal is generated. One much solution has been described by Sezginer et al. [43]. The algorithm by which segments are generated is depicted schematically in Figure 2.74, which shows part of a polygon and the corresponding evaluation points. The points are placed significantly denser than would be required for a standard OPC correction. As in previous discussions of OPC, a cost function is used that depends on deviations between the desired and actual contours. The derivative of the cost function with respect to segment movements indicates the amount and direction of segment movement to minimize the cost function and thus achieve printed contours that better match the target contours desired. Figure 2.74 shows a hypothetical example; vectors attached to the evaluation points indicate the necessary movements. The orientation of these vectors may change from an outward movement to an inward movement along an existing segment. The crossover point from negative to positive direction provides the optimum location for introducing a new segment, and the

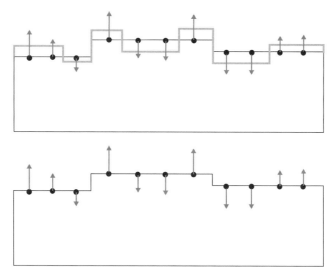

Figure 2.74 Schematic depiction of the algorithm for generating segments.

existing segment is split up at the crossover point. Based on this algorithm, executed in an iterative fashion, an optimized segmentation can be generated. Segment generation occurs only when needed, and thus excessive segmentation can be avoided.

Other approaches to limiting polygon segmentation to the necessary minimum are based primarily on design intent. They are part of a more general concept referred to as *design-aware OPC* [53,54]. The overall goal is to limit post-OPC data sizes, fracture counts, and OPC run times based on design intent. Two implementations of this concept have been described in the literature. One of them modifies the tolerance requirements for OPC based on a timing and power analysis of the circuit. In this case, additional design information is made available for optimizing the manufacturing process. A more detailed description of this approach is given below. The other implementation is a true DFM case in the sense that shot count information is transferred into the design domain through the use of empirical models.

In the context of OPC, tolerance represents the extent to which actual contours may deviate from the contour desired. Tolerance information may be used to control the density of segments. It also allows control over OPC run times by reducing the number of OPC iterations. As one example we describe how the timing impact of various gate- and metal-level variations is converted to tolerance information usable in OPC. The process starts with the creation of several layout versions of the same library cell. In essence, these represent the worst-case wafer contours that would occur as a result of the planned relaxed tolerances for the OPC algorithm. The different versions are created by adding to the drawn width of the poly a bias representative of the planned relaxed tolerance. The biasing may be applied in locally nonuniform fashion that mimics an under- or overexposed image. In addition, different biases can be applied depending on whether or not the poly shape represents an active gate or simply a poly connection. Larger biases may be applied to poly connections than to active gates. Finally, the magnitude of the biasing may be varied as well. The resulting cells are evaluated for their timing behavior using extensive SPICE modeling to create timing tables as a function of input slew rate and output capacitance. Based on these characteristics the cells may be used during synthesis and provide input for static timing analysis of the final circuit. In OPC the varying tolerances allow different treatments for the different cells. Larger tolerances may be used to limit segmentation. Executing OPC on the cells with relaxed tolerances provides sizable shot count reductions of up to 40% for poly when allowing dimensional tolerances of 30%. Above this level the benefits level off. At the same time the timing impact for the same level of tolerance relaxation was about 3%.

A significantly simpler albeit reasonably efficient implementations of this approach is to apply the most refined segmentation for those portions of the gate mask that intersect active areas (and thus form an active transistor) as opposed to those regions where poly connections fall over field oxide and thus do not require tight line-width control.

Providing shot-count information at the design stage (i.e., before OPC has been executed) has been proposed by Cote et al. [54]. and is referred to as *mask cost comparison* (MCC). MCC provides a model of the shot count as a function of tolerances based in the design stage. Using these models the design team can make the appropriate trade-offs between mask cost and timing considerations. To generate the mask cost model, OPC is executed on a set of standard library elements. The shot counts of the resulting post-OPC layout are recorded as a function of OPC tolerances. A set of simple layout characteristics, such as the length of the perimeter (a measure of the outside dimension of the cell), average shape-to-shape distance, and poly vertex count, provide the necessary parameters for an empirical model.

One of the issues faced by mask manufacturing is that only post-OPC shapes are transferred to the mask shop. Under these circumstances the only choice available to the mask manufacturer is to reproduce, as closely as possible, the template provided to them. Segment reduction strategies are implemented primarily at the OPC stage, and thus design information is used prior to transferring the data to the mask house. Mask inspection, on the other hand, is a manufacturing step that would greatly benefit from additional information in an effort to reduce mask costs. Mask inspection is the step whereby the quality of the mask with respect variations in critical dimensions and the occurrence and severity of mask defects is assessed. Mask defects are deviations of the actual image as it appears after mask manufacturing on the mask relative to the post-OPC GDS supplied. Most commonly, these defects are the result of small particles that have dropped onto the reticle during the manufacturing process or the result of imperfections in the resist coatings or mask blanks. In general, such defects can be tolerated only if they do not affect significantly, the image as these defects will be reproduced on every chip printed on the wafer. Thus, the presence of defects on a reticle results in either an attempt to repair the mask or a mask rebuild, both of which are quite expensive, and it is thus desirable to have high confidence that the defects will actually result in nonfunctional chips.

Mask inspection tools typically qualify defects based on obvious criteria such as defect sizes, defect type (clear or opaque), and the location of the defect on the mask. Similar to the wafer inspection process, review tools are available that provide further information by assessing the impact the defect has on the resulting images. Another important criterion for dispositioning defects is the impact of a defect on circuit performance. For example, defects that occur on dummy fill patterns have little if any impact on the device performance and thus should not be a reason to reject the reticle. Additional, more sophisticated approaches have been proposed. Similar to the OPC case, tolerance information derived, for example, from timing information, information on the location of critical paths, or even information as simple as the differentiation between active gates and poly wiring on the gate level could be used. Based on this information, stringent defect-dispositioning criteria are used only with circuits that have small tolerances; more relaxed criteria can be applied to other regions. As the tightest tolerances are typically required for only a small portion of a chip, these approaches would greatly facilitate a reduction in mask costs.

REFERENCES

1. ASML presentation to Deutsch Bank Investment.
2. C. Spence et al., Mask data volume: historical perspective and future requirements, *Proc. SPIE*, vol. 6281, 2006.
3. H. Martinsson et al., Current status of optical maskless lithography, *JEM3*, vol. 4, p. 3, 2005.
4. L. Pain et al., Electron beam direct write lithography flexibility for ASIC manufacturing: an opportunity for cost reduction, *Proc. SPIE*, vol. 5751, p. 35, 2005.
5. T. Miura et al., Nikon EPL tool: the latest development status and results, *Proc. SPIE*, vol. 5751, p. 477, 2005.
6. M. Melliar-Smith, Lithography beyond 32 nm: a role for Imprint, Plenary paper, *Proc. SPIE*, vol. 6520, 2007.
7. N. Harned et al., EUV lithography with the Alpha demo tool: status and challenges, *Proc. SPIE*, vol. 6517, paper 651706, 2007.
8. B. Lin, Marching of the lithographic horses: electron, ion and photon—past present and future, *Proc. SPIE*, vol. 6520, paper 652001, 2007.
9. Y. Zhang et al., Mask cost analysis via write time analysis, *Proc. SPIE*, vol. 5756, p. 313, 2005.

10. B. Lin, The $k3$ coefficient in nonparaxial λ/NA scaling equations for resolution, depth of focus and immersion lithography, http://www.spie.org, doi: 10.1117/1.1445798.

11. A. E. Rosenbluth et al., Optimum mask and source patterns to print a given shape, *J. Microlithogr. Microfabrication Microsyst.*, vol. 1, no. 1, pp. 13–30, Apr. 2002.

12. H. Fukuda et al., Spatial filtering for depth of focus and resolution enhancement in optical lithography, *J. Vac. Sci. Technol. B*, vol. 9, no. 6, pp. 3113–3116, 1991.

13. E. Wolf and Y. Li, Conditions for the validity of the Debye integral representation of focused fields, *Opt. Commun.*, vol. 39, no. 4, pp. 205–210, Oct. 1981.

14. ASML presentation at ISS Halfmoon Bay, 2007.

15. International Technology Roadmap for Semiconductors, http://www.itrs.net/reports.html.

16. A. Wong, *Optical Imaging in Projection Microlithography*, SPIE Press, Bellingham, WA, 2005.

17. A. Maharowala, Meeting critical gate linewidth control needs at the 65 nm node, *Proc. SPIE*, vol. 6156, paper 6156OM, 2006.

18. Y. Sumiyoshi et al., Analysis of precise CD control for the 45 nm node and beyond, *Proc. SPIE*, vol. 5754, p. 204, 2005.

19. M. Dusa et al., Pitch doubling through dual patterning lithography challenges in integration and litho budgets, *Proc. SPIE*, vol. 6520, paper 65200G, 2007.

20. C. Brodsky et al., Lithography budget analysis at the process module level, *Proc. SPIE*, vol. 6154, paper 61543Y, 2006.

21. S. Hector et al., Evaluation of the critical dimension control requirements in the ITRS using statistical simulation and error budget, *Proc. SPIE*, vol. 5377, p. 555, 2004.

22. W. Koestler et al., CD control: Whose turn is it? *Proc. SPIE*, vol. 5754, p. 790, 2005.

23. A. Nackaerts et al., Lithography and yield sensitivity analysis of SRAM scaling for the 32 nm node, *Proc. SPIE*, vol. 6521, paper 65210N, 2007.

24. U.S. patent 7,124,396, Alternating phase shift mask compliant design.

25. L. Liebmann et al., High performance circuit design for the RET-enabled 65 nm technology node, in *Design and Process Integration for Microelectronic Manufacturing II*, L. Liebmann, Ed., *Proc. SPIE*, vol. 5379, p. 20, 2004.

26. J. Wang et al., Performance optimization for gridded-layout standard cells, 24th Annual BACUS Symposium, *Proc. SPIE*, vol. 5567, p. 107, 2004.

27. N. Cobb, Flexible sparse and dense OPC algorithms, *Proc. SPIE*, vol. 5853, p. 693, 2005.

28. N. Cobb, Dense OPC and verification for the 45 nm node, *Proc. SPIE*, vol. 6154, paper 61540I-1, 2006.

29. Y. Cao et al., Optimized hardware and software for fast, full simulations, *Proc. SPIE*, vol. 5754, p. 407, 2005.

30. B. Wong, Ed., *Physical Design in the NanoCMOS Era*, Wiley-Interscience, Hoboken, NJ, 2005, Chap. 3.

31. A. Balasinski, A methodology to analyze circuit impact of process related MOSFET geometry, *Proc. SPIE*, vol. 5379, p. 91, 2004.

32. F. E. Gennari et al., A pattern matching system for linking TCAD and EDA, ISQED 2004.

33. L. Capodieci, From proximity correction to lithography-driven physical design (1996–2006): 10 years of resolution enhancement technology and the roadmap enablers for the next decade, *Proc. SPIE*, vol. 6154, paper (6154)1-1, 2006.

34. D. Perry et al., Model based approach for design verification and co-optimization of catastrophic and parametric related defects due to systematic manufacturing variations, *Proc. SPIE*, vol. 6521, 2007.

35. S. Kotani et al., Development of a hotspot fixer, *Proc. SPIE*, vol. 6156-14, 2006.

36. S. Kobayashi et al., Automated hot-spot fixing system applied for metal layers of 65 nm logic devices, *Proc. SPIE*, vol. 6283-101, 2006.

37. E. Roseboom et al., Automated full chip hotspot detection and removal flow for interconnect layers of cell based designs, *Proc. SPIE*, vol. 6521, paper 65210C, 2007.

38. J. Kibarian, Design for manufacturability in the nanometer era: system implementation and silicon results, Session 14, Low power wireless and advanced integration, International Solid-State Circuits Conference, San Jose, CA, 2005.

39. J. Mitra et al., RADAR RET aware detailed routing using fast lithography simulation, *Proceedings of the Design Automation Conference*, ACM Press, New York, 2004.

40. T. Kong et al., Model assisted routing for improved lithographic robustness, *Proc. SPIE*, vol. 6521, paper 65210D, 2007.

41. M. Terry et al., Process and interlayer aware OPC for the 32 nm node, *Proc. SPIE*, vol. 6520, paper 65200S, 2007.

42. P. Yu et al., Process variation aware OPC with variational lithography modeling, *Proceedings of the Design Automation Conference*, ACM Press, New York, 2006.

43. A. Sezginer et al., Sequential PPC and process window aware mask synthesis, *Proc. SPIE*, vol. 6156, p. 340, 2006.

44. R. Socha, Contact hole reticle optimization by using interference mapping technology, *Proc. SPIE*, vol. 5446, p. 516, 2004.

45. S. Shang et al., Model based insertion and optimization of assist features in contact layers, *Proc. SPIE*, vol. 5446, p. 516, 2004.

46. F. Zach et al., Across field CD improvement for critical layer imaging: new applications for layout correction and optimization, *Proc. SPIE*, vol. 6156, paper 61560I, 2006.

47. H. M. Shieh, C. L. Byrne, and M. A. Fiddy, Image reconstruction: a unifying model for resolution enhancement and data extrapolation—Tutorial, *J. Opt. Soc. Am. A*, vol. 23, no. 2, pp. 258–266, Feb. 2006.

48. T. M. Habashy and R. Mittra, Review of some inverse methods in electromagnetics, *J. Opt. Soc. Am. A*, vol. 4, no. 1, pp. 281–291, Jan. 1987.

49. K. M. Nashold and B. E. A. Saleh, Image construction through diffraction-limited high-contrast imaging systems: an iterative approach, *J. Opt. Soc. Am. A*, vol. 2, no. 5, pp. 635–643, May 1985.

50. A. Klatchko and P. Pirogovsky, Conformal mapping in microlithography, 24th BACUS Symposium on Photomask Technology, *Proc. SPIE*, vol. 5567, pp. 291–300, Oct. 2004.

51. D. Abrams, D. Peng, and S. Osher, Method for time-evolving rectilinear contours representing photo masks, U.S. patent 7,124,394 B1, Oct. 2006.

52. A. Yehia et al., *Proc. SPIE*, vol. 6520, paper 65203Y-1, 2007.

53. P. Gupta et al., Modeling OPC complexities for design for manufacturing, *Proc. SPIE*, vol. 5992, paper 59921W, 2005.

54. M. Cote et al., Mask cost reduction and yield optimization using design intent, *Proc. SPIE*, vol. 5756, pp. 389–386, 2005.

3

INTERACTION OF LAYOUT WITH TRANSISTOR PERFORMANCE AND STRESS ENGINEERING TECHNIQUES

3.1 INTRODUCTION

The semiconductor industry has a love–hate relationship with mechanical stress. For the first 40+ years it was hate only. Stress was considered to be the cause of defects, and industry tried to avoid it by any means [1–4]. Around 2000 the picture changed: Many semiconductor manufacturers began intentionally introducing stress to boost circuit performance [5–8]. So by now it's mostly love, but you always try to avoid stress-induced defects.

Intentional stress introduction into transistors is usually referred to as *stress engineering*. It has been demonstrated to almost double the transistor drive current [9], so if you don't use stress engineering, your circuits are at a huge competitive disadvantage. That is why all leading-edge companies have embraced it.

The first evidence of stress-enhanced hole and electron mobilities was reported at Bell Laboratories in 1954 [10]. It was not used in the industry until the mid-1990s, when IBM started exploring biaxially stretched thin silicon layers on top of thick relaxed SiGe layers [5]. This technique demonstrated a great performance boost for very wide nMOSFETs ($W > 10\,\mu m$). However, there was no significant pMOSFET improvement, and the nMOSFET enhancement disappeared for the narrow channel widths that are typical of logic and

Nano-CMOS Design for Manufacturability: Robust Circuit and Physical Design for Sub-65 nm Technology Nodes
By Ban Wong, Franz Zach, Victor Moroz, Anurag Mittal, Greg Starr, and Andrew Kahng
Copyright © 2009 John Wiley & Sons, Inc.

memory circuits ($L < W < 10L$). A refined version of this technique, applied to silicon-on-insulator (SOI) technology, became popular soon after 2000 [11].

The major breakthrough in introducing stress engineering into semiconductor manufacturing was reported by Intel [7]. A 15% performance enhancement was achieved for nMOSFETs using a strained nitride contact etch-stop layer (CESL), and similar performance enhancement for pMOSFETs was obtained using an embedded SiGe source–drain. Since then, many new stress engineering techniques, as well as refinements of the original stress engineering techniques, have been proposed, further increasing transistor performance.

Besides the obvious benefits, stress engineering introduces several new variability mechanisms that add to the existing transistor performance variations. Most of this new variability is systematic rather than random and can be traced to neighborhood effects for transistors inserted into different layout environments. In this chapter we discuss in detail the physical mechanisms, application, and side effects of stress engineering.

3.2 IMPACT OF STRESS ON TRANSISTOR PERFORMANCE

Mechanical stress affects transistor performance in many ways. During the manufacturing process, stress alters diffusion and segregation of the dopants in the channel and source–drain. As the wafer goes through process steps, stress in the transistors keeps changing whenever new layers are deposited or patterned, the wafer heats up or cools down, and the dopants are introduced by implantation and diffusion. Once the manufacturing process is over, the geometry of the transistor no longer changes, and its temperature changes in a relatively narrow range. This makes the residual stress in the transistors stable during circuit operation.

The stress σ applied to a semiconductor crystal lattice distorts it, causing it to shrink, stretch, or warp. The amount of distortion is characterized by the strain, ε. As long as the lattice remains defectless, the crystal behavior is elastic, with strain being directly proportional to stress:

$$\sigma = E\varepsilon \tag{3.1}$$

where E is Young's modulus. Amorphous materials have isotropic mechanical properties which can be described by a single scalar value of Young's modulus. Crystalline semiconductors have anisotropic mechanical properties that can be described by a tensor. Mechanical strength of silicon varies by about 30%, from $E = 135\,\text{GPa}$ along the weakest (100) crystallographic direction to $E = 187\,\text{GPa}$ along the strongest (111) crystallographic direction.

If the amount of strain exceeds the mechanical strength of the crystal lattice, crystal defects form to relax the stress. In silicon, this happens at strain levels higher than about 1%. Typical defects are dislocations or slip planes:

Table 3.1   Mechanical properties of typical materials used in the
semiconductor industry[a]

Material	Young's Modulus (GPa)	Poisson Ratio
Silicon	165	0.28
Silicon oxide, SiO_2	66	0.20
Silicon nitride, Si_3N_4	[b]	0.28
Germanium	141	0.27
Gallium arsenide	85	0.31
Polysilicon	165	0.28
Aluminum	86	0.30
Nickel	186	0.34
Nickel silicide	115	0.33
Tungsten	366	0.30
Copper	130	0.34
Hafnium oxide	57	0.20

[a]For anisotropic crystals, averaged properties are shown.
[b]Young's modules depends on the process: up to 350 for high deposition temperatures, and below 190 for the lower deposition temperatures that are more typical for current processes.

either missing or extra planes of the highest density [i.e., usually (111) planes]. Once defects form, the elastic relationship (3.1) is no longer valid, but in the semiconductor industry such defects are avoided because they usually increase junction leakage and are undesirable. Therefore, for practical purposes we can assume that semiconductors follow elastic behavior (3.1).

Table 3.1 lists Young's moduli for materials that are widely used in the semiconductor industry. According to equation (3.1), for the same force applied to a material that is characterized by stress σ, the amount of lattice distortion that is characterized by strain ε is inversely proportional to Young's modulus. For example, the same stress applied to silicon and to oxide introduces almost three times greater strain in the oxide than in silicon.

Strain alters the band structure of semiconductors and insulators such as silicon and oxide. Under strain, some of the valleys move up, whereas others move down. Both the tensile and compressive stresses narrow the bandgap. Typically, tensile strain lowers the conduction band, and the compressive strain elevates the valence band. Roughly, the bands move about 50 mV per gigapascal of hydrostatic pressure, but the exact band structure depends on the particular stress pattern. In the following sections we describe various aspects of the stress impact on transistor performance.

3.2.1   Electron Mobility

The first comprehensive study of the impact of stress on semiconductor conductivity was performed at Bell Laboratories in 1954 [10]. The effect that was discovered in that study is known as *piezoresistance*. Bulk conductivity was

Table 3.2   Impact of 1-GPa stress on electron mobility based on the bulk piezoresistance effect[a]

Stress Component	Tensile Stress (%)	Compressive Stress (%)
Longitudinal	+30	−30
Transverse	+20	−20
Vertical	−50	+50

[a]The numbers show the percentage of mobility change compared to stress-free silicon. The standard wafer surface orientation of (100) is used along with the standard current direction along <110>.

measured for silicon slabs of n- and p-type. Stress was applied in longitudinal (i.e., along the current flow), transverse (i.e., across the current flow), and vertical (i.e., out of the plane) directions. Some of the applied stress patterns, such as tensile longitudinal, tensile transverse, and compressive vertical, increased the conductivity of n-type silicon. The opposite stress patterns—compressive longitudinal, compressive transverse, and tensile vertical—reduced the conductivity of n-type silicon. Let's refer to the stress patterns that increase conductivity as beneficial and to the opposite stress patterns as harmful.

The beneficial stress patterns lower the conduction valley with light electrons (i.e., with lower effective mass) into the bandgap. This gives more light electrons and fewer heavy electrons than with relaxed silicon. The light electrons have higher mobility and therefore increase the conductivity of silicon under beneficial stress. On the contrary, the harmful stress patterns lower the conduction valley with heavy electrons and reduce the effective electron mobility.

Table 3.2 shows the impact on electron mobility of stress applied in different directions. The results obtained at Bell Laboratories [10] apply to electron mobility in lightly doped n-type silicon. The MOSFET transistors have channels within about 3 nm of the silicon surface. Therefore, surface effects can be potentially strong in the MOSFETs and can affect or even overshadow the bulk stress effect. Analysis of the MOSFET response to stress shows that it is very similar to the behavior of bulk piezoresistance effect. Potentially, the effect of stress on band structure can be different in the vicinity of an interface. Stress can also affect electron scattering at the interface, although no direct evidence of this has been found so far.

3.2.2   Hole Mobility

The holes exhibit a similar mechanism of mobility response to the stress as electrons. The first comprehensive study of the hole piezoresistance effect came form Bell Laboratories [10]. The main difference from the situation as applied to electrons is that here what matters is the band structure of the valence band rather than the conduction band. Figure 3.1 illustrates holes in

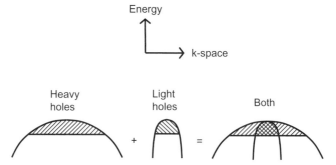

Figure 3.1   Valence band structure in stress-free silicon. The population of free holes contains a mixture of light holes with high mobility and heavy holes with low mobility.

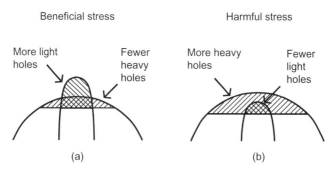

Figure 3.2   Valence band structure of silicon under (a) beneficial and (b) harmful stress. Stress-induced shifts in the energy bands lead to repopulation between light and heavy holes.

the valence band located in two valleys with different radii of curvature. The effective mass of the hole is proportional to the curvature radius of the valley it is in. The mobility is inversely proportional to the effective mass. In stress-free silicon, the population of holes is split between the two valleys so that the overall mobility gets averaged between the light and heavy holes.

Beneficial stress patterns for the holes are different from those for electrons. Once a beneficial stress pattern is applied, it raises the valence valley with light holes into the bandgap, as shown in Figure 3.2a. This repopulation puts more holes into the valley with lower hole effective mass and therefore higher mobility. The harmful stress patterns raise the valence valley with heavy holes as shown in Figure 3.2b and repopulate holes into the valley with high effective mass and low mobility.

The impact of stress on hole mobility is shown in Table 3.3. Comparison of Tables 3.2 and 3.3 shows that:

- The electrons are most sensitive to vertical stress, whereas the holes are least sensitive to vertical stress.

Table 3.3   Impact of 1-GPa stress on hole mobility based on the bulk piezoresistance effect[a]

Stress Component	Tensile Stress (%)	Compressive Stress (%)
Longitudinal	−70	+70
Transverse	+70	−70
Vertical	+1	−1

[a]The numbers show the percentage of mobility change compared to stress-free silicon. The standard wafer surface orientation of (100) is used along with the standard current direction along <110>.

- The holes are more sensitive to lateral stress than are the electrons.
- The only common stress that increases hole and electron mobilities simultaneously is the tensile transverse stress.

So far we have addressed electron and hole mobilities at room temperature in lightly doped silicon and at relatively low stress levels. The impact of temperature, high doping levels, and high stress levels is discussed in the following sections.

3.2.3   Threshold Voltage

Threshold voltage exhibits fairly low direct sensitivity to stress. It depends on a particular stress pattern, but rarely exceeds 10 mV in a typical MOSFET, due to the stress present in its channel. The tensile and compressive stresses both lower the threshold, due to their impact on semiconductor band structure. Threshold voltage often depends strongly on the particular layout of the transistor, such as the channel width W and the size and shape of the source and drain areas. Let's look at the W dependence.

3.2.3.1   Variation of Threshold Across the Channel Width

Figure 3.3 shows the top view of an nMOSFET with a nominal poly gate length of 47 nm obtained using a scanning tunneling microscope. Several interesting effects can be seen on that image:

1. The polysilicon gate shape exhibits line-edge roughness (LER), which is an inherent property of photoresist. This roughness has a random nature, with an amplitude (across the poly stripe) of about ±5 nm and a period (along the poly stripe) of about 30 nm. For the channels that are much wider than the LER period, $W \geq 30$ nm (which will be true until the 32-nm technology node), LER does not affect the transistor drive (i.e., on) current, but degrades the off current. For channels that are comparable or narrower than the LER period, LER becomes introducing a

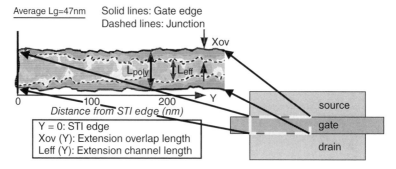

Figure 3.3 Scanning tunneling microscope image of the nMOSFET channel. (From ref. 12.)

systematic shift in channel length, making some transistors shorter and others longer, and therefore affects both on and off currents.

2. There are atomistic-level fluctuations due to the finite number of dopant atoms in the channel and source/drain. These fluctuations are random and affect threshold voltage and both on and off currents. The channel is doped with boron, which serves as an acceptor, and the source and drain are doped with arsenic, which serves as a donor.

3. There is a systematic trend in the effective channel length L_{eff} across the channel, with L_{eff} measured from the source–channel junction to the channel–drain junction. Next to the shallow trench isolation (STI), L_{eff} is about 43 nm. Away from the STI, toward the channel center, L_{eff} shrinks to about 17 nm.

4. There is a systematic shift in the channel doping, with boron concentration decreasing away from STI.

Counting boron atoms as a function of the distance from STI (i.e., quantifying effect 4) produces the trends depicted in Figure 3.4. Also shown is a boron profile with nitrogen co-doping. Apparently, the introduction of nitrogen suppresses most of the systematic boron variation.

Figure 3.5 shows the extension overlap,

$$L_{\text{ext}} = \frac{L_{\text{poly}} - L_{\text{eff}}}{2} \tag{3.2}$$

extracted from scanning tunneling microscope images (quantifying effect 3). It is clearly increasing away from STI toward the channel center. The implication of trend 3 is that the effective channel length is a function of channel width. This effect is neither expected nor desired, and happens outside the scope of lithography-related effects, and therefore cannot be fixed using optical proximity correction (OPC) techniques. This effect has not been

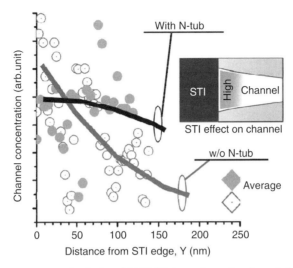

Figure 3.4   Boron concentration in the nMOSFET channel as a function of distance from STI, extracted from the scanning tunneling microscope image.

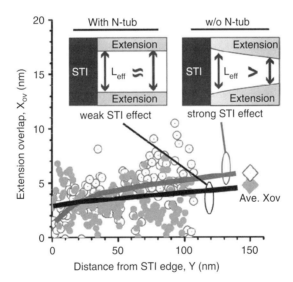

Figure 3.5   Source/drain extension overlap in nMOSFET as a distance from the STI, extracted from scanning tunneling microscope image.

studied extensively so far, but was reported, for example, by Fukutome et al. [12] and Hook et al. [13].

The implication of trend 4 is that the threshold voltage is a function of channel width. The dependence of threshold on the channel width is a well-characterized effect that is determined by several independent physical mechanisms [14–16]. Measuring the threshold for different W's gives you values that are integrated across the channel. However, variations of boron doping (and therefore threshold voltage) across the channel have not yet been well characterized.

Let's look at the possible mechanisms behind effects 3 and 4. Stress-dependent dopant diffusivity is often blamed for such effects. The dopant diffusion happens at high temperatures during the spike anneal, which has reached temperatures of about 1050 °C for the last several technology nodes. The peak temperature is expected to scale down to 1000 °C or even somewhat lower when used in combination with a millisecond anneal. During the spike anneal, the main and often the only stress source is thermal mismatch stress induced by STI. At these temperatures, the STI-induced stress in the transistor channel is either zero or weak tensile, within +50 GPa. This weak tensile stress would modify the arsenic and boron diffusivities by less than 5%, which is definitely not enough to explain the observed variations depicted in Figures 3.3 to 3.5. Moreover, tensile stress would slow down arsenic and speed up boron, but apparently both arsenic and boron diffusivities are suppressed near STI. Figures 3.4 and 3.5 demonstrate that a nitrogen implant effectively reduces both effects 3 and 4, although the low-dose nitrogen implant used in this study definitely does not generate any significant stress. Therefore, these effects are not driven by stress.

An alternative explanation involves point defect recombination at the silicon–STI interfaces. The point defects, specifically excess silicon self-interstitials, usually denoted as I's, are introduced every time an implant step is performed in the process flow. The excess interstitials travel around and help dopants such as arsenic and boron to diffuse faster, a phenomenon referred to as *transient-enhanced diffusion* (TED): transient because the excess interstitials eventually recombine at the silicon surfaces, and dopant diffusivities reduce back to their thermal equilibrium values.

For a typical spike anneal and silicon–STI interface, the recombination length of interstitials is about 100 nm, which is comparable to the effects observed in Figures 3.3 to 3.5. Interstitials quickly recombine within the proximity of the silicon–STI interface and therefore do not participate in TED. This can explain the fact that both arsenic and boron diffusivities decrease as you get closer to STI.

The presence of nitrogen is known to suppress I recombination at silicon–oxide interfaces [17], and STI is essentially a deposited oxide. So in samples with a nitrogen implant, the silicon–STI interface loses its ability to quickly remove excess interstitials, and therefore the strong spatial STI proximity effect disappears. Now let's look at the impact of the size and shape of source–drain areas on threshold voltage.

3.2.3.2 Impact of Source–Drain Layout on Threshold Voltage

It has been observed that threshold voltage could be very sensitive to the distance from the poly gate to the STI (Figure 3.6). In this particular case, the threshold decreases as the poly-to-STI distance shrinks. For different pro-

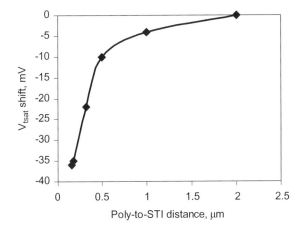

Figure 3.6 Threshold voltage shift as a function of the poly-to-STI distance. (Data from ref. 18.)

cesses, it can be flat or even change the sign and increase with shrinking poly-to-STI distance. For the same process, the trends for nMOSFET and pMOSFET can be either the same or opposite.

The apparent sensitivity of threshold voltage to poly-to-STI distance is often interpreted incorrectly as a stress effect [19]. The STI-induced residual stress in the transistor channel is fairly low, usually below 300 MPa. Such stress would lower the threshold by only a few millivolts as the poly-to-STI distance shrinks to the minimum allowed. The threshold dependencies observed for poly-to-STI distance can be much stronger—typically about 20 mV and sometimes over 50 mV. Depending on the specific process flow, the threshold could either increase or decrease with poly-to-STI distance. Both the magnitude and the various signs of the effect indicate that it cannot be related directly to the residual STI stress.

One potential indirect mechanism of stress impact on threshold voltage is the stress-dependent dopant diffusion in the channel. However, as we mentioned earlier and will look at in detail later in the chapter, STI stress disappears at the high temperatures at which the diffusion takes place. Therefore, this effect cannot be attributed to stress either.

The dominant mechanism at play here is layout-dependent transient enhanced diffusion. Let's take a close look at this effect. Figure 3.7 shows typical layouts for isolated and nested transistors. In an isolated transistor, there is just one poly gate inside the diffusion area. On one side of the poly gate is the transistor's source. On the other side is the transistor's drain. The nested transistors contain several poly gates per diffusion area, with three gates shown in Figure 3.7b.

Figure 3.8 shows the source, drain, and channel profiles in a nested 80-nm nMOSFET. All doping profiles are symmetric with respect to the source and

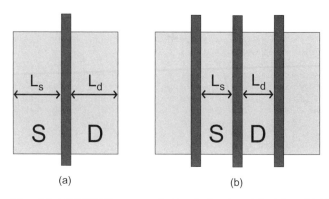

Figure 3.7 (a) Isolated MOSFET is surrounded by shallow trench isolation; (b) MOSFET with nested gates has neighbors' poly gates behind its source (S) and drain (D).

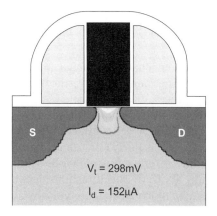

$V_t = 298mV$

$I_d = 152\mu A$

Figure 3.8 Nested nMOSFET with low V_t and strong I_d. Nested transistor with $L_s = L_d = 220\,nm$ is surrounded by similar transistors on the left and on the right, without STI nearby.

drain, and the linear threshold voltage is close to the nominal value of 300 mV. Reflective boundary conditions are used on the left and right sides of the structure, which means that this transistor is repeated to the left and the right, as shown in Figure 3.9.

Figure 3.10 shows an isolated asymmetric transistor simulated using the same process flow as that used for the nested device, but with STI behind the source and drain. The linear threshold voltage is about 50 mV higher than for the nested transistor, and the linear drain current is 15% lower. The asymmetry of the channel doping profile is caused by the asymmetric L_s and L_d. The difference in performance between the nested and isolated transistors is due to the different recombination pattern for the implant damage introduced by the source and drain implants (Figure 3.11).

The proximity of STI to the channel in the isolated transistor enhances damage recombination and therefore reduces the amount of transient-

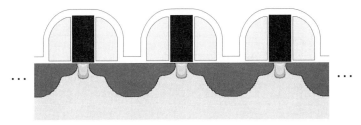

Figure 3.9   Cross section of a structure with nested transistors.

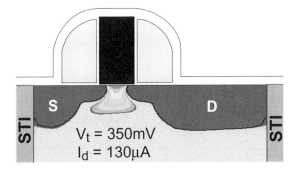

Figure 3.10   Isolated, asymmetric nMOSFET with small source and drain sizes, leading to higher V_t and weaker I_d. Here, $L_s = 120\,$nm and $L_d = 240\,$nm.

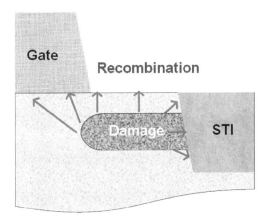

Figure 3.11   Recombination pattern for damage introduced by source and drain implants.

enhanced diffusion experienced by the channel and halo implants. Typical channel doping profiles after the halo implant and after the entire process flow are shown in Figure 3.12. The as-implanted profile has a peak at a depth of several tens of nanometers from the channel surface. After annealing, the peak is close to the channel surface.

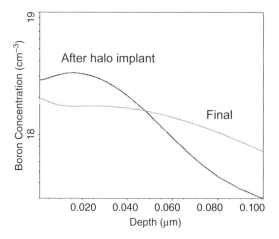

Figure 3.12   Vertical boron profile in an nMOSFET channel.

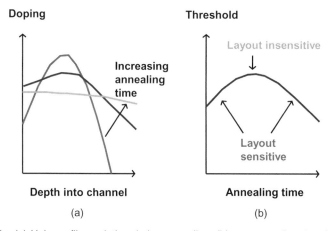

Figure 3.13   (a) Halo profile evolution during annealing; (b) corresponding threshold voltage evolution.

Figure 3.13 illustrates halo profile evolution during annealing. At the initial annealing stage, surface concentration goes up, but eventually this trend reverses and surface concentration goes down with annealing time. Moving STI closer to the channel would correspond to the reduced annealing time, and vice versa, because both the thermal budget and the STI proximity affect the amount of TED experienced by the channel dopant.

For a specific annealing thermal budget, it is possible to adjust the halo implant energy and tilt angle such that a transistor with minimum L_s and L_d values corresponds to the point just left of the peak on Figure 3.13b. All transistors with larger L_s and L_d values as well as the nested gate transistors will be to the right of the transistor with minimum L_s and L_d values. Because the threshold voltage is almost independent of the amount of TED in this range of thermal budget, the final device and circuit performance would exhibit less sensitivity to layout.

3.2.3.3 Well Proximity Effect

The bulk CMOS process flow includes the formation of deep retrograde n- and p-wells to degrade properties of the parasitic thyristors and protect transistors from latching up [20,21]. The nMOSFET is built on top of the p-well, whereas the pMOSFET is built on top of the n-well. The formation of a deep retrograde well requires a high-energy implant step that places the peak of implanted ions deep under the channel surface. Whenever a p-well is formed for the nMOSFETs, the pMOSFETS are covered by a thick photoresist, and vice versa. The location of the photoresist edges are determined by the well masks.

The photoresist captures the high-energy ions and prevents them from penetrating the wrong well. The ions get scattered by the atoms that comprise photoresist, and eventually come to rest before they can reach silicon underneath. However, at the edge of the photoresist mask, some of the ions scatter out of the photoresist through its wall, as illustrated in Figure 3.14. The ions that get scattered out have lower energy, due to the scattering that they experienced inside the photoresist. Therefore, they get deposited at a shallower than intended depth in silicon that is adjacent to the photoresist edge. The shallow dopants near the channel surface unintentionally alter the threshold voltage, sometimes as much as >100 mV [22].

One sure way to eliminate this effect is to move the transistors far enough from the well masks such that the scattered ions do not reach the transistor channels. Practically, this cannot be done, as the lateral scattering range is on

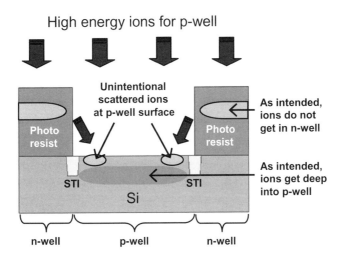

Figure 3.14   Well proximity effect due to ion scattering off the photoresist mask.

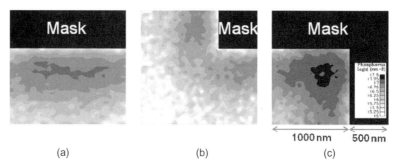

←———— 1000 nm ————→ ←———— 500 nm ————→

(a) (b) (c)

Figure 3.15 Scattering of phosphorus implanted at 400 keV over a silicon surface next to (a) a straight photoresist mask, (b) a mask with a convex corner, and (c) a mask with a concave corner.

the order of 1 μm, and having a 2-μm-wide margin between the nMOSFET and the pMOSFET would be an unacceptable waste of chip area. Instead, this effect is being tolerated by being accounted for using appropriate models. The models make the threshold a function of the distance from the well mask to the channel [19]. First, heuristic models that are based on the inherently noisy measured data assumed monotonic reduction and eventual saturation of the threshold voltage with increasing distance to the well mask.

More sophisticated models based on predictive Monte Carlo simulation of the ion-scattering events revealed nonmonotonic threshold behavior [23]. Figure 3.15 illustrates spatial distributions of the scattered ions around straight well masks as well as around the concave and convex mask corners. The modeled threshold shifts for each transistor due to the well proximity effect according to its particular layout environment are written as instance parameters (DELVTO) in the SPICE netlist. The circuit simulation is done using these threshold shifts to account for particular properties of each transistor.

Figure 3.16 illustrates WPE distribution in a library cell. Both nMOSFET and pMOSFET thresholds increase due to WPE, with positive threshold shifts in the lower part of the library cell that contains the nMOSFETs and negative shifts in pMOSFETs due to the negative nominal threshold there. The maximum threshold shift is 47 mV for the nMOSFET and −87 mV for the pMOSFET, with both extremes happening near concave mask corners.

3.2.4 Junction Leakage

Junction leakage is the current that flows through reverse-biased p-n junction. In MOSFETs, this happens at the drain junction, which is under reverse bias whenever the drain bias is applied. Junction leakage contributes to the off current of the transistor, along with several other leakage mechanisms, such

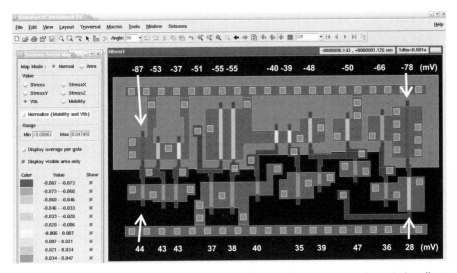

Figure 3.16   Variations of threshold voltage in a library cell due to the well proximity effect. (Calculated using *Seismos* [23].)

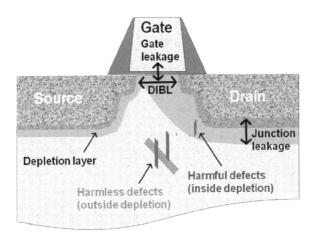

Figure 3.17   The depletion region in a transistor determines the leakage current of the drain junction.

as gate tunneling current, band-to-band tunneling, and the punch-through current associated with drain-induced barrier lowering (DIBL).

Let's look at the possible mechanisms of junction leakage. Figure 3.17 shows a cross section of a transistor. The junction depletion region is narrow on the source side because the source junction has zero bias. The depletion is much wider on the drain side due to the drain bias. The depletion region is depleted of free carriers by definition. Any free electrons that happen to be within the depletion region will be driven by an electric field toward the n-type

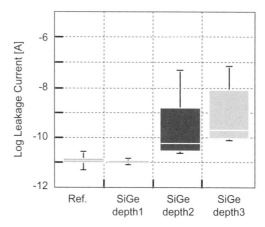

Figure 3.18 Measured leakage current of a drain junction. Reference shows silicon S/D, and all other transistors include eSiGe S/D, with SiGe depth increasing from depth1 to depth3. Depth 1 is less than the depth of drain junction, and depths 2 and 3 are larger than the junction depth.

side and contribute to the junction leakage. Similarly, any free holes will go toward the p-type side.

In a silicon lattice that is completely free of any crystal defects, the number of thermally generated free carriers is determined by the bandgap, E_g [24]. As we already know, any stress applied to the silicon lattice would narrow its bandgap. Therefore, any stress increases junction leakage. If you recall that the industry is relying heavily on stress engineering to squeeze performance gain out of transistors, you can see that there is reason to worry here. The presence of high concentrations of impurities in silicon, such as Ge in SiGe, also narrows the bandgap and therefore increases the leakage current.

The ratio of leakage current in strained silicon J_σ to stress-free silicon J_0 can be expressed as

$$J_\sigma/J_0 = \exp(\Delta E_g/2kT) \tag{3.3}$$

If a pMOSFET junction is located within a strained SiGe drain with 20% Ge, the bandgap narrowing would be about 80 mV due to the Ge presence, and another 90 mV due to the compressive stress within SiGe [25]. The combined ΔE_g of 170 mV leads to a 30-fold increase in leakage, which is similar to the measured values shown in Figure 3.18. This is why everybody keeps drain junctions outside the strained SiGe.

Whenever stress becomes unbearably high for the silicon lattice, it relaxes the stress by introducing crystal defects such as dislocations. If such defects are located away from the junction depletion region, they help to reduce the leakage current by partially or completely relaxing the stress and therefore increasing the bandgap. However, if such defects happen to be within the depletion region, they often introduce deep levels in the bandgap. A midgap

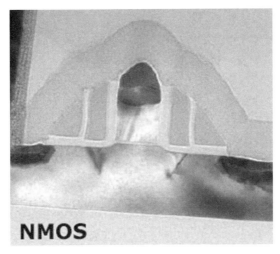

Figure 3.19   TEM image of a 50-nm SOI transistor from an AMD Athlon 64 2700 chip, showing dislocations under the spacers that are apparently harmless. (Courtesy of Chipworks.)

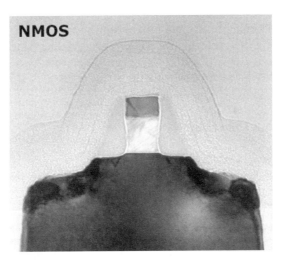

Figure 3.20   TEM image of a 42-nm transistor from Intel's Presler chip, showing dislocations under the spacers that are apparently harmless. (Courtesy of Chipworks.)

level increases electron–hole pair generation tremendously, as it effectively shrinks the bandgap in half. That is why dislocations are avoided and usually are not observed, with some notable exceptions. For example, Figures 3.19, 3.20, and 3.21 show TEM images of the cross sections of AMD, Intel, and Matsushita transistors from their 90- and 65-nm technology nodes, which show consistent dislocation formation away from the drain depletion regions and are therefore harmless.

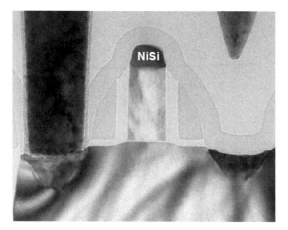

Figure 3.21 TEM image of a 55-nm transistor from Matsushita's DVD SOC chip, showing dislocations under the silicide that are apparently harmless. (Courtesy of Chipworks.)

3.2.5 High Stress Levels

The industry is learning to live with increasing stress levels within transistors to keep enhancing the performance. This raises two questions:

1. Will mobility keep increasing as the stress level goes up?
2. How much stress is too much?

We address these questions next.

3.2.5.1 Nonlinear Mobility Versus Stress

The piezoresistance effect was discovered and well characterized for low stress levels. The mobility response to stress was found to be linear. What happens when we push stress into the GPa range? Based on the physical mechanisms behind the piezoresistance effect, mobility enhancement is expected to saturate as soon as the repopulation of carriers from the heavy valley into the light valley is finished. This is exactly what is observed with electrons. Figure 3.22 shows electron mobility enhancement saturating at about 75%. There is no firm evidence of larger stress-induced enhancements of the electron mobility. In case a strong harmful stress pattern is applied to an nMOSFET, we should expect the mobility degradation to saturate similarly at a −75% level (i.e., the absolute mobility would be one-fourth of its value in stress-free silicon).

The holes exhibit surprisingly high mobility gains, well beyond the value following repopulation. Figure 3.23 shows that it almost triples as the compressive longitudinal channel stress approaches 2 GPa. At this point, it is not quite clear why this happens. We can see the following two possibilities. The most

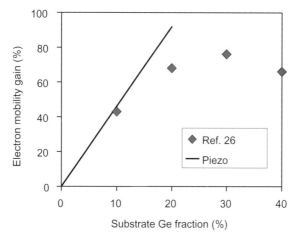

Figure 3.22 Enhanced electron mobility under high stress levels for biaxial tensile stress in thin silicon film on top of a thick relaxed SiGe. The piezoresistance model works well at low stress levels but does not describe enhancement saturation observed at higher stress levels, above about 1 GPa. (Data from ref. 26.)

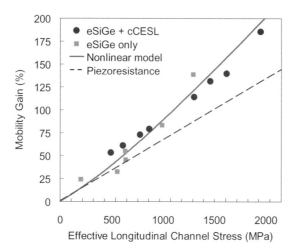

Figure 3.23 Enhanced hole mobility under high stress levels for a stress pattern with dominant longitudinal component. The classic linear piezoresistance model works well at low stress levels but fails to describe superlinear mobility versus stress trend at high stress levels. (From ref. 9.)

likely mechanism is warping of the light hole valley such that its curvature keeps decreasing under higher stress, and therefore the effective hole mass keeps shrinking. An alternative (or perhaps a complementary) possible mechanism is stress-induced reduction of hole scattering by the silicon–oxide interface, which is located within 2 nm from the inversion layer. The tripling of hole mobility was observed in a pMOSFET that has all the current flowing inherently close to the silicon surface, which is different from the purely bulk piezoresistance effect discovered back in 1954. An experiment with similarly high stress applied to a p-type bulk resistor will help to determine which physical mechanism is responsible for such favorable behavior of the hole mobility with respect to stress.

For all practical purposes, we know that the current in all MOSFET variations, including finFETs, is always in close proximity to the surface, just as in the strained pMOSFET described by Nouri et al. [25]. Therefore, the hole mobility there will exhibit pretty much linear response to stress, even at high stress levels. One of the side effects of this section is that the standard bulk piezoresistance model is surprisingly good at describing mobility response to stress except for the electrons at very high stress levels. Therefore, we will continue to use the piezoresistance model in this chapter while keeping that exception in mind.

3.2.5.2 Defect Formation

As we have seen, there are no good reasons to keep increasing the nMOSFET stress level beyond the point where electron mobility versus stress dependence saturates. However, we saw that the hole mobility keeps increasing upon seeing any beneficial stress patterns that were thrown at it. We will probably see higher stresses applied to pMOSFETs in the next few technology nodes. The question then becomes: What will happen first: hole mobility saturation with respect to stress or formation of harmful defects?

It is hard to say at this point, but it appears likely that pMOSFET stress engineering will be limited by the defects. Figure 3.24 shows a stacking fault in the middle of the channel. The longitudinal compressive stress level in the channel of this transistor is about 2.1 GPa, obtained by a combination of SiGe source/drain and a 100-nm-thick CESL with 2.5 GPa compressive stress. We believe that this defect is a {111} disk of missing lattice silicon atoms (essentially, vacancies). The missing atoms are squeezed out by the high compressive stress. This probably happened during the high-temperature spike anneal that was performed after the SiGe epitaxy.

At high temperatures (and the peak temperature at spike anneal usually exceeds 1000 °C), the lattice atoms are able to move around and if necessary, to move away from an overstressed area to relieve some of that stress. Many other stress engineering approaches begin after the high-temperature anneals and therefore might be able to introduce somewhat higher stress levels without harming the transistor. At room temperature, silicon lattice can withstand stress as high as several gigapascal before being destroyed. So we will probably see major pMOSFET stress engineering done toward the end of the front-end process flow to avoid the combination of high stress and high temperature.

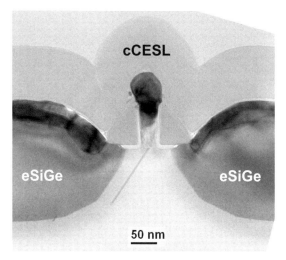

Figure 3.24   TEM image of a 45-nm pMOSFET with 20% germanium eSiGe source–drain and CESL with 2.5 GPa compressive stress. The {111} stacking fault in the channel is considered to be of a vacancy origin (i.e., a missing plane of lattice silicon atoms).

3.2.6   Crystal Orientations

All quantitative response of mobility to stress that we discussed so far applies to the standard orientations of the wafer and the channel direction. For CMOS circuits, standard wafer orientation since the 1970s is (100). The MOSFET channels are oriented along the <110> crystal direction. Let's refer to this combination as (100)/<110>. This combination provides the highest electron mobility in stress-free silicon and one of the highest sensitivities of the electron mobility to stress. For the holes, the picture is different: Much higher stress-free hole mobility as well as its sensitivity to stress can be achieved in (110) wafers. However, choosing this wafer orientation would degrade the nMOSFET performance, and having hybrid orientations (different for an nMOSFET and a pMOSFET) is too expensive, at least for bulk CMOS, but might become practical for silicon-on-insulator (SOI) circuits in the near future [27].

Figure 3.25 shows the response of electron mobility to stress for two wafer orientations: (100) on the left and (110) on the right. For each wafer orientation, all possible channel directions are shown (with a 360° transistor rotation on the wafer surface). The mobility response is based on the bulk piezoresistance coefficients and is shown for the two lateral stress components: longitudinal Π_l and transverse Π_t. The response to the vertical stress component is not shown, as it is independent of the wafer rotation.

The idea behind these circular plots is taken from an article by Kanda et al. [28]. Here is how to read these plots. If mobility increases with compressive stress, it is shown in the upper half of the circle. Therefore, the curves shown in the lower half of the circle mean that mobility increases in tensile

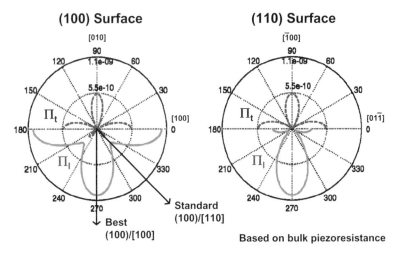

Figure 3.25   Response of electron mobility to longitudinal and transverse stress components for different crystal orientations of the wafer surface and the channel direction.

stress. The inside circle shows a 55% mobility change compared to stress-free silicon, and the outside circle shows a 110% mobility change. For example, let's take the standard (100)/<110> case and see what happens when we apply 1 GPa of tensile stress to the lateral directions. For the longitudinal stress, the electron mobility will increase by about 30%. For the transverse direction, it will increase by about 20%. Biaxial tensile stress of 1 GPa would yield 30% + 20% = 50% mobility gain. You can see that a wafer that is rotated 45° exhibits higher electron mobility sensitivity to longitudinal stress, whereas the transverse component flips the sign such that it degrades in tensile stress. A biaxial tensile stress of 1 GPa would yield the same 100% − 50% = 50% mobility gain.

Figure 3.26 shows the response of hole mobility to stress, also based on the bulk piezo effect. A compressive longitudinal stress of 1 GPa applied to the standard oriented (100)/<110> transistor improves the hole mobility by about 70%. Any biaxial stress (either tensile or compressive) does not provide significant hole mobility change, as the two components cancel each other out. The best combination for a pMOSFET is (110)/<111>, as shown on the right side. It provides both the highest stress-free mobility and the highest mobility sensitivity to stress.

You can see that the best combination for nMOSFET (100)/<100> makes the pMOSFET virtually insensitive to stress. Many companies used this combination at the 65-nm technology node to achieve the following three goals:

1. Maximize nMOSFET gain under longitudinal stress induced by tensile CESL.

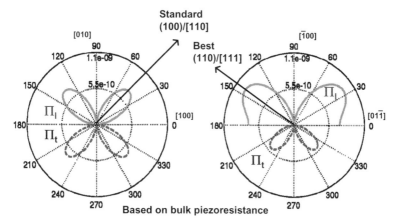

Figure 3.26   Response of hole mobility to longitudinal and transverse stress components for different crystal orientations of the wafer surface and the channel direction.

2. Increase pMOSFET performance by up to 20% (due to the anisotropic hole velocity saturation), which is due to the crystal orientation only, and almost independent of stress.
3. Keep the manufacturing cost low by using only one tensile CESL over both n- and p-type transistors. Use of a single CESL is cheaper than use of a dual stress liner (DSL) and increases nMOSFET performance while not degrading the pMOSFET, due to its low stress sensitivity for this orientation.

At the 45-nm node, most semiconductor manufacturers return to the standard (100)/<110> orientations to use more sophisticated stress engineering for both n- and p-transistors.

Figure 3.27 shows the mobility gain for electrons under 1 GPa longitudinal tensile stress and for holes under 1 GPa longitudinal compressive stress. The highest gain is obtained at different crystal orientations for the holes and electrons, but promises twofold electron mobility and fourfold hole mobility, making pMOSFETs as strong as nMOSFETs. If achieved, this would make the layout of n- and p-type transistors in CMOS symmetric, instead of the pMOSFETs being 1.5 times wider than the nMOSFETs before the stress engineering era. Even if complete parity of n- and p-type transistors is not achieved, we should expect to see the pMOSFETs getting closer to the historically stronger nMOSFETs, just because the holes exhibit the ability to go faster under stress than the electrons.

3.2.7   Uniaxial, Biaxial, and Arbitrary Stress Patterns

It is impossible to obtain a strictly uniaxial (i.e., nonzero stress along only one direction) or biaxial (i.e., identical nonzero stresses along only two orthogonal

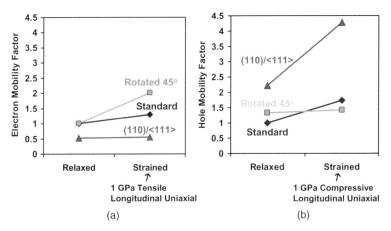

Figure 3.27 Mobility gain for (a) electrons under 1 GPa of longitudinal tensile stress and (b) holes under 1 GPa of longitudinal compressive stress under $E_{eff} = 0.5$ MV/cm.

directions) stress pattern in circuits. All three stress components will always be nonzero, although it is possible to obtain stress patterns with dominant stress components. For example, if you manage to apply force in one direction only (and none of the existing stress engineering techniques can do that), the Poisson's ratio effect will kick in and create noticeable stress in the other two directions, on the order of 20 to 30% of the intended stress. Moreover, specific transistor patterns and their neighborhood would further alter all three stress components within the transistor's channel.

By nature of silicon mechanical properties, stress gradients span several hundreds of nanometers. One consequence of this is that in short channel transistors, with L below about 100 nm, stress distribution within the channel is fairly uniform. On the contrary, in a long channel transistor, say 1 μm long, stress distribution can be very nonuniform, and can even change sign as you go along the channel from source to drain. Even for transistors that have the same short channel length, the channel width makes a big difference in the channel stress. Specifically, the Poisson's ratio effect creates transverse stress of different signs in narrow versus wide transistors [29].

Wafer bending is a popular approach to studying the impact of stress on transistors. Wafer bending is usually expected to induce a purely uniaxial stress in the channel [30]. In fact, it does create uniform stress on a large scale. On a transistor scale, the stress remains uniform only if there are no patterns, such as in a bare silicon wafer or a wafer covered with layers of uniform thickness. However, as soon as you introduce some patterns, as in a transistor, the wafer bending stress becomes very nonuniform, as illustrated in Figure 3.28. This means that even such a seemingly simple technique as wafer bending requires careful stress analysis for proper interpretation of the results.

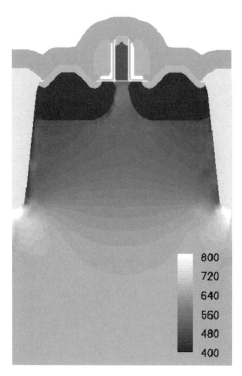

Figure 3.28   Distribution of longitudinal stress component in megapascal in a transistor sub-jected to wafer bending.

3.2.8   Stress Gradients

Stress is typically highest at stress sources, and decays as you move away from a source. The nonuniform stress distributions imply that there are always stress gradients. So far we were looking at the effects related to local stress, considering it uniform within some vicinity. Stress gradients lead to several additional effects, one being that the diffusing impurity atoms experience drift due to the stress gradient. This applies only to the stress present in silicon during high-temperature annealing steps when the impurity diffusion is significant. The amount of drift is determined by the gradient in conduction and valence bands induced by the stress gradient. The direction of the impurity drift is determined by whether a specific mobile impurity-defect pair tracks the conduction band or the valence band [31]. Another important effect associated with the stress gradient is the drift of electrons and holes in the built-in field induced in conduction and valence bands by the stress gradient. Such a carrier drift can speed up or slow down transistor operation, depending on its sign and location.

Bandgap engineering has been used successfully for many years in hetero-junction transistors, where different semiconductors or variable mole fractions of the semiconductor components affect the bandgap. Stress gradients can be used for the same purpose. For example, conduction band lowering from the source to the channel creates a drift of carriers that increases the drain current. The flip side of stress gradients is that they are dangerous in terms of defect formation. Fast transition from stretched to compressed matter makes it prone to the defects that compensate the gradient. The most important practical example of this effect is the vacancy drift from tensile to compressive metal, causing void formation and failure of the interconnects and vias.

3.2.9   Effects of Temperature and High Dopant Concentrations

Temperature affects the distribution of carrier energy, usually described by Fermi–Dirac or Boltzmann statistics [24]. At zero temperature, there are no free carriers. The higher the temperature, the more free carriers are there, and at higher energies. This leads to the following implication for stress-enhanced mobility. Stress engineering lowers the valley with lower effective carrier mass, so that more carriers repopulate into it. At higher temperatures the number of free carriers and their energies rise such that they fill up the desirable valley and then spill into the other valleys. This defeats the purpose of stress engineering by spreading the carriers across good and bad valleys. Lower temperature, on the contrary, helps stress engineering to keep the carriers in the right places.

The effect of dopant concentration is similar to the effect of temperature. Higher doping means that there are more carriers, which are spread across a wider energy spectrum. Quantitatively, these effects are described by simple expressions based on Fermi integrals [28]. It is important to note that the doping effect applies to majority carriers only. For MOSFETs, that means carriers inside the source and drain. Channel doping does not directly affect stress-enhanced mobility of the minority carriers in the channel. The indirect effect is mobility degradation in a high normal electric field that does increase with channel doping. This effect is much weaker than the one that affects majority carriers.

3.2.10   Stress Effects in Nonsilicon Semiconductors

The original piezoelectric effect was discovered in silicon and germanium [10]. However, the semiconductor industry then moved almost exclusively into silicon, which later became the target of most stress engineering studies. Now there is a renewed interest in nonsilicon semiconductors, so we should expect to see more reports on stress effects in these materials.

Qualitatively, the impact of stress on nonsilicon semiconductors is similar to what happens in silicon. Let's look at some differences, though. One important difference is that silicon is the most rigid of semiconductors of interest. That means that the same applied force leads to a larger strain in nonsilicon than in silicon semiconductors. For example, silicon has a Young's modulus of 165 GPa, whereas GaAs has a Young's modulus of only 85 GPa. The same stress corresponds to twice as large a strain in GaAs than in silicon. The reason this is important is that it is strain that determines the band structure of a semiconductor. The flip side of a lower Young's modulus is that the material is not as strong and suffers from defects and cracks at a lower stress level.

3.3 STRESS PROPAGATION

Consider that some stress is generated by a stress source. How far will it propagate from the source? The answer depends on the size and shape of the stress source and on the mechanical properties of the surrounding material. A point defect, which can be an extra atom (i.e., an interstitial), or a missing atom (i.e., a vacancy) or an impurity atom, creates stress that propagates only about 1 nm into the surrounding silicon lattice. Large stress sources can create stress that extends as far as 2 μm into the silicon.

3.3.1 Stress Propagation for Various Stress Source Geometries

For simple geometries, there are simple solutions in terms of stress distributions [32]. Consider stress sources that take one of the simple shapes. An infinite plane introduces stress that decays as $1/r$, where r is the distance from the plane. An infinitely long line introduces stress that decays as $1/r^2$. An infinitely small point introduces stress that decays as $1/r^3$. The displacements u_i are related to the diagonal components of strain ε_{ii} as [32]

$$\varepsilon_{xx} = \frac{\partial u_x}{\partial x} \quad \varepsilon_{yy} = \frac{\partial u_y}{\partial y} \quad \varepsilon_{zz} = \frac{\partial u_z}{\partial z} \tag{3.4}$$

This means that the decay of displacements and the strain for simple stress sources behave as shown in Table 3.4.

Figure 3.29 illustrates the decay of displacements in silicon calculated around a pointlike stress source 0.2 nm in diameter, an infinitely long line with a 0.2-nm cross-sectional diameter, and a 0.2-nm-thick plate with a 30-nm diameter. It is clear from this figure that for a platelike surface 30 nm in diameter, within about 30 nm of the plate it looks like an infinitely large wall, and its displacement exhibits almost no decay. However, for distances greater than 30 nm, it begins to look more like a point, and the stress decay switches to

Table 3.4 Stress penetration and decay around stress sources of several simple shapes

Shape of the Stress Source	Strain	Displacement
Point	$1/r^3$	$1/r^2$
Line	$1/r^2$	$1/r$
Plane	$1/r$	$\ln r$

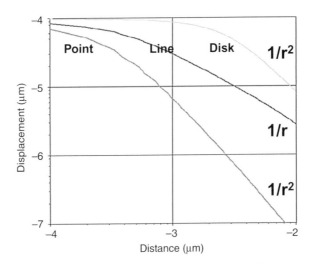

Figure 3.29   Stress propagation around defects of different sizes. Displacement is shown as a function of distance from the stress source. The point stress source is a sphere 0.2 nm in diameter. The line stress source is an infinitely long line with a 0.2-nm cross-sectional diameter. The disk stress source is 30 nm in diameter and 0.2 nm thick.

something that is similar to the $1/r^2$ trend that is expected for a point stress source. Similar behavior can be seen for the other shapes of the stress source. In silicon technology there are no such simple shapes. The actual shapes, however complex, usually fall somewhere in between the simple cases. Moreover, with finite sizes for the stress sources, the stress decay is different close to the source from decay far away from the source.

The simple solutions mentioned here refer to stress propagation in an isotropic matter. All crystal lattices have anisotropic mechanical properties, and the impact of stress on transistor behavior requires considering individual stress components rather than hydrostatic pressure. Moreover, stress is very sensitive to the exact shape of stress sources and of other layers and parts that make up transistors. Therefore, accurate three-dimensional stress distribution in transistors can be obtained only using numerical stress analysis of structures with exact geometry and the appropriate stress sources. Later in the chapter we discuss several simplified stress approximation techniques that do a reasonably good job of describing stress distribution along the silicon surface based on the results of three-dimensional stress analysis.

3.3.2   Stress Propagation Through STI and Other Barriers

Whenever stress goes across an interface between two materials, the strain remains continuous, but stress has a discontinuity that is proportional to the ratio of the Young's moduli of the two materials. Stress is higher in more rigid materials and lower in softer materials. Figure 3.30 illustrates how stress goes

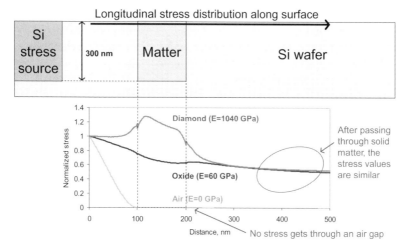

Figure 3.30 Stress propagation through different layers. For example, a diamond, an oxide, or an air gap are considered in place of the "matter" layer, which is 300 nm deep and 100 nm wide, a typical size for an STI.

through several materials with very different mechanical properties. As we move inside silicon away from the stress source, it decays according to silicon's mechanical properties. As we enter the softer oxide, the stress drops, but as we reenter silicon on the other side of the oxide layer, stress rises and continues as if the oxide were not there.

If, instead of the oxide, we track stress penetration through a diamond, we see stress increasing inside the diamond due to its higher rigidity, but then dropping back upon reentering silicon as if the diamond were not there. Therefore, when looking at the layout stress proximity effects, we can assume that stress propagates unchanged through layers such as STI. A notable exception from this rule is an air gap. As shown in Figure 3.30, the stress relaxes completely at the free silicon–air boundary and nothing penetrates past the air gap.

During wafer processing, some air gaps are encountered temporarily. For example, you etch a trench in silicon before filling it with STI for isolation or with other materials for a capacitor. Whatever stress was present in the silicon before such etching, it relaxes next to the air gap. Whatever stresses appear in the wafer after the trench is filled with solid-state matter, they will remain in silicon as the temporary air gap is gone.

3.3.3 Free Boundaries

A surface of solid-state material exposed to the air is called a *free boundary* from the point of view of mechanical analysis. Any forces in the direction normal to such boundaries can easily relax, as the material is free to expand or shrink in that direction. The tangential stress components remain, but get affected by the Poisson's ratio effects. Usually, free boundaries are horizontal surfaces of the top layer that covers the wafer. That assists relaxation of the vertical stress component in these layers but keeps the horizontal stress components intact. In some cases during process flow, some of the vertical surfaces temporarily become free boundaries. Usually, this happens after etching or patterning of the STI, poly gates, and so on. Whenever such a vertical free boundary is next to a strained layer such as SiGe, it relaxes horizontal stress components in the strained layer [33,34].

One interesting example of the importance of free boundaries is the stress memorization technique (SMT). It will be dealt with in detail later in the chapter, but here we would like to emphasize that it only happens when the transistor is covered with a rigid nitride film during high-temperature anneal. The nitride provides a constraint that changes boundary conditions for force rebalancing and makes the SMT effect possible. Without the nitride, free boundaries make the stress disappear.

3.4   STRESS SOURCES

There are several ways to classify stress sources in transistors. One distinction is whether the stress source is intentional (introduced for the purpose of stress engineering) or unintentional (essentially, a side effect of the wafer processing). The unintentional stress is not always harmful and can be beneficial, depending on the transistor geometry. Another distinction is whether the stress is induced temporarily during the wafer processing and disappears by the end of the process or survives the process and persists in the circuit during its operation. The temporary stress can leave a permanent impression on the transistor by affecting dopant diffusion and layer growth. The residual stress alters the bandgap, leakage, threshold voltage, and carrier mobility in the transistors. In this section we review stress sources that have been invented so far.

3.4.1   Thermal Mismatch: STI and Silicide

Fundamentally, all materials expand when heated and shrink when cooled down. There are some notorious exceptions to this general rule associated with phase transitions, but they are irrelevant for this discussion. The mechanism behind volume expansion with rising temperature is that the atomic vibrations increase with temperature so that each atom pushes its neighbors farther away, elongating the bonds that hold them together. The volume expansion ΔV is proportional to the temperature change ΔT and the thermal expansion coefficient α:

$$\Delta V = \alpha \Delta T \qquad (3.5)$$

Table 3.5   Thermal expansion coefficients of typical materials used in the semiconductor industry

Material	Thermal Expansion Coefficient, $1/K$
Silicon	3.05×10^{-6}
Silicon oxide, SiO_2	1.0×10^{-6}
Silicon nitride, Si_3N_4	3.0×10^{-6}
Germanium	3.05×10^{-6}
Gallium arsenide	6.05×10^{-6}
Polysilicon	3.05×10^{-6}
Aluminum	2.44×10^{-5}
Nickel	1.25×10^{-5}
Nickel silicide	$9.5e \times 10^{-6}$
Tungsten	$4.6e \times 10^{-6}$
Copper	1.65×10^{-5}
Hafnium oxide	1.31×10^{-6}

As long as the materials behave elastically, the volume expansions of the adjacent materials create thermal mismatch stress, which is proportional to the volume mismatch. Most of the time, materials do behave elastically in silicon processing, and later in the section we look at some of the important nonelastic effects.

The fundamental trend among different material types is that insulators are the least sensitive to temperature, and conductors have the most thermal expansion and contraction. Semiconductors are, as expected, somewhere in between. Table 3.5 lists thermal expansion coefficients of the major materials used in semiconductor manufacturing.

STI is traditionally filled with oxide using TEOS (tetraethyl orthosilicate) deposition. The deposition is performed at elevated temperatures, typically in the range 700 to 800°C. During high-temperature deposition, the oxide is stress-free and fills the shallow trench in silicon without difficulty. During subsequent wafer cooling to room temperature, the oxide does not shrink as much as the silicon. As a result, the silicon trench becomes too small for the oxide that filled it. Therefore, the squeezed oxide pushes in all directions. Due to the free surface on the top, the top oxide surface expands upward, drastically reducing the vertical stress. However, laterally, it squeezes the adjacent silicon. The silicon next to the STI receives the highest compressive stress, typically several hundreds of megapascal. As you move farther into silicon, the compressive stress decays, disappearing in about 2 μm from the stress source. The impact of compressive STI stress on transistor performance through enhanced hole mobility and degraded electron mobility was noted in the semiconductor industry in the 1990s [35].

Usually, there are several high-temperature annealing steps in CMOS process flow after STI formation. Here is what happens during such annealing steps. During wafer heating, the oxide expands, but the silicon expands even more. Upon reaching the initial STI deposition temperature, the stress returns to zero. If the peak temperature is higher than the STI deposition temperature, the stress sign reverses, because faster silicon thermal expansion makes the trench too large for the oxide. Therefore, when the dopant diffuses at highest temperatures, both STI and the adjacent silicon are at a low tensile stress. If the peak temperature exceeds 950 °C, the oxide liquefies and becomes viscous. Viscous flow of the amorphous oxide relaxes the stress within oxide as well as in the adjacent silicon. Subsequent cooling of the wafer to room temperature creates thermal mismatch stress starting at zero stress at 950 °C.

There are several additional stress mechanisms associated with oxide, such as:

- Buildup of high tensile stress during STI densification (when water evaporates from deposited TEOS) and the oxide volume shrinks by 5 to 10%
- Compressive oxide stress due to the 126% volume expansion of silicon being converted to oxide during thermal oxidation
- Tensile oxide stress when an oxide grown at lower temperatures (below about 950 °C) relaxes and gets less dense during subsequent annealing at high temperatures

These stress mechanisms introduce additional stress into the oxide and surrounding silicon temporarily, but most of this stress disappears by the end of the process flow and the dominant remaining oxide-related stress is due to the thermal mismatch.

The STI depth is about 300 nm for the 65-nm technology node, and it exhibits extremely slow scaling due to the specifics of latch-up protection. The lateral size of STI (let's refer to it as STI width) when it was introduced in the 1990s was over 500 nm, which translates into an STI aspect ratio of 300/500 = 0.6. The STI width strictly follows layout scaling of 0.7-fold per technology node to allow higher transistor densities. At the 45-nm technology node, the STI width is expected to be about 70 nm, which translates into an STI aspect ratio of 300/70 = 4.3. It is difficult to fill a trench with such a high aspect ratio using TEOS. An SOG (spin-on-glass) oxide is considered for filling such trenches for the 45-nm node and beyond. It is a softer material that creates tensile rather than compressive stress. Usually, the bottom of the STI is filled with tensile SOG and the top part is filled with conventional compressive TEOS. This makes for a complicated stress pattern, explained in detail by Ota et al. [36]. Briefly, transistors located close to such double-layer STI experience short-range compressive stress from the top part of the STI, whereas remote transistors experience longer-range tensile stress from the bottom of the STI. Therefore, by moving the transistor relative to STI, you can see its mobility both increase and reduce as the STI-induced stress changes sign.

Silicide is a good conductor and usually generates tensile stress inside the silicide as well as in the adjacent silicon. Being shallower than the STI (a typical silicide thickness is about 15 nm for the 65-nm node), it induces relatively short-range stress that affects the nearest transistor channel, but hardly affects neighbors.

3.4.2 Lattice Mismatch: eSiGe and Si:C

Germanium and carbon are elements of group IV of the periodic table. When mixed with silicon, they take substitutional positions and do not introduce charges, because like silicon they use the four valence electrons to form bonds with the silicon lattice. The atomic radii of carbon, silicon, and germanium are 0.77, 1.17, and 1.22 Å, respectively. The germanium lattice is 4.2% larger than the silicon lattice, and the diamond lattice size is 45% smaller than silicon.

Each substitutional germanium atom creates small compressive stress in silicon lattice, and each substitutional carbon atom creates a large tensile stress in silicon lattice. The amount of stress and the stress pattern depend on the specific geometry and require numerical three-dimensional stress analysis for typical transistors with complicated shapes. For some simple geometries, there are simple solutions. For example, a strained SiGe or Si:C layer grown epitaxially on top of a planar silicon wafer has zero vertical stress and the following amount of stress in all lateral directions:

$$\sigma = \frac{E\varepsilon}{1-\nu} \tag{3.6}$$

where $E = 165\,GPa$ is Young's modulus of silicon averaged among different crystal orientations, ε is strain, and $\nu = 0.28$ is the silicon Poisson's ratio.

Strain is defined as a relative size distortion. *Zero strain* means that silicon lattice has a nominal size. A positive strain of 0.01 means that the lattice is stretched by 1%. A negative strain of −0.01 means that the lattice shrank by 1%. For the lattice mismatch stress we have

$$\varepsilon = \varepsilon_0 \theta \tag{3.7}$$

where ε_0 is the maximum strain between 100% silicon and 100% nonsilicon (i.e., germanium or carbon) lattices. For germanium, $\varepsilon_0 = -0.042$, and for carbon, $\varepsilon_0 = 0.45$. The θ is a mole fraction of germanium or carbon in Si. Plugging in $\theta = 0.1$ for $Si_{0.9}Ge_{0.1}$ with 10% germanium, we get compressive lateral stress $\sigma = -976\,MPa$. For a $Si_{0.99}C_{0.01}$ with 1% carbon, we get tensile lateral stress $\sigma = 1046\,MPa$.

The fact that each germanium atom introduces relatively low stress into silicon lattice makes it possible to mix germanium and silicon in arbitrary fractions and still have a perfect lattice with all germanium and silicon atoms

Table 3.6 Critical thickness of the SiGe layer for various germanium contents

Germanium Content (%)	Critical Thickness (μm)
10	1.0
15	0.3
20	0.15
30	0.05
40	0.02

in substitutional lattice positions. If SiGe is epitaxially grown, the lattice of the underlying silicon wafer is reproduced. If germanium is implanted, an anneal helps germanium atoms to get into substitutional positions. Usually, a germanium implant amorphizes silicon and the subsequent solid-phase epitaxial recrystallization works similar to the epitaxial layer growth.

There is a concept of critical SiGe layer thickness beyond which misfit dislocations form at the Si–SiGe interface to relax the lattice misfit stress [37]. This happens for thick SiGe layers and/or high-germanium mole fractions. Table 3.6 lists critical SiGe layer thicknesses for different germanium content. The concept of critical layer thickness applies to a blanket layer grown over the entire planar wafer. If SiGe is broken into small islands that serve as source and drain for pMOSFETs, which is how it is done in a typical layout, the SiGe thickness can be somewhat thicker than critical without generating misfit dislocations. The layout-specific critical SiGe thickness can be calculated by estimating the accumulated strain energy for each SiGe island.

Figure 3.31 shows a cross section of Intel's 50-nm pMOSFET for the 45-nm technology node. The diamond-shaped SiGe source–drain regions all have the same small size about 100 nm in the lateral direction. The 100-nm-thick SiGe with 30% germanium is twice as thick as the critical layer thickness and is bound to generate misfit dislocation network whenever a large SiGe lateral size is encountered. Therefore, all SiGe regions have to be chopped into small islands to avoid the defect formation. The small lateral SiGe sizes observed in Figure 3.31 help to keep the strain energy low and avoid misfit dislocations.

Another interesting feature of the diamond-shaped SiGe regions depicted in Figure 3.31 is that the Si–SiGe interface follows (111) planes. This minimizes shear stress projected onto (111) dislocation slip planes and helps to suppress formation of dislocations in adjacent silicon. In fact, the TEM image does not exhibit any dislocations, as opposed to previous generations of Intel's embedded SiGe source–drain, which had similar SiGe thickness and even lower germanium content, but a rounded rather than a diamond SiGe shape.

Each carbon atom generates about 10 times higher stress than each germanium atom. This makes it difficult to incorporate large carbon fractions as

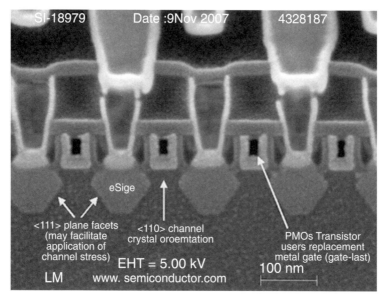

Figure 3.31 TEM image of the 50-nm pMOSFETs with nested gates from Intel's 45-nm technology node.

substitutional atoms. Silicon with more than about 0.5% carbon becomes unstable, with carbon atoms clustering at high-temperature anneals. Carbon–carbon and carbon–interstitial clusters are stress free and form readily during annealing, especially with implant-introduced extra interstitials. This complicates using Si:C in stress engineering, and explains why it is still not used in production at the 65-nm technology node, whereas SiGe is widely used. Research is ongoing and there are hopeful results reported for the Si:C using an advanced millisecond annealing technique [38].

3.4.3 Layer Growth

Most layer deposition steps introduce minor unintentional stress that does not significantly affect transistor performance. Thermal oxidation of silicon is one of the few process steps that can introduce significant stress into the adjacent silicon and even generate dislocations at the silicon–SiO_2 interface. Oxidation-induced dislocations were a major concern in the 1980s and early 1990s, when transistors were isolated using a thick thermal oxide LOCOS. Now, isolation is done using STI, and the thickness of thermally grown oxide is getting smaller with every subsequent process generation. Therefore, oxidation-induced dislocations are no longer a major concern.

When oxygen reacts with silicon to form SiO_2 [39], the resulting oxide volume is 126% larger than the volume of silicon consumed. If the silicon surface is planar, the volume expansion is relaxed due to the top oxide surface being a free boundary. However, either concave or convex corners constrain volume expansion and lead to a buildup of compressive stress in the oxide [40].

Thermal oxidation of silicon is usually performed multiple times during the process flow, for different purposes. Each time it is performed, the adjacent silicon gets stressed. Most of this stress is temporary, though. Some disappears when the oxide is etched away, which happens to the sacrificial oxidations; some relaxes, due to the viscous oxide flow at high-temperature annealing steps. By the end of the process flow, the residual oxide stress is dominated by the thermal mismatch discussed earlier in this chapter.

Silicide growth is also associated with significant volume expansion, sometimes more than 200%. However, silicide is known for not generating much as-grown stress nor even responding to it [33,40]. Therefore, similar to the oxide, although for a different reason, residual silicide stress is dominated by the thermal mismatch stress.

3.4.4 Intrinsic Stress: CESL and DSL

Nitride contact etch-stop layers (CESLs) have been used in semiconductor manufacturing for decades to help mitigate contact misalignment problems. Once stress was identified as a vehicle for improving transistor performance, strained CESL layers appeared right away. The CESL deposition is done at the end of front-end process (i.e., transistor formation) and before the back-end process (i.e., interconnects). There are no high-temperature steps after CESL deposition, which means that there is no danger of losing stress due to possible structural changes in the nitride. It also means that the CESL stress does not affect dopant diffusion. The reason that CESL introduces some stress into the channel is due purely to the transistor geometry. A flat strained nitride layer does not introduce any noticeable stress into the underlying silicon. A strained nitride layer that is "disrupted" by going over a gate stack introduces significant longitudinal and vertical stress components into the channel.

Consider a compressive CESL that is used to improve pMOSFET performance. Figure 3.32 illustrates the distribution of the longitudinal stress component. The horizontal parts of compressive CESL are trying to expand and stretch the silicon that lies immediately under CESL. The horizontal CESL to the right of the gate stack pulls to the left as it tries to expand and sees disruption by the gate stack as an opportunity to relax some of its stress. That pulling compresses silicon in the channel (i.e., under the gate stack) and stretches the drain (i.e., immediately under the horizontal part of CESL). Similarly, the horizontal part of CESL to the left of the gate stack also compresses silicon in the channel and stretches silicon in the source, immediately under the CESL. As you can see from electron piezo coefficients in Table 3.3, longitudinal compressive stress enhances hole mobility, which is why compressive CESL works for pMOSFET.

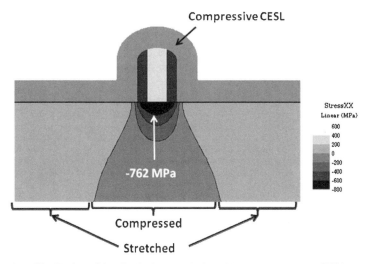

Figure 3.32 Distribution of longitudinal stress induced by a compressive CESL in a stand-alone pMOSFET.

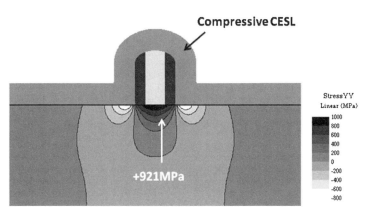

Figure 3.33 Distribution of vertical stress induced by a compressive CESL in a stand-alone pMOSFET.

The vertical stress induced by CESL is illustrated in Figure 3.33. The two vertical parts of compressive CESL on either side of the gate stack are trying to stretch, and by doing so, pull the gate stack up away from the channel. That creates tensile vertical stress in the channel, which does not affect hole mobility. For the nMOSFETs, compressive CESL would degrade its performance, and therefore a tensile CESL is used there instead. It creates a stress pattern opposite to that of compressive CESL: tensile longitudinal and compressive vertical channel stress. Both of these stress components enhance electron mobility, as can be seen from Table 3.2. Currently, compressive CESL can

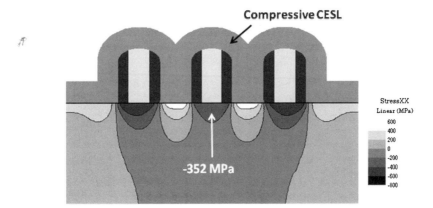

Figure 3.34 Distribution of longitudinal stress induced by a compressive CESL in a pMOSFET with two neighbors.

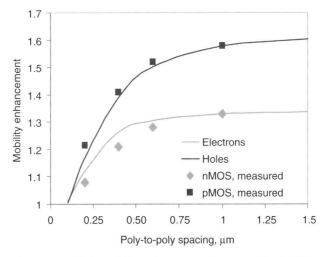

Figure 3.35 Electron and hole mobility enhancement due to strained CESL as a function of the poly-to-poly spacing. (Simulation performed using *Seismos* [23]. Measured data from ref. 41.)

reach built-in stress up to $-3\,\mathrm{GPa}$, whereas tensile CESL can reach stress up to about $1.8\,\mathrm{GPa}$. Different built-in nitride stress is achieved by manipulating deposition conditions to form preferentially N–H or Si–H bonds.

So far we have been looking at a stand-alone transistor with strained CESL. For transistors placed on the chip and surrounded by neighbors, the CESL effect will be different. Transistors placed to the left and to the right of the current transistor as shown in Figure 3.34 disrupt the CESL and weaken its longitudinal stress. A typical trend for mobility enhancement as a function of poly-to-poly distance is shown in Figure 3.35. You can see that a shorter poly pitch degrades mobility and increases the sensitivity of mobility to layout variations. A longer poly pitch enhances mobility and suppresses layout sensitivity, but of course increases the chip area. These competing trends make selection of the optimum poly pitch a trade-off.

Whenever both tensile and compressive CESL are used on the same chip, which is an increasingly popular option called DSL (dual stress liner), the location of the boundary between the two nitrides makes a difference in the performance of both transistor types. Usually, keeping the boundary between tensile and compressive CESL as close as practical to pMOSFETs and as remote as practical from nMOSFETs improves the performance of both transistor types. For the fastest circuit performance and minimum layout area, the place-and-route tool should take into account the impact of neighborhood on DSL-induced stress and find the best trade-off.

3.4.5 Stress Memorization Technique

The stress memorization technique (SMT) is a good example of a process that is not well understood but nevertheless is being embraced by the industry for the 45-nm technology node because of its clear benefits. Here is how it is done. After the amorphizing source/drain implants, a nitride layer is deposited over the gate stacks, and the high-temperature spike anneal is performed with this nitride on. After annealing, the nitride is removed, but the nMOSFET performance increases noticeably. What is new in this approach is that the nitride covers the wafer during annealing.

The built-in stress in the nitride used in SMT does not matter, but there is a direct correlation of SMT effect with the nitride density [42]. Several consecutive SMT steps have been reported to accumulate mobility improvements, with the beneficial stress coming from the poly gate as well as from the source and drain [43]. SMT is known to improve nMOSFET and degrade pMOSFET, so it is applied only to the nMOSFET. Stress is believed to be responsible for nMOSFET improvements, whereas harmful stress and increased series resistance are viewed as possible pMOSFET degradation mechanisms.

Lack of understanding of the exact physical mechanisms responsible for the SMT effect translates into lack of a quantitative model. Having tensile stress upon epitaxial recrystallization of the amorphized source and drain can explain tensile longitudinal stress in the channel, with nitride playing a crucial role. Namely, it puts a constraint on the usually free top surface of the source and drain and does not allow the swelled amorphized silicon to shrink during epitaxial recrystallization. It is not known at this time how SMT puts compressive stress into the poly gate.

3.4.6 Stress Measurement Techniques

Several stress measurement techniques have been invented so far. All of them measure strain, or the length of the lattice bonds averaged over some volume. A good review and comparison of several popular stress measurement

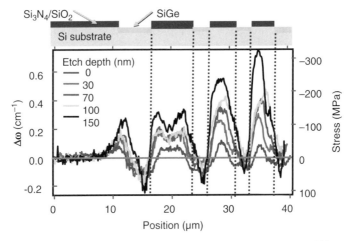

Figure 3.36   Silicon compressive stress induced by SiGe layers of different widths and depths as measured by micro-Raman. (From ref. 25.)

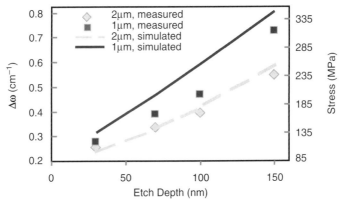

Figure 3.37   Simulated and measured silicon stress induced by SiGe layers of different widths and depths. (From ref. 25.)

techniques is provided in an article by Senez et al. [44]. The main problems of these techniques are that they have trouble resolving small areas and therefore work well on larger patterns, but cannot resolve individual transistors, and that they need complicated interpretation, which in turn requires some assumptions about the stress pattern.

An approach that has been working well so far is to apply one of the measurement techniques to relatively large patterns that they can easily resolve and to calibrate a stress model based on these data. Then apply the calibrated model to the smaller patterns of interest. Here is an example of successful application of this approach. The simulated eSiGe stress was verified using micro-Raman stress measurements for the long silicon and SiGe stripes of different widths shown in Figure 3.36. Figure 3.37 shows good agreement of simulated and measured stress for the 2-μm-wide silicon stripes, but systematically lower measured stress values for the narrower 1-μm-wide silicon stripes. This is expected, as micro-Raman technique cannot resolve features below about 2 μm. Nominal mechanical properties were used for silicon, SiGe, and other materials, and no parameter tuning was done for these simulations.

The model that was verified on patterns that are larger than 2 μm was then applied to the 40-nm pMOSFETs with an eSiGe source–drain [45]. The stress calculated for the actual complex pMOSFET geometry and the stress-enhanced hole mobility that was extracted from electrical measurements create a mobility versus stress curve that is consistent with wafer-bending experiments for pMOSFETs as well as with the classic piezoresistance experiments for the bulk resistance. In addition to the conventional stress measurement techniques, there are new approaches that promise to achieve higher resolution, like the nano-Raman method [46].

3.4.7 Stress Simulation Techniques

The problem that we are trying to solve here is to find the distribution of strain and stress in the wafer, which contains transistors and other active and passive structures with rather complicated shapes. The strain and stress are caused by mechanical forces that arise from several different physical mechanisms. The exact solution of such problem exists only for a limited number of structures with simple geometries. The transistors are hopelessly complex for a simple analytic closed-form solution. Therefore, numerical stress analysis is typically used to solve such problems.

Numerical stress analysis leads to a rigorous numerical solution to the fundamental stress equations and provides results that are accurate within the numerical error, which is on the order of 1%. An example of typical general-purpose stress simulation software is ANSYS [47], an example of typical stress simulation software that is specialized for semiconductor process is FAMMOS [48]. A specialized tool such as FAMMOS has a database of related material properties, can read layout information from GDS files, and can build transistors by simulating a sequence of process steps. Detailed descriptions of the equations and solution algorithms involved may be found in articles by Moroz et al. [33] and Senez et al. [49].

Besides solving the appropriate equations accurately, it is very important to have an adequate formulation of the problem. Here is one example. Consider deposition of a layer with built-in stress, such as SiGe, or a strained CESL. To model stress induced by such a strained layer, one can use the following intuitively apparent one-shot approach:

1. Deposit the new layer.
2. Introduce the appropriate built-in stress by making all three stress components the same.

3. By solving the stress equations, rebalance the forces to find the stress distribution.

This approach will give an accurate solution only if the surface of the structure is planar. For an arbitrary geometry, including planar and non-planar surfaces, the following multistep approach gives accurate stress distributions:

1. Deposit a small fraction of the layer thickness, approximately one-tenth of the layer thickness.
2. Introduce the appropriate built-in stress by making all three stress components the same in the newly deposited thin layer.
3. By solving the stress equations, rebalance the forces to find the stress distribution.
4. Keep returning to step 1 until the entire layer thickness is achieved.

From a physics point of view, the multistep approach means that force rebalancing happens during growth and deposition rather than waiting for the entire deposition to finish. In fact, the rebalancing is expected to happen even on a smaller atomic scale, as each atom gets deposited on the surface. The atom immediately feels the surrounding forces and only needs to move a fraction of an angstrom to accomplish force rebalancing. Because the move is so small, it does not require thermal activation to jump over any barriers and therefore happens instantly on the scale of deposition step duration. The multistep approach does not need to go as far as atomic scale, as subdividing the deposition into 10 to 20 substeps is usually enough to get accurate results. Figure 3.38 compares results of the one-shot and multistep approaches and shows that the one-shot approach does not have the right range nor the right amount of stress to describe the measured mobility versus poly spacing trend.

Another important factor to consider is whether to use full-blown three-dimensional analysis or reduce it to the simplified two-dimensional analysis. The two-dimensional analysis is easier to set up and faster to solve, but it can only be used for really wide transistors, with a channel width of at least several microns. Figure 3.39 shows longitudinal and transverse stress components in the channel of a pMOSFET with an eSiGe source/drain as a function of channel width W. It is clear that the compressive transverse stress reduces with shrinking W, and at small channel widths even reverses sign and becomes tensile.

The mechanism behind the transverse stress switching from compressive at large W to tensile at small W has to do with the Poisson's ratio effect. If a piece of a material is stretched in one direction, in response to that it will shrink in a plane that is normal to the stretching direction. This is referred to as the *Poisson effect* and is determined by the value of Poisson's ratio for each

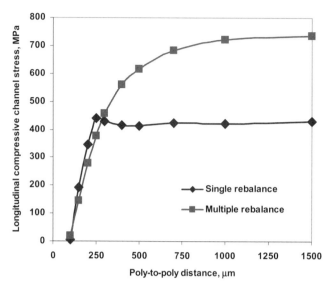

Figure 3.38 Comparison of longitudinal CESL-induced stress as a function of poly-to-poly distance calculated using single and multiple stress rebalances.

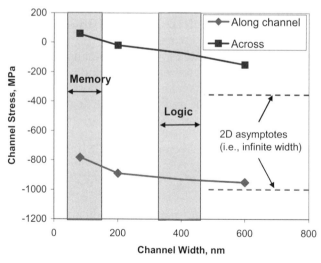

Figure 3.39 Longitudinal and transverse stress components in the channel of a pMOSFET with eSiGe source–drain as a function of channel width.

material. The Poisson effect induces stress of the opposite sign in the material that has free boundaries in the directions that are normal to the main force. However, a material that is mechanically restricted in the normal directions will respond differently, like changing the sign of the transverse stress, as in the case of the narrow channel above.

Typical pMOSFET channel widths are well below 1 μm in the 65- and 45-nm nodes, and therefore all practically important transistors do require three-dimensional stress analysis to obtain accurate stress distributions. Only extra-wide test transistors can be handled in two dimensions.

The main problem with rigorous numerical stress analysis is that it cannot handle large layouts and is limited to about 10 transistors at a time. That is why several simplified semiempirical approaches are proposed to handle large layouts, potentially with full chip coverage. One simplified stress estimation approach is the length-of-diffusion (LOD) model [19]. It assumes that STI introduces compressive stress in the longitudinal direction that decays as 1/distance from STI into the diffusion area. This model is used widely in the industry to handle STI stress effects for the 130-, 90-, and 65-nm nodes. Another simplified stress estimation approach is presented in *Seismos* [23]. In addition to the longitudinal LOD effect, it accounts for STI stress generated in the transverse direction. It goes beyond the single diffusion region and accounts for the STI stress generated by all neighbors within a certain search radius, which is typically 2 μm. In addition to STI stress, it accounts for CESL, DSL, eSiGe, contact, and SMT stresses.

3.5 INTRODUCING STRESS INTO TRANSISTORS

Many methods have been suggested for putting intentional stress into transistors. Some of these methods used widely in the industry and are considered an enabler of further progress in technology. The most popular methods used to some extent in the 90-nm node and to a larger extent in the 65- and 45-nm nodes are:

- Tensile CESL for nMOS and compressive CESL for pMOS
- eSiGe for pMOS
- SMT for nMOS

Some additional methods that might gain popularity in the near future are:

- Globally prestrained thin silicon film (for SOI only)
- Using different (hybrid) orientations for pMOS and nMOS to maximize stress engineering gain
- Si:C source–drain for nMOS
- Tensile STI for both nMOS and pMOS
- Tensile contacts for nMOS with recessed source–drain

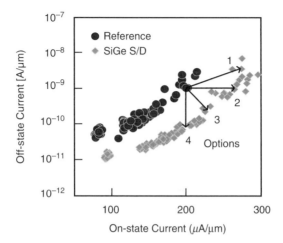

Figure 3.40 Stress-enhanced mobility can be used to improve driving current or to suppress OFF current, or to do a little bit of both.

When performing stress engineering, it is important to target the stress components that are most sensitive for a given carrier type and surface and channel crystal orientations. The reason for this is to maximize the beneficial channel stress while keeping the other stress components low enough that they do not introduce harmful defects. Another important aspect of stress engineering is deciding what to do with the stress-related performance gain. Stress-enhanced mobility increases I_{on} and I_{off} simultaneously; usually, I_{off} enhancement is larger than I_{on} enhancement. However, that still leaves you some room for tailoring the exact transistor performance.

Consider the example shown in Figure 3.40. The reference process exhibits $I_{on} = 200\,\mu A/\mu m$ at $I_{off} = 1\,nA/\mu m$. If we take a transistor with this performance and add eSiGe stress engineering (option 1), its I_{on} increases by 37% and I_{off} triples to $3\,nA/\mu m$. However, that is not the only possible option. By adjusting the process we can keep the I_{off} unchanged and still achieve a respectable 33% I_{on} gain (option 2). Option 3 demonstrates splitting the benefits and achieving both 15% I_{on} gain as well as a fivefold I_{off} reduction. Option 4 demonstrates a 14-fold drop in I_{off} at a fixed I_{on} value. Different options might be preferred for different circuit applications, such as low power versus high performance.

Stress engineering is a young field and we see it evolving quickly. It is likely that new stress engineering methods will continue to appear.

3.5.1 Stress Evolution During Process Flow

The process flow affects differently stresses generated by thermal mismatch (e.g., STI) and stresses generated by built-in stress (e.g., CESL, eSiGe). The STI stress is zero at the STI deposition temperature (which is around 800 to

900 °C). Once the wafer cools down to room temperature, compressive stress inside STI appears to be due to thermal mismatch. There is tensile stress in silicon below STI, but it does not affect anything, as usually no devices are formed under the STI. Around the STI, the silicon surface is compressed in lateral directions.

As the wafer goes through multiple thermal cycles, it is heated repeatedly to a variety of peak temperatures, and cooled down repeatedly to room temperature. As soon as the wafer reaches about 900 °C, the thermal mismatch stress in STI and surrounding silicon disappears. It reappears upon subsequent wafer cooling, with some stress increase if the wafer spends enough time above 950 °C due to the viscous stress relaxation in oxide at high temperatures.

Built-in stress such as eSiGe easily survives high-temperature steps but can be affected unintentionally by the oxide etching steps. Consider the transistor depicted in Figure 3.28. On one side (facing the channel) the swollen eSiGe layers are pushing the silicon channel. On the other side (facing away from the transistor), they are pushing the STI. If later in the process flow some of the STI oxide gets etched away during one of the sacrificial oxide etching steps, the eSiGe loses STI support there, and its vertical boundaries that face away from the transistor become free boundaries. That allows the eSiGe to expand laterally into the space once occupied by the STI. This, in turn, relaxes some of the eSiGe stress and reduces the beneficial pMOSFET channel stress.

Strained CESL has a similar problem. The contact holes that are etched in CESL perforate it and relax part of the CESL-induced stress. More contacts, placed closer to the channel, improve source–drain conductivity but simultaneously degrade the CESL-enhanced carrier mobility. Therefore, the number and location of the contacts become the trade-offs of layout optimization.

3.5.2   Stress Relaxation Mechanisms

In this chapter we analyzed the following stress relaxation mechanisms:

- Viscoelastic
- Free boundary
- Poisson ratio effects
- Defect formation

Viscoelastic flow can happen in amorphous materials only and in silicon processes is usually observed in the oxide above about 950 °C. The practical implication of this stress relaxation at high temperature is that you get significant compressive oxide stress, due to thermal mismatch while the wafer is cooling down to room temperature. The free boundaries redistribute some of the normal stress components into tangential stress components due to the Poisson effect and relax the remaining normal stress, as the material is free to expand or contract in the direction normal to a free boundary. Either crystalline or amorphous materials exhibit this relaxation mechanism.

The Poisson effect introduces transverse stress in response to a force applied in a certain direction. It does not make stress go away (i.e., it does not relax it), just redistributes it. The effect complicates interpretation of measured stress and interpretation of wafer-bending experiments, but can easily be calculated using rigorous three-dimensional numerical stress analysis with the appropriate boundary conditions. Either crystalline or amorphous materials exhibit this stress redistribution mechanism.

The defects form in the silicon lattice in response to overly high stress. They can be beneficial as long as they are placed away from troublesome areas such as a junction depletion region. Due to the difficulty controlling defect formation and placement, they are usually avoided altogether. From the mechanical analysis point of view, defect formation is the mechanism responsible for the nonelastic behavior of crystals such as silicon.

3.5.3 Combining Several Stress Sources

As long as the material exhibits elastic behavior, stress generated by different sources adds up as a linear superposition. This statement comes from the mature branch of science known as mechanical engineering and is, of course, confirmed in the few experiments performed in the young field of semiconductor stress engineering [9,11]. Figure 3.41 illustrates the distribution of stress-enhanced hole and electron mobilities in a library cell for a process containing STI, eSiGe, and DSL stress engineering. The lowest mobility enhancement,

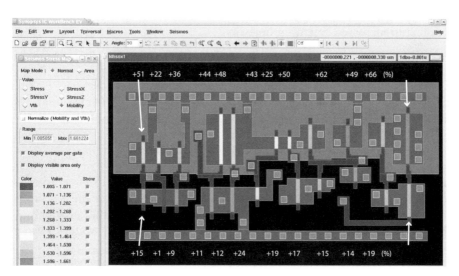

Figure 3.41 Distribution of stress-enhanced hole and electron mobilities in a 45-nm library cell that has STI, eSiGe, and DSL stress engineering.

1% (compared to the relaxed silicon), is observed in the nMOSFET that is second from the left. The highest enhancement of 66% is observed in the rightmost pMOSFET. The overall mobility range is 65%, which corresponds to a layout-induced systematic variation of driving current of about 33%.

3.5.4 Stress-Engineered Memory Retention

The main focus of stress engineering has been on enhancing carrier mobility. Here we discuss stress engineering as applied to improving charge retention in flash memory. In a flash memory cell, the charge is stored on a floating polysilicon gate. The reason it is called floating is because it is completely surrounded by oxide and is not connected to metal conductors. The main mechanism used by electrons trapped on the floating gate to escape is thermionic emission, going over the conduction band barrier between the poly and the oxide. The thermionic emission current J is an exponential function of the barrier height E_b:

$$J = A_R T^2 \exp\left(\frac{-E_b}{kT}\right) \tag{3.8}$$

where A_R is the Richardson constant, which is proportional to the effective electron mass, T is the temperature, and k is the Boltzmann constant.

The poly-to-oxide barrier is very high, so flash memory has good charge retention. To improve the retention time further, the following stress engineering technique has been suggested [50]. If tensile stress is somehow applied to the floating poly gate, it would increase the activation energy for thermal detrapping and would simultaneously increase the effective electron mass, leading to increased retention time.

3.5.5 Layout-Induced Variations

Stress engineering has brought significant transistor performance improvements and is considered imperative for the future of semiconductors. The flip side of stress engineering is that it introduces additional random and systematic performance variations, most of which are related to the layout. Stress-related random performance variations can be caused by mask misalignment. For example, whenever a gate mask is misaligned with respect to the diffusion mask, it changes the poly-to-STI distance. That in turn changes the STI stress as well as the eSiGe stress. Within a single chip, the same variation affects all transistors, but it appears random at the die-to-die and wafer-to-wafer levels.

Even if there is no mask misalignment, there are several mechanisms of systematic transistor performance variation due to the different layout neighborhood for transistors that have the same L and W. For example, different distances to neighboring transistors and different poly-to-STI distances would affect DSL, STI, and eSiGe stresses. Let's investigate the impact of layout on several simple circuits with stress engineering.

3.5.5.1 CMOS Inverter

For a simple CMOS inverter, let's estimate the effects of the layout density on its performance. Figure 3.42 shows two CMOS inverters that are scaled for the 45-nm technology node. One is a stand-alone inverter surrounded by large STI areas to the left and right. Another is surrounded by similar inverters separated by the minimum possible STI spacings.

Figure 3.43 shows the distribution calculated for stress-induced changes in the electron and hole mobilities for the two inverters. In some locations, the

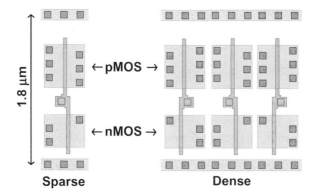

Figure 3.42   Identical 45-nm CMOS inverters in different layout environments.

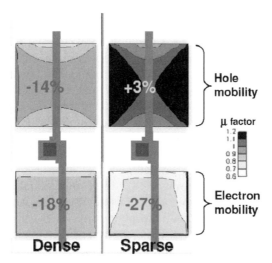

Figure 3.43   Stress-induced changes in carrier mobility for two identical 45-nm CMOS inverters in different layout environments.

mobility is degraded by up to −40% (corresponding to the white color and mobility multiplication factor of 0.6) relative to the stress-free transistor, whereas in other locations the mobility is enhanced by up to +20% (corresponding to the black color and multiplication factor of 1.2). Overall, darker colors show stress-enhanced mobility, and lighter colors show stress-degraded mobility.

For transistors in the sparse layout environment, STI introduces compressive stress from all lateral directions. The hole mobility increases with compressive longitudinal stress along the channel [10], which is illustrated by the dark areas on the left and right sides of the pMOSFET in the upper right corner of Figure 3.43. In contrast, transverse compressive stress across the channel degrades the hole mobility [10], shown as the light gray areas at the top and bottom of the same pMOSFET. The average stress-induced mobility change in the channel of the pMOSFET is +3%, compared to the hole mobility in a stress-free pMOSFET. In practice, any transistor with a large diffusion area (i.e., both the channel width and the LOD are large) is essentially stress-free.

The stress-induced hole mobility distribution in a pMOSFET placed in a dense layout environment is shown in the upper left corner of Figure 3.43. There is much less STI located behind its source and drain, and this thin STI area introduces much less beneficial longitudinal stress into the transistor channel. Meanwhile, the harmful transverse compressive stress is the same as in the case of sparse layout, because the amount of STI in the vertical direction did not change. This leads to significant degradation of the pMOSFET performance, with the average stress-induced mobility change of −14%.

The electron mobility degrades with any compressive stress in the longitudinal and transverse directions [10]. In the sparse layout there is more STI surrounding the nMOSFET, which leads to stronger compressive stress and therefore a larger mobility degradation of −27%, compared to a degradation of "only" −18% for an nMOSFET in a dense layout environment.

Looking at the CMOS inverter as a whole, we see that the nMOSFET degrades by 9% when going from a dense to a sparse layout density. At the same time, the pMOSFET improves by 17%. This shifts the nMOS/pMOS balance by 9% + 17% = 26%. This is summarized in Figure 3.44, which includes the results of a similar analysis done for pMOSFETs with embedded SiGe source and drain regions (S/D).

The inclusion of an embedded SiGe S/D stressor in a pMOSFET design modifies the results above in three major ways: (1) there is significant enhancement in the pMOSFET performance; (2) the conventional trend for pMOSFETs gaining performance when going from dense to sparse layout density is reversed; and (3) the impact of layout density on transistor performance becomes much larger because of the embedded SiGe stressor.

So far, we have been looking only at the impact of stress on carrier mobility. Table 3.7 shows how modification of the low-field mobility is translated into changes in the on-current (I_{on}), the off-current (I_{off}), and the inverter delay.

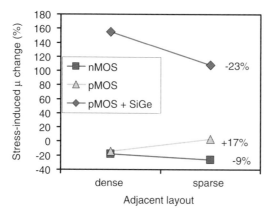

Figure 3.44 Impact of layout density on stress-induced variation in the transistor performance.

Table 3.7   Impact of layout density on transistor and inverter performance

| | Percent Change (Dense vs. Sparse) | | | |
| | STI-Only Stress | | Combined STI and SiGe Stresses | |
Parameter	nMOS	pMOS	nMOS	pMOS
I_{on}	8.2	−10.7	6.0	16.2
I_{off}	10.4	−14	7.0	16.9
Delay	−7.8 (fall)	12.3 (rise)	−5.5 (fall)	−13.9 (rise)

These device characteristics were computed using energy balance device simulation to consider the impact of stress on both low-field mobility and velocity overshoot. The changes in inverter delays correlate well to the changes in I_{on} and exceed 10%, which can be significant enough to affect the proper functionality of a circuit that contains such variations.

3.5.5.2 Ring Oscillator

Figure 3.45 demonstrates how the two inverters with both STI and SiGe stress sources compare when included in a three-stage ring oscillator as computed using mixed-mode device simulation. The ring oscillator that is surrounded by a dense layout neighborhood performs significantly faster, getting ahead by 3.5 ps in just three cycles. This happens because the pMOSFETs in the dense layout experience less relaxation of SiGe-induced stress, due to the soft STI. These transistors have a higher level of beneficial compressive longitudinal stress in the channel and therefore a higher hole mobility.

3.5.5.3 Library Cell

We now consider a library cell with a layout that is more complicated than just a simple rectangular shape. Figure 3.46 shows a library cell that is scaled

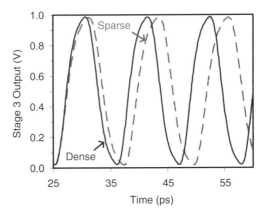

Figure 3.45   Impact of layout density on the performance of a three-stage ring oscillator with STI and SiGe stress sources.

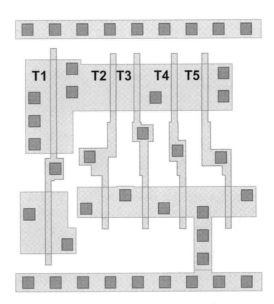

Figure 3.46   Layout of a 45-nm library cell that includes five p-channel transistors with SiGe S/D.

for the 45-nm technology node and includes five pMOSFETs with SiGe S/D. Transistors T2 through T5 have the same channel width of 330 nm. The channel width of the leftmost transistor, T1, is twice as wide as that of the other pMOS-FETs. This means that its driving current, I_{on}, is expected to be twofold higher than for the other transistors. The nonregular layout of the pMOS diffusion areas introduces nonuniform stress in the channels of the five transistors. The

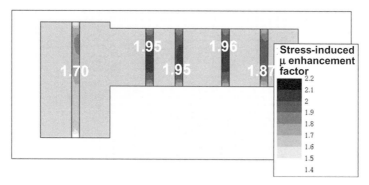

Figure 3.47   Stress-induced mobility variation in five pMOSFETs with SiGe S/D.

nonuniform stress distribution, in turn, affects the channel mobility as shown on Figure 3.47.

The three transistors in the middle, T2 through T4, are located inside a large SiGe area and exhibit fairly similar mobility enhancements of +95% compared to that of the stress-free silicon. The transistors on the sides of the SiGe area, however, are subjected to less beneficial compressive longitudinal stress and therefore exhibit lower mobility enhancement. As a result, the wide transistor, T1, is weaker by about 15% than intended by its W/L channel size. An unexpected performance variation of this magnitude might degrade circuit behavior or possibly even cause a circuit failure.

Optimization of the library cell layout with the goal of minimizing unexpected variations and maximizing performance is described in Chapter 6.

3.5.6   Bulk Transistors versus SOI and FinFET

The main focus in this chapter has been on the neighborhood effects in bulk CMOS transistors, which have dominated the semiconductor industry for the last several decades.

Silicon-on-insulator (SOI) circuits are used in several niche applications [11,51]. All neighborhood effects that affect bulk transistors also apply to SOI transistors. The TED-related effects are somewhat weaker in partially depleted SOI (PDSOI) transistors currently in production compared to bulk transistors. These effects are expected to become weaker yet in fully depleted SOI (FDSOI) transistors, which might replace PDSOI in the future [52]. An additional stress proximity effect has been reported for SOI that is not observed in bulk transistors [53]. It affects SOI transistors with small channel width or small poly-to-STI distance, due to the thermal oxidation lifting SOI film at SOI–STI boundaries.

Multigate transistors such as finFETs harvest higher driving current per unit wafer area and simultaneously exhibit better leakage behavior [54]. However,

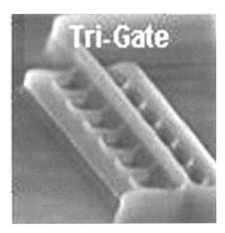

Figure 3.48   SEM image of a finFET-like trigate transistor with multiple channels. (Courtesy of Intel Developer Forum.)

several process integration issues and higher manufacturing costs will prevent their use in production until at least the 22-nm node. From the stress engineering point of view, finFETs have the following differences from bulk transistors:

- The tall and narrow silicon "fin" can serve as a lever for clever application of beneficial forces.
- Once the fins are patterned, their sides become free boundaries and relax the stress, so stress has to be introduced later in the process flow.
- The sides of the fin have different crystal orientations from its top, which affects both stress-free carrier mobility and stress-enhanced carrier mobility.
- There is no STI, and therefore no STI-induced stress or associated layout sensitivity.

What distinguishes FDSOI and finFETs from bulk transistors is that they connect several channels (or "fingers") into a single transistor (Figure 3.48). This makes the structure more regular, because instead of different channel widths, it is just a different number of fingers, with each finger having the same geometry. This more regular structure reduces layout-induced stress variations and therefore makes circuit behavior more predictable.

Another feature shared by FDSOI and finFETs is that their channels are undoped, which means that there are no random dopant fluctuations or WPE effect. The threshold voltage is determined by channel geometry rather than channel doping. Therefore, channel geometry (i.e., width and height of the fingers) variations contribute to transistor variability. In this case, the layout sensitivity happens through optical proximity effects as well as etched micro loading effects [55].

REFERENCES

1. N. D. Theodore et al., A transmission electron microscopy study of process-induced defects in submicron SOI structures, *J. Electrochem. Soc.*, vol. 139, no. 1, pp. 290–296, 1992.

2. H. L. Tsai, Structural integrity and crystal defect generation in VLSI/ULSI devices, *J. Electrochem. Soc.*, vol. 137, no. 1, pp. 309–313, 1990.

3. S. M. Hu, Stress from isolation trenches in silicon substrates, *J. Appl. Phys.*, vol. 67, no. 2, pp. 1092–1101, 1990.

4. P. Smeys et al., Influence of process-induced stress on device characteristics and its impact on scaled device performance, *IEEE Trans. Electron Devices*, vol. 46, no. 6, pp. 1245–1252, 1999.

5. K. Rim et al., Transconductance enhancement in deep submicron strained Si n-MOSFETs, *IEDM Tech. Dig.*, pp. 707–710, 1998.

6. S. Ito et al., Mechanical stress effect of etch-stop nitride and its impact on deep submicron transistor design, *IEDM Tech. Dig.*, p. 247, 2000.

7. T. Ghani et al., A 90 nm high volume manufacturing logic technology featuring novel 45 nm gate length strained silicon CMOS transistors, *IEDM Tech Dig.*, pp. 978–980, 2003.

8. S. Pidin et al., A novel strain-enhanced CMOS architecture using selectively deposited high-tensile and high-compressive silicon nitride films, *IEDM Tech. Dig.*, pp. 213–216, 2004.

9. L. Washington et al., pMOSFET with 200% mobility enhancement induced by multiple stressors, *IEEE Electron Device Lett.*, vol. 27, no. 6, pp. 511–513, 2006.

10. C. S. Smith, Piezoresistance effect in germanium and silicon, *Phys. Rev.*, vol. 94, no. 1, pp. 42–49, 1954.

11. M. Horstmann et al., Integration and optimization of embedded-SiGe, compressive, and tensile stressed liner films, and stress memorization in advanced SOI CMOS technologies, *IEDM Tech. Dig.*, pp. 243–246, 2005.

12. H. Fukutome et al., Direct measurement of effects of shallow-trench isolation on carrier profiles in sub-50 nm n-MOSFETs, *VLSI Technol. Dig.*, pp. 140–141, 2005.

13. T. B. Hook et al., The dependence of channel length on channel width in narrow-channel CMOS devices for 0.35–0.13 µm technologies, *IEEE Trans. Electron Devices*, vol. 21, no. 2, pp. 85–87, 2000.

14. P. P. Wang, Device characteristics of short-channel and narrow-width MOSFET's, *IEEE Trans. Electron Devices*, vol. 25, pp. 779–786, 1978.

15. S. Kim et al., Explanation of anomalous narrow width effect for nMOSFET with LOCOS/NSL isolation by compressive stress, *Proc. SISPAD*, pp. 189–191, 1997.

16. J.-W. Jung et al., Dependence of subthreshold hump and reverse narrow channel effect on the gate length by suppression of transient enhanced diffusion at trench isolation edge, *Jpn. J. Appl. Phys.*, vol. 39, pp. 2136–2140, 2000.

17. M. E. Rubin et al., Interface interstitial recombination rate and the reverse short channel effect, *Mater. Res. Soc. Symp. Proc.*, vol. 568, pp. 213–218, 1999.

18. H. Tsuno et al., Advanced analysis and modeling of MOSFET characteristic fluctuation caused by layout variation, *VLSI Technol. Dig.*, pp. 204–205, 2007.

19. M. V. Dunga et al., BSIM4.6.0 MOSFET model, http://www-device.eecs.berkeley. edu/~bsim3/BSIM4/BSIM460/doc/BSIM460_Manual.pdf, 2006.

20. W. Morris et al., Buried layer/connecting layer high energy implantation for improvedCMOS latch-up, *IIT Proc.*, pp. 796–799, 1996.

21. J. O. Borland et al., LOCOS vs. shallow trench isolation latch-up using MeV implantation for well formation down to 0.18 μm design rules, *IIT Proc.*, pp. 67–70, 1998.

22. I. Polishchuk et al., CMOS Vt-control improvement through implant lateral scatter elimination, *ISSM Proc.*, pp. 193–196, 2005.

23. *Seismos User's Manual v. 2007.09*, Synopsys, Mountain View, CA, 2007.

24. S. M. Sze, *Physics of Semiconductor Devices*, Wiley, New York, 1981, p. 880.

25. F. Nouri et al., A systematic study of trade-offs in engineering a locally strained pMOSFET, *IEDM Tech. Dig.*, pp. 1055–1058, 2004.

26. J. Welser et al., Strain dependence of the performance enhancement in strained-Si n-MOSFETs, *IEDM Tech. Dig.*, pp. 373–376, 1994.

27. M. Yang et al., High performance CMOS fabricated on hybrid substrate with different crystal orientations, *IEDM Tech. Dig.*, pp. 423–426, 2003.

28. Y. Kanda, A graphical representation of the piezoresistance coefficients in silicon, *IEEE Trans. Electron Devices*, vol. 29, no. 1, pp. 64–70, 1982.

29. V. Moroz et al., Analyzing strained-silicon options for stress-engineering transistors, *Solid State Technol.*, July 2004.

30. S. E. Thompson et al., Key differences for process-induced uniaxial versus substrate-induced biaxial stressed Si and Ge channel MOSFETs, *IEDM Tech. Dig.*, pp. 221–224, 2004.

31. P. Pichler, *Intrinsic Point Defects, Impurities, and Their Diffusion In Silicon*, Springer, New York, 2004, p. 554.

32. S. Timoshenko, *Theory of Elasticity*, McGraw-Hill, New York, 1970, p. 608.

33. V. Moroz et al., Modeling the impact of stress on silicon processes and devices, *Mater. Sci. Semicond. Proc.*, vol. 6, pp. 27–36, 2003.

34. S. M. Cea et al., Front end stress modeling for advanced logic technologies, *IEDM Tech. Dig.*, pp. 963–966, 2004.

35. R. A. Bianchi et al., Accurate modeling of trench isolation induced mechanical stress effects on MOSFET electrical performance, *IEDM Tech. Dig.*, pp. 117–120, 2002.

36. K. Ota et al., Stress controlled shallow trench isolation technology to suppress the novel anti-isotropic impurity diffusion for 45 nm-node high-performance CMOSFETs, *VLSI Tech. Dig.*, pp. 138–139, 2005.

37. R. People, Physics and applications of GexSi1-x/Si strained-layer heterostructures, *J. Quantum Electron.*, vol. 22, no. 9, pp. 1696–1710, 1986.

38. Y. Cho et al., Experimental and theoretical analysis of dopant diffusion and C evolution in high-C Si : C epi layers: optimization of Si : C source and drain formed by post-epi implant and activation anneal, *IEDM Tech. Dig.*, pp. 959–962, 2007.

39. B. E. Deal and A. S. Grove, General relationship for the thermal oxidation of silicon, *J. Appl. Phys.*, vol. 36, no. 12, pp. 3770–3778, 1965.

40. D.-B. Kao et al., Two-dimensional thermal oxidation of silicon: II. Modeling stress effects in wet oxides, *IEEE Trans. Electron Devices*, vol. 35, no. 1, pp. 25–37, 1988.

41. K.V. Loiko et al., Multi-layer model for stressor film deposition, *SISPAD Proc.*, pp. 123–126, 2006.

42. C. Ortolland et al., Stress memorization technique (SMT) optimization for 45 nm CMOS, *VLSI Tech. Dig.*, pp. 78–79, 2006.

43. A. Wei et al., Multiple stress memorization in advanced SOI CMOS technologies, *VLSI Tech. Dig.*, pp. 216–217, 2007.

44. V. Senez et al., Strain determination in silicon microstructures by combined convergent beam electron diffraction, process simulation, and micro-Raman spectroscopy, *J. Appl. Phys.*, vol. 94, no. 9, pp. 5574–5583, 2003.

45. L. Smith et al., Exploring the limits of stress enhanced hole mobility, *IEEE Electron Device Lett.*, vol. 26, no. 9, pp. 652–654, 2005.

46. E. J. Ayars et al., Fundamental differences between micro- and nano-Raman spectroscopy, *J. Microsc.*, vol. 202, pp. 142–147, 2001.

47. *ANSYS Mechanical User's Manual, v. 11.0*, ANSYS, Canonsburg, PA, 2007.

48. *Fammos User's Manual, v. 2007.06*, Synopsys, Mountain View, CA, 2007.

49. V. Senez et al., Two-dimensional simulation of local oxidation of silicon: calibrated viscoelastic flow analysis, *IEEE Trans. Electron Devices*, vol. 43, no. 5, pp. 720–731, 1996.

50. R. Arghavani et al., Strain engineering in non-volatile memories, *Semiconductor International*, 2006.

51. G. K. Celler and S. Cristovoleanu, Frontiers of silicon-on-insulator, *J. Appl. Phys.*, vol. 93, no. 9, pp. 4955–4978, 2003.

52. V. Moroz and I. Martin-Bragado, Physical modeling of defects, dopant activation and diffusion in aggressively scaled Si, SiGe and SOI devices: atomistic and continuum approaches, *Mater. Res. Soc. Symp. Proc.*, vol. 912, pp. 179–190, 2006.

53. P. S. Fechner and E. E. Vogt, Oxidation induced stress effects on hole mobility as a function of transistor geometry in a 0.15 µm dual gate oxide CMOS SOI process, *SOI Conf. Proc.*, pp. 163–165, 2005.

54. X. Huang et al., Sub 50-nm finFET: PMOS, *IEDM Tech. Dig.*, pp. 67–70, 1999.

55. C. J. Mogab, The loading effect in plasma etching, *J. Electrochem. Soc.*, vol. 124, pp. 1262–1268, 1977.

II

DESIGN SOLUTIONS

4

SIGNAL AND POWER INTEGRITY

4.1 INTRODUCTION

Chip design in the 45- and 32-nm nodes brings significant problems related to interconnect design. The interconnect hierarchy is shown in Figure 4.1, and a representative set of 45-nm design rules is presented in Table 4.1. As the technology scales, interconnect delay has been increasing in absolute terms of picoseconds per millimeter, and since the gate delay has been decreasing, apparent delay referenced to gate delays has been increasing dramatically. This has required more sophisticated methods of compensating for the delay.

Increased process complexity related to subwavelength lithography, chemical--mechanical polishing, and the implementation of low-k dielectrics leads to higher variability of resistance and capacitance. Electromigration (EM) reliability has been degraded due to process changes requiring layout and circuit changes. Design for manufacturability (DFM) requires more stringent design rules and extra cycles of post-layout iteration to correct for manufacturing-related issues. While statistical static timing analyzers are of limited value in the optimization of place and route, a statistical SPICE model would be helpful for analog design. Tighter wire spacing increases coupling and crosstalk-related noise requiring new layout and circuit techniques. Power supply integrity for performance through reduced IR drop and for reliability are affected by manufacturing difficulties at finer nodes. Circuit techniques to

Nano-CMOS Design for Manufacturability: Robust Circuit and Physical Design for Sub-65 nm Technology Nodes
By Ban Wong, Franz Zach, Victor Moroz, Anurag Mittal, Greg Starr, and Andrew Kahng
Copyright © 2009 John Wiley & Sons, Inc.

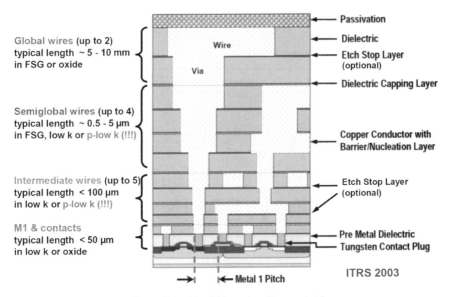

Figure 4.1 Metal hierarchy. (From ref. 1.)

Table 4.1 Representative 45-nm BEOL design rules: width, spacing or pitch thickness, AR

	ITRS 2004 hp 65 manuf. 2007	ITRS 2004 hp 45 manuf. 2010	Toshiba 45 nm[a]	Fujitsu 45 nm[b]	NEC 45 nm[c]
Metal 1	152 (pitch) 1.7 AR	108 (pitch) 1.8 AR	65/65	65/65	70/70
Vias (1×, intermediate)	1.6 AR	1.6 AR	75 × 75	70/70	70 diameter
Intermediate wires (1×)	195 (pitch) 1.8 AR	135 (pitch) 1.8 AR	80/70	70/70	70/70
Vias (2×, semiglobal-1)	—	—	140 × 140	140/140	
Semiglobal-1 wires (2×)	290 (pitch)	205 (pitch)	140/140	140/140	
Vias (4×, semiglobal-2)	—	—	280 × 280		
Semiglobal-2 wires (4×)	—	—	280 × 280		
Vias (global wires)	2.0 AR	2.1 AR	600 × 600		
Global wires (>10×)	1560 (pitch) 2.2 AR	1350 (pitch) 2.3 AR	1000/1000		
Pad metallization	—	—	—	Al pad	—

Source: Ref. 1.

[a]M. Iwai et al., *VLSI 2004*, p. 12.
[b]S. Sugiura et al., *IITC 2005*, p. 15.
[c]M. Tada et al., *VLSI 2005*, p. 18.

characterize performance loss as a function of dc and transient supply droop are described. Methods of reducing power supply noise and improving performance with decoupling capacitors are described.

4.2 INTERCONNECT RESISTANCE, CAPACITANCE, AND INDUCTANCE

Scaling to advanced process nodes has a significant impact on interconnect delay and power system integrity. In this section we describe the impact of process scaling on wire and via resistance, capacitance, and inductance and the resulting effect on interconnect delay and crosstalk. Further complicating the situation is the worsening variability of resistance and capacitance with the ensuing consequence of variation of interconnect delay. Design for manufacturability (DFM) measures must be taken to combat this problem. Scaling and variability also take their toll on power supply systems. With higher current demands at faster edge rates, lower supply impedance is required. This can be achieved primarily by low package inductance and by on-chip decoupling capacitance.

4.2.1 Process Scaling and Interconnect Fabrication

The economics of smaller die size and more die per wafer drives process scaling to finer features. The principal technology enabling this scaling is lithography. Current production subwavelength lithography produces half-pitch geometries of 45 nm using 193-nm-wavelength light (deep UV). Table 4.2 shows the expected progress of process scaling on metal pitch. The ever-smaller dimensions pose challenges to manufacturing, especially lithography and back-end-of-line fabrication. Layout and mask design are required to meet increasingly restrictive design rules to enable production yield.

Beginning at 45-nm and going beyond to 32-nm technology, immersion lithography will be needed. Resolution is determined by the wavelength λ and numerical aperture (NA) of a system [2]:

$$R(\text{half-pitch}) = \frac{k_1 \lambda}{\text{NA}} \tag{4.1}$$

Modern air-based lithography has an NA limit of 0.93. The NA is proportional to the refractive index of the medium. For higher NA, immersion in water

Table 4.2 Lithography Roadmap

Half-Pitch (nm)	65	45	32	22	16	11
Year	2005	2007	2009	2011	2013	2015

Source: Ref. 2.

(index of refraction = 1.44) gives an NA value of 1.33. Higher-index liquids, up to 1.64, are being researched. Double-exposure methods increase resolution for one-dimensional patterns by exposing alternate features at a larger spacing. The wafer is exposed by two different masks but does not leave the scanner. Double patterning enables a significant increase in resolution at greater expense. The wafer is exposed, leaves the scanner for developing, etching, and so on, and then is exposed again with a second mask. Fundamentally, the method results in a subtraction of the two patterns, enabling smaller dimensions, but has an extreme requirement for alignment or mask overlay of 1 to 3 nm compared to the current 45-nm technology alignment of 4 to 8 nm. See Chapter 2 for further details on the lithography-related aspects of DFM.

As shown in Figure 4.1, the width and height of the wires increases with higher layer number. Metal 1, in low-k dielectric or silicon dioxide, is the finest but highest resistance and is used for short-range connectivity. The intermediate layers, typically twice the thickness and height of M1, are used for interconnect in the range of about 100 μm. The semiglobal wires are typically four times thicker and wider than M1 and are used up to several millimeters in length. The top one or two levels of global interconnect can be up to 10 times thicker and wider than M1 and are typically used for power and ground distribution.

The metal lines and vias are fabricated using the dual-Damascene chemical--mechanical polishing (CMP) process. CMP involves the chemical reaction of slurry with copper forming a passivation layer, followed by mechanical removal of this layer. The method involves depositing the dielectric layers, etching the vias and trenches (for the wires), then electroplating the copper and subsequently using CMP to remove the copper down to the dielectric layer. Dual-Damascene etches holes for both vias and trenches before metal deposition, either vias first or trenches first. Problems with CMP include delamination of the hard mask used for feature definition, CMP scratches caused by abrasive particles in the slurry, thinning of the metal, dielectric erosion, edge erosion, trenching, dishing, and corrosion. These problems are illustrated in Figure 4.2. Of these, dishing and thinning in particular are determined by interconnect pattern density and reduce interconnect thickness significantly. A recent advance uses electropolishing (ECMP) followed by low-pressure CMP. The removal rate depends on the current applied and is independent of the down force, enabling abrasive-free copper removal, resulting in less dishing and thinning. The reduced down force also limits damage to the fragile low-k dielectric. Since out-diffusion of the copper would poison the transistors, this requires that the vias and trenches are lined with cladding to prevent outdiffusion. The cladding has a minimum thickness, causing the usable cross-sectional area of the wire or via to decrease faster than the scaling rate. Also, the copper is capped with a different metal (e.g., cobalt) to improve electromigration reliability, but this metal has a higher resistivity value than that of copper.

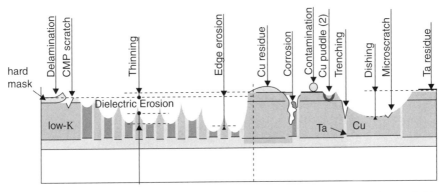

Figure 4.2   Chemical–mechanical polishing issues affecting metal thickness. (From ref. 2.)

The need to reduce capacitance has led to the development of various low-k dielectrics. Intermetallic dielectrics (IMD) at the upper levels have used SiO_2, $k = 4.0$; fluoridated silico-glass (FSG), $k = 3.7$; SiLK, $k = 2.65$; and porous SiCOH, $k < 2.5$. Interlayer dielectrics have used black diamond (SiOC), $k = 2.5$ to 2.9. Porous dielectrics are created by removal of a sacrificial material, leaving a porosity of 0.6 to 0.1, with pores typically 2 to 5 nm in size. A primary limitation of porous low-k materials is their mechanical weakness, making CMP and wirebonding difficult. Reducing the low-k benefit is carbon depletion at the sidewalls caused by etching and resist strip, resulting in higher k values. Furthermore, the porous materials need to be sealed to reduce leakage between lines. A recent move to air-gap IMD holds the promise of effective k-values as low as 1.5.

4.2.2   Impact of Process Scaling on Resistance and Capacitance

As copper interconnect lines have been scaled to increasingly smaller dimensions, the resistance per unit length has increased due to reduced cross section. For a given scaling factor of k, the thickness and width of the wire each scale by k, and the cross-sectional area therefore scales with k^2. Also, since the copper cladding and capping layers cannot be reduced proportionally, the effective area decreases faster than k^2, increasing the resistance substantially. But in addition to that, the effective bulk resistivity has also increased, due to four separate factors. The first is conduction electron scattering caused by grain boundaries. Figure 4.3 shows that as the line width decreases, the grains become smaller also, causing electron scattering from the grain boundaries. In addition, there is scattering from the surface of the metal line, which becomes proportionally greater as the line width scales down. Figure 4.4 shows a conceptual three-dimensional diagram of the metal line, with scattering from the surface as well as the grain boundaries. The combined effect of these two characteristics is illustrated dramatically by Figure 4.5, which shows more than

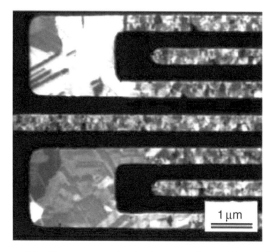

Figure 4.3   Decreased grain size in narrow lines. (From ref. 1.)

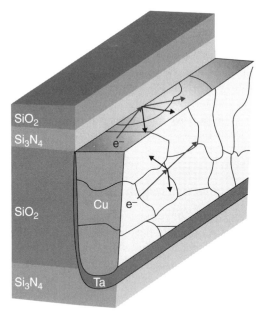

Figure 4.4   Narrow metal line showing electron scattering from the surface as well as from the grain boundaries. (From ref. 1.)

three times as much resistivity as would be expected from the line dimensions alone. A third effect is due to impurity scattering in the bulk of the metal caused by aluminum and manganese being added to the copper to improve electromigration (EM) robustness. The fourth is due to increased scattering due to a photolithographic effect called *line edge roughness* (LER), which

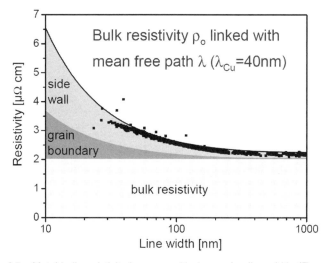

Figure 4.5   Metal bulk resistivity increase with decreasing line width. (From ref. 1.)

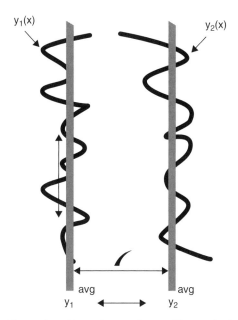

Figure 4.6   Line-edge roughness. (From ref. 2.)

refers to the irregular texture on the edge and sidewall of a patterned feature (Figure 4.6). It is generated in the resist during photolithographic exposure and transferred into the interlayer dielectric etch and subsequently, to the electroplated copper line. This roughness should be less than 10 nm for a 50-nm line width. The increase in resistivity due to LER is shown in Figure 4.7.

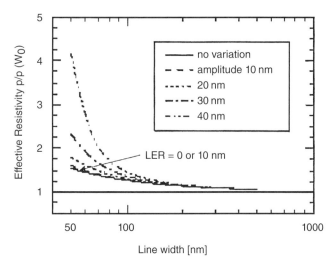

Figure 4.7   Resistivity increase due to line-edge roughness. (From ref. 3.)

Using low-k and ultralow-k dielectrics helps reduce the total capacitance, but the problems of these dielectrics causes some of the low-k advantage to be lost. For example, the cladding or barrier material needed to enclose the copper such as SiN or SiC is a higher k. The trench walls in porous dielectrics during cleaning attain a higher k value. Since to first order, the capacitance between two lines is constant with respect to constant-aspect-ratio scaling, the use of lower-k materials does lead to some reduction in capacitance, but not in proportion to the lower k.

4.2.3 Scaling and Reliability

There are several aspects to the impact of scaling on reliability. Copper diffuses readily into dielectric materials and silicon. In the dielectric it can cause shorts between metal lines, and in silicon it can degrade transistors. It needs to be encapsulated with barrier metals such as Ta or TaN or dielectrics such as SiC or SiN. These have the disadvantages of having either lower conductivity than copper or higher k than the low-k dielectric. As the process scales to smaller dimensions, the thickness of these layers does not decrease proportionally, leading to higher effective k or higher resistance of the metal line. Electromigration in copper is primarily a surface diffusion function, with voids forming and coalescing until the line resistance increases to an unsustainable value or causes an opening. Since as the lines scale, the ratio of surface to volume increases, and a greater EM problem is caused. Capping the copper with other metals (e.g., cobalt) can improve EM, but it increases the effective resistance.

As shown in Table 4.3, current densities are expected to increase with further process scaling, requiring significant improvements in fundamental

Table 4.3   ITRS interconnect reliability requirements with scaling

Total Interconnect Length[a] (m/cm^2)	Failure Rate of Interconnects[b] (FITs/m·cm$^2 \times 10^3$)	J_{max}[c] (A/cm^2)
1019	4.9	8.91×10^5
1212	4.1	1.37×10^6
1439	3.5	2.08×10^6
1712	2.9	3.08×10^6
2000	2.5	3.88×10^6
2222	2.3	5.15×10^6
2500	2	6.18×10^6
2857	1.8	6.46×10^6
3125	1.6	8.08×10^6

[a]Metal 1 and five intermediate levels, active wiring only [1].
[b]Excluding global levels [2].
[c]Intermediate wire at 105°C.

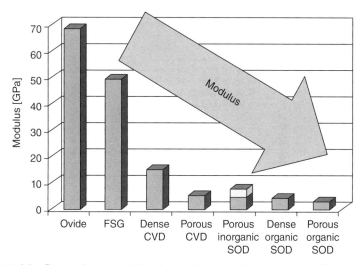

Figure 4.8   Decreasing strength (modulus) with newer low-k dielectrics. (From ref. 1.)

EM performance. Process changes in the metal are likely to further increase interconnect metal resistivity, such as doping the metal with impurities such as aluminum and manganese, but these cause more electron scattering and higher resistivity.

In addition to the EM issues mentioned above, there are increasing issues with the dielectric. While the move to low-k dielectrics has enabled some mitigation of the reduction in signal propagation speed with scaling, the rest of the news is all bad. First, the strength or modulus of the low-k and porous dielectrics is greatly reduced with respect to SiO_2 (Figure 4.8). This exacerbates difficulties with regard to CMP. In addition, it makes wirebonding problems

- Low modulus - absorbs ultrasonic energy
- Low fracture toughness - easily cracked
- Low adhesion - delamination
- CTE mismatch - delamination
- Moisture adsorption - delamination

  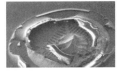

Figure 4.9   Wirebond issues with weak low-*k* dielectric. (From ref. 4.)

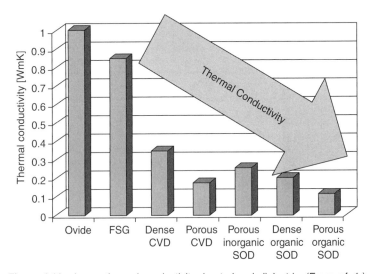

Figure 4.10   Lower thermal conductivity due to low-*k* dielectric. (From ref. 1.)

(Figure 4.9). Cracking and delamination failures in bonding are major issues. Wirebond design rules require metal structures below the pad to increase the strength to withstand shear stresses during wirebonding. The second problem is with lower thermal conductivity (Figure 4.10). Heat dissipated by current flowing in interconnect wires causes higher temperatures, exacerbating electromigration reliability. The third problem is that the low-*k* dielectrics have much higher temperature coefficient of expansion (Figure 4.11), causing tensile stress on the metal system. This tensile stress causes the metal to migrate, called *stress migration*, causing voids, especially near the vias. This leads to "via popping" failures, which may not appear until after packaging and test.

As the space between metal lines scales down, the electric fields increase, causing time-dependent dielectric breakdown (TDDB). Reliability depends on the quality of the edge seals and passivation during and after assembly.

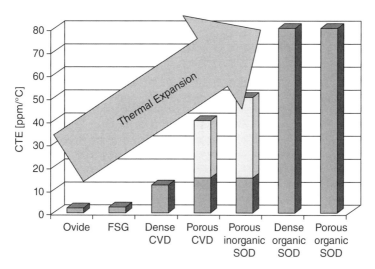

Figure 4.11 Higher thermal coefficient of expansion, causing tensile stress in the metal system. (From ref. 1.)

4.2.4 Interconnect Delay, Energy, and Scaling Implications

RC descriptions [5] of signal delay are often based on simple π- or T-models of parasitic resistance and capacitance and use Elmore delay:

$$D = R_{\text{gate}}(C_{\text{wire}} + C_{\text{par}} + C_{\text{load}}) + R_{\text{wire}}(0.5R_{\text{wire}} + C_{\text{load}}) \qquad (4.2)$$

or approximately, for long wires,

$$D = k(\text{FO4}) + 0.5R_{\text{wire}}C_{\text{wire}} = k(\text{FO4}) + 0.5\left(R_{\text{wire/mm}}C_{\text{wire/mm}}\right)L^2 \qquad (4.3)$$

which is quadratic in wire length (mm). The factor FO4 refers to the delay of an inverter driving a load equal to four times its input capacitance and is a prime measure of the transistor-level performance of a process. In a typical 65-nm process, for example, the FO4 delay is about 16 ps. The factor *k* is a fitting parameter dependent on the relative strength of the driver. For long wires, the resistance of the wires tends to shield the driver somewhat from the sum of the interconnect and load capacitance, allowing a faster rise time than when considering the total capacitance without the resistance as a load on the driver.

As noted earlier, the resistance per millimeter of wires increases rapidly with scaling, exacerbating this delay. The capacitance per millimeter is roughly constant, since the height of the wire and the spacing to the neighboring wire usually scale in a similar fashion. The reason that it seems like the interconnect delay is getting worse with every generation is that, yes, the absolute unrepeated delay is increasing (Table 4.4), but since the gate delays have been getting much faster, delay by this measure is increasing even faster (Table 4.5).

Table 4.4   Increase in resistance, slow decrease in capacitance, and 20- to 30-fold increase in absolute delay

Parameter for Unrepeated Wires	Technology Node (nm)					
	65	45	32	22	15	10
Resistance (Ω/mm), optimistic to conservative	135	240 350	470 725	1000 1560	2150 3420	4000 6950
Capacitance (pF/mm), optimistic to conservative	0.36	0.34 0.35	0.31 0.33	0.285 0.325	0.260 0.325	0.240 0.315
Delay, 0.5 RC (ps/mm/mm), optimistic to conservative	24	41 61	73 120	140 255	280 560	480 1090

Table 4.5   Increase in resistance, slow decrease in capacitance, and 100- to 200-fold increase in delay in terms of FO4 inverter delay

Parameter for Unrepeated Wires	Technology Node (nm)					
	65	45	32	22	15	10
Resistance (Ω/mm), optimistic to conservative	135	240 350	470 725	1000 1560	2150 3420	4000 6950
Capacitance (pF/mm), optimistic to conservative	0.36	0.34 0.35	0.31 0.33	0.285 0.325	0.260 0.325	0.240 0.315
Delay, 0.5 RC (FO4/mm·mm), optimistic to conservative	1.6	3.6 5.4	9.1 15	25 46	75 150	175 400

Source: Ref. 5.

Also, any given functional block has shrunk by the scaling factor, to some extent mitigating the apparent interconnect delay increase due to the shorter interconnect length, and since the gates will get much faster, the speed of the block as a whole can increase. For global wires which have to go the same distance as before, however, the delay increase is indeed large.

For example, in a 65-nm process, wire delay is approximately 2 to 3 FO4/mm². Circuit solutions focus on the use of repeaters, either buffers or inverters. The delay of a signal line with optimally sized repeaters placed at optimum spacing can be described approximately by

$$D = 1.65[(FO4) R_{wire} C_{wire}]^{1/2} \tag{4.4}$$

where the constant depends on repeater sizing and spacing as well as the distribution of resistance and capacitance. Correct optimization depends on accurate extraction of wire resistance and capacitance. Optimization also depends on the cost function being optimized. For single wires, the objective is usually minimum delay, but this comes at a cost in incremental energy and area. Rather than minimizing delay alone, optimizing the energy $\times$ delay product gives a better result in terms of power with minimal extra delay. A typical solution in 65 nm would be 1.6-mm spacing using transistors of 60 times minimum size. A drawback to repeaters is that they must use many vias to get from the (usually) higher-level interconnect down to the substrate, disrupting routing and taking space for the repeaters. Also, the optimum spacing may put the desired location for the repeaters into a macro, which may not be possible, hence a different repeater spacing would be required.

For high-performance chips with high clock frequencies and large die size, the distance the signal needs to travel may encounter a delay greater than a clock cycle, requiring pipelining of the interconnect with flip-flops along the way.

Capacitive crosstalk between signal wires is becoming more critical with scaling for longer wires due to resistive shielding. The capacitive coupling factor is given by

$$\frac{C_c}{C_g + C_c M} \tag{4.5}$$

and is further determined by the Miller or switching factor M, which is 2 if the wires are switching in opposite directions, 1 if the attacker is not switching, and 0 if the wires are switching in the same direction. For timing analysis, windows of the clock cycle need to be determined when the victim line is switching, which is when it is most sensitive to crosstalk. If the attacker switches when the victim line is held by its driver strongly to ground or V_{dd}, the resulting disturbance to the quiet line is usually not enough to cause a false transition or signal delay. This needs to be confirmed by parasitic extraction and circuit simulation. Guidelines need to be established to confirm that crosstalk to quiet lines will not cause false transitions.

Crosstalk can be reduced by increasing spacing to neighboring lines or by inserting grounded shield lines, both of which take extra area. The more common approach is to use repeaters, which can be either inverters (simpler) or buffers (less crowbar current but slower). Figure 4.12 shows two approaches [5] to reducing crosstalk using repeaters. The first uses noise cancellation by inserting inverters from the attacker line to the victim line through a capacitor to counter the induced charge through the parasitic line–line capacitance. The second is to stagger inverters on each line so that half of the line–line capacitance is in phase and the other half is out of phase, which cancels out for any signal combination.

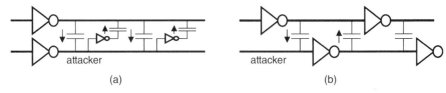

Figure 4.12 Reducing capacitive crosstalk using (a) noise cancellation and (b) staggered repeaters. (From S. Naffziger, Interconnect on GHz CPUs, International Solid-State Circuits Conference, 2001.)

The optimum design of interconnect depends on the bandwidth needed. Note that the bandwidth of the wire can be increased by using a wider (or thicker at higher levels) wire. However, in terms of bandwidth per unit cross-sectional area (i.e., layout space limited), more bandwidth can be achieved by using multiple (repeatered) wires. The wire between repeaters should be set such that the wire transitions to 95% of V_{dd}, roughly 11 FO4 delays.

The energy of a repeatered wire is

$$E = 1.35 C_{wire} V_{dd} V_{swing} \qquad (4.6)$$

or about 0.6 pJ/bit per millimeter (= 0.6 mW/Gbps per millimeter). To send this signal across a 10-mm chip, this is 6 mW/Gbps. With scaling, the power devoted to driving the interconnect portion of the total chip power is proportional to the total number of wires for a chip of the same size and number of metal layers, which will increase by about 50% per layer per generation (scaling = 0.7×).

Optimizing for a low energy-delay product instead of just optimizing for minimum delay has a high energy cost. A better strategy is to minimize energy × delay. For example, for 10% less performance, energy savings are on the order of 30%. In a 65-nm process, suggested spacings are 1.6 mm and transistor sizing is 60 times the minimum-size transistor. Representative numbers of repeaters for the entire chip are 20,000 in the 180-nm Itanium and 16,000 in the 130-nm Ultrasparc. A further option is to spread the logical function along the path, thus also using the repeaters to perform logic rather than having the logical processing all done in one place.

The architectural implication of this is one of the primary factors driving microprocessors to multiple cores. Each core shrinks with scaling, and although its speed does not suffer, across-chip global communication (core–core) slows down considerably.

4.3 INDUCTANCE EFFECTS ON AN INTERCONNECT

4.3.1 Emergence of Inductance as a Limiting Characteristic on Signal Integrity

With the faster switching times of transistors and the low resistance of copper interconnects, longer on-chip wires exhibit transmission-line behavior. Whereas

RC interconnect modeling earlier gave adequate estimations of delay and crosstalk, inductance must now be taken into account. In addition to processors running in the multiple-gigahertz range, the trend to multicore means that there are more buses running between the cores for longer distances. Power–frequency throttling of individual cores and localized clock gating within a single core cause high dI/dt and resulting high noise values and droop in the value of V_{dd}. This means that accurate modeling of inductance in signal wires and power grids is becoming more important than ever.

The need for accurate delay estimation driven by higher clock frequencies is complicating static timing analysis. The influence of magnetic fields on circuit performance at higher speeds requires *RLC* modeling. Inductive effects have a much longer range than capacitive coupling, requiring estimation of the interaction between many more circuit elements over a much larger distance beyond the nearest neighbor. In addition to parallel coupling, inductive coupling has forward and reverse propagation as well, extending over a large distance (Figure 4.13). Inductive coupling needs to be taken into account for crosstalk between circuit elements at high frequencies.

An example given by Beattie and Pileggi [7] involves a 32 b bus with one line switching in a quiet bus and all the other lines remaining static, hence without crosstalk. The difference between an *RC* and an *RLC* simulation is shown in Figure 4.14 for near- and far-end responses. For the simple three-wire example shown in Figure 4.15, the difference between *RC* and *RLC* simulation of the far-end response for out-of-phase switching (−1 1 −1) and in-phase switching (1 1 1) is shown in Figure 4.16, where a rising signal is represented as 1 and a falling signal as −1. This illustrates that whereas the worst-case switching for *RC*-coupled lines is when they are switching in opposite directions, for inductively coupled lines the worst case is for lines switching in the same direction, due to the mutual inductive coupling, which leads to a large error for the *RC* simulation compared to the full *RLC* simulation.

At what point does inductance need to be taken into consideration? The important factor is when the inductive impedance, ωL, is a significant fraction of the resistance, *R*, per millimeter of the line segment. Estimating the inductance of a line is a strong function of the (usually unknown) return path, but a representative value ranges from 0.6 to 1.3 nH/mm [9]. We can calculate the

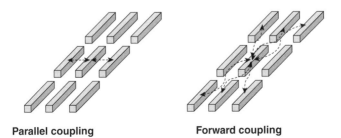

Parallel coupling   **Forward coupling**

Figure 4.13   Inductive parallel and forward coupling. (From ref. 6.)

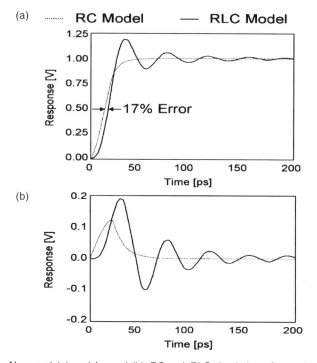

Figure 4.14 Near-end (a) and far-end (b) *RC* and *RLC* simulation of one signal in a static bus. (From ref. 1.)

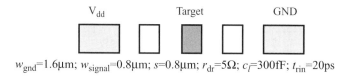

Figure 4.15 Three-wire crosstalk example simulated in Figure 4.16. (From ref. 2.)

minimum frequency at which the inductive impedance is 10% or more of the resistance, which gives $f \geq R \times 60\,\text{MHz}$. For example, the top-level metal at $10\,\Omega/\text{mm}$ gives $f = 600\,\text{MHz}$, where inductive effects first become noticeable. For a lower-level metal at which $R = 100\,\Omega/\text{mm}$, this is $6\,\text{GHz}$.

Because the nature of the inductance and its return paths determines the circuit design methods employed to mitigate inductance effects, in the next section we summarize the characteristics of the parasitic inductance and the simulation and extraction requirements.

4.3.2 Inductance Simulation and Extraction

Inductance is a loop characteristic defined by

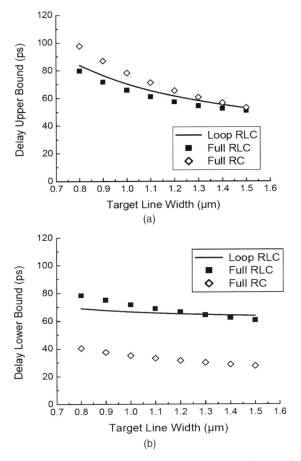

Figure 4.16   Impact of *RC* versus *RLC* simulation with (a) out-of-phase switching (–1 1 –1) and (b) in-phase switching (1 1 1).

$$L_{loop} = L_{self} + L_{return} - 2L_m \qquad (4.7)$$

The problem is that the return paths and mutual inductances can be far away and be composed of many separate line segments. The total number of loops is on the order of $O(N)^2$ [7], where N is the number of line segments, so identifying all possible loops and their inductive coefficients is not possible. According to Pileggi [6], the number of forward coupling elements is much greater than the number of parallel coupling elements, shown in Figure 4.13. Discarding some partial mutual inductances which couple two other segments indirectly can cause stability problems when simulating the inductance matrix. The effectiveness of the return path is also affected strongly by the resistance and distribution of that path, so in trying to reduce the total inductance, as much emphasis needs to be put on the return path as on the initial line segment itself. Calculating the return path is a chicken-and-egg problem, because we need to know the currents in the paths to calculate the total inductance, but we need the inductance to calculate the currents. To get around this problem we use the method of partial inductances.

Partial inductance is defined [7] as the magnetic flux created by the current of one segment through the virtual loop that another segment forms with infinity (Figure 4.17). The total loop inductance is the sum of the partial self- and mutual inductances of the segments that form all loops in the system. This leads to huge complexity because of the mutual couplings between all wire segments. By defining each segment as forming its own return loop with infinity, partial inductances are used to represent the eventual loop interactions without prior knowledge of the actual current loops.

To enable modeling of integrated-circuit (IC) interconnects using common circuit simulation tools, capacitive and partial inductive elements were combined with resistors to form an analytical method called *partial element equivalent circuits* (PEEC). The conductors are divided into filaments (Figure 4.18) to capture the nonuniform distribution of current in the conductors and to model one aspect of the frequency dependence of inductance. Full simulation is prohibitive, but there are approximate methods for many situations, leading to useful analytical expressions useful for circuit design and timing extraction.

A widely used approximation of the IC current-return paths is to assume that the signal nets are shorted at the far end to a ground, or that a return line is placed nearby so that the entire current-return path is known without capacitance in the model. This enables full-wave tools such as FastHenry to be used efficiently for long wires. However, to take account of the displacement currents in lateral coupling parasitic capacitances, the long wires need to be divided into shorter segments.

An important frequency dependence of inductance is the *line proximity effect* [7,11] or current crowding. Current will flow along the path of least resistance, and at low frequencies, impedance is dominated by resistance, causing the current to spread out over the cross section of the conductor. As the frequency increases, the inductive component of the impedance, $R + j\omega L$, starts to dominate. To minimize the impedance, the loop size must now be

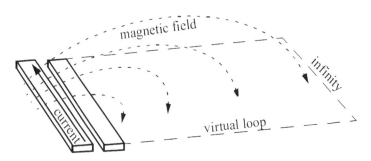

Figure 4.17   Loop definition of partial inductance. (From ref. 7.)

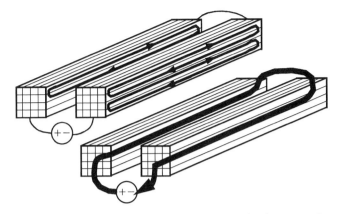

Figure 4.18 Dividing conductors into filaments to capture the frequency-dependent non-uniform current distribution. (From ref. 10.)

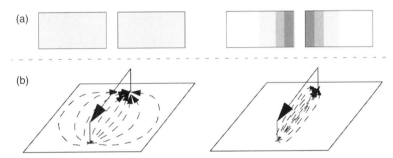

Figure 4.19 Line proximity effect showing the constriction of current flow at high frequencies on the right side for (a) neighboring lines and (b) the ground plane. (From ref. 7.)

minimized to reduce the loop inductance L. Therefore, the current tends to return closer to the signal line. The line proximity effect is shown in Figure 4.19. Also, at higher frequencies, as shown, the return current distribution in the ground plane is more constricted. The line proximity effect will cause increased resistance at higher frequencies:

$$R_{\text{eff}}(f) = R_{\text{dc}}\left[1 + \frac{1}{2}\left(\frac{f\mu_0 W^2}{R_{\text{sheet}}P}\right)2\right] \qquad (4.8)$$

where μ_0 is the permeability of free space, W is the conductor width, and P is the pitch.

The *loop proximity effect* leads to higher return resistance and a decrease in loop inductance at high frequencies, due to the fact that at higher frequencies the current returns in the nearest-neighbor wires only, as shown in Figure 4.20. The frequency dependency can be approximated by [11]

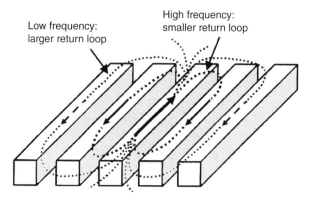

Figure 4.20   Loop proximity effect. The return loops shrink with increased frequency. (From ref. 11.)

$$\text{loop factor} = \frac{1 - p_1 e^{-fp2}}{2} \tag{4.9}$$

where the factor ½ comes from the fact that at very high frequency most currents will only return through the nearest two neighboring wires. The two fitting parameters p_1 and p_2 are extracted from full-wave simulations.

In contrast to the previously mentioned proximity effects caused by inductance, the *skin effect* is caused by the attenuation of electromagnetic waves in conducting material. At higher frequencies, the current flows closer to the surface. *Skin depth* is defined as the distance at which the field drops to $1/e$ of its surface magnitude and is given by [7]

$$\delta = (\pi f \mu \sigma)^{-0.5} \tag{4.10}$$

where f is the signal frequency and σ is the conductivity. For example, at 1 GHz in copper, the skin depth is 1.2 μm. The effect of skin depth is to increase the frequency-dependent resistance and to decrease inductance (due to the exclusion of magnetic field from the core of the conductor). This is modeled by subdividing the conductor into filaments that distribute the current nonuniformly throughout the cross section as shown in Figure 4.18. In general, this applies primarily to conductors with thickness greater than a skin depth, usually power and ground conductors at the top layer of metal.

4.3.3   Inductance-Influenced (*RLC*) Signal Delay

Traditional *RC* analysis assumes zero impedance for the return path in the power–ground network, but at higher frequencies the impedance of the return path is a significant part of the total impedance [12]. Even at 1 MHz, the effect of increasing the number of return paths decreases the effective resistance by about 14% up to 20 pairs of power–ground return paths, where the power–ground pairs are situated on both sides of the signal line. In an example layout [13] with one signal line surrounded by 17 return paths, between 100 MHz and 10 GHz, the inductance falls from 2.9 nH to 2.3 nH for this 3-mm line, due to the loop proximity effect. Above 10 GHz, it is necessary to discretize the conductors into filaments, which will cause a further increase in the frequency-dependent resistance and a slight decrease in effective inductance, both due to proximity and skin effects.

Using an extraction algorithm [14] that defines a window enclosing the extracted nets with the nearest power–ground return paths, and using FastHenry, a 2.82-mm wire of width 1.2 μm has a partial inductance of 4.67 nH and loop inductance of 5.7 nH. The mutual inductance with one of its neighbor lines is 2.26 nH. Using analytical expressions valid for the case of two parallel lines [12] yields

$$L = \frac{\mu_0 l}{2\pi}\left[\ln\left(\frac{2l}{w+t}\right)+\frac{0.2235(w+t)}{l}+\frac{1}{2}\right] \quad (4.11)$$

$$M = \frac{\mu_0 l}{2\pi}\left[\ln\left(\frac{2l}{s}\right)-1+\frac{s}{l}\right] \quad (4.12)$$

showing that the self-inductance is on the order of $l \cdot \ln[2l/d]$ and the mutual inductance is on the order of $l \cdot \ln(2l/d)$, where l is the line segment length, d is the spacing between the line segments, and w and t are the line width and thickness.

4.3.4 Single-Line Delay Considerations

Optimization strategies depend on the cost function being optimized, which is often different in different situations. For single signal lines the primary objective is usually minimum delay and minimum crosstalk. Design approaches to minimize inductance can also make the extraction of parasitics more tractable. Minimizing inductance can be done by providing return paths through lateral parallel power and ground shield lines as close as possible to the signal path to minimize the enclosed area. A further method of minimizing and isolating the inductance of signal lines is to provide metal planes above and below the signal line, as was done in the Alpha microprocessors, which is costly in terms of area and complexity.

The provision of power and ground shield lines must take account of the driver of the line and its connection to its local power and ground. The ground for V_{dd} is not necessarily the ground for drivers [6]. These power and ground lines must be closely connected for minimum return-path inductance [15] (Figure 4.21). When the driver is switching in the positive direction, the p-channel transistor is connected to the V_{dd} line, which will provide the return

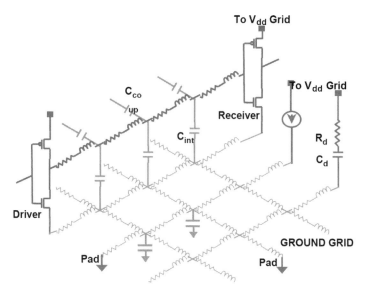

Figure 4.21   Drivers connected to power and ground return paths.

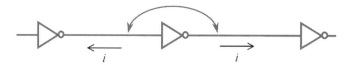

Figure 4.22   Repeaters showing cancellation of forward inductive coupling in a single line. (From ref. 16.)

path. The n-channel transistor is connected to the parallel V_{ss} line, but it is turned off, so the return-path current cannot flow in this loop. This problem can be alleviated somewhat by providing good decoupling capacitance between power and ground as close as possible to the driver, enabling some return-path current to flow through the V_{ss} line. It can be seen from this description that the inductance calculation is dependent on the state of the driver and must be calculated separately for each state since the return paths are different.

Referring to Figure 4.13, the number of forward inductive coupling elements is much greater than the number of parallel ones. The coupling of forward elements of the same line to each line segment causes an increase in inductance of that segment. This is noted in most inductive approximations as a supralinear dependence of inductance on line length, usually in the form $l \cdot \ln(l)$, where l is the line length. In this case of a single line, its inductance can be reduced by inserting inverting repeaters at optimum distances (Figure 4.22), causing the direction of current flow to reverse with each inverter. Now the mutual inductance between these forward elements works to reduce the total inductance and hence the total delay. This must be balanced with the cost of the repeater delay. This must now be coupled with the optimization of RC delay and repeater spacing to arrive at an integrated solution for both types of delay reduction. Each solution will be unique, depending on the signal line's return path environment and the frequency of operation. On top of all this, the optimization must take account of the power dissipation of the repeaters. For a long single line in the critical path, it may be worthwhile to optimize for delay alone.

4.3.5   Delay and Crosstalk for Buses

Modern processors running at clock frequencies of several gigahertz combined with die sizes up to 20×20 mm have buses that require consideration of inductance for optimization for delay, area, and throughput. Following Pileggi [10], in inductance extraction, numerous mutual inductance terms are of negligible magnitude. This means that only couplings between segments that are closer to each other than a given window size are considered. Such sparsification approaches can lead to indefinite partial inductance matrices, hence unstable equivalent-circuit models. The minimum window size for stable matrices increases with the length of the bus for any given bus width, and for lengths of 20 times the width or greater, the entire bus must be enclosed in the analysis window.

Compared to capacitive coupling, inductive coupling behavior leads to opposite polarity noise. For this reason, the switching behavior of the bus needs to be taken into account when calculating the delay. For the example from Cao et al. [8] of three signal wires between power and ground shield wires, the delay was shown in Figure 4.16 as a function of switching patterns $(-1\ 1\ -1)$ and $(1\ 1\ 1)$ where 1 indicates a rising edge and -1 indicates a falling edge. In-phase switching $(1\ 1\ 1)$ results in the largest inductive coupling, but also the lowest capacitive coupling, in contrast with the opposite situation in out-of-phase switching $(-1\ 1\ -1)$. The $(1\ 1\ 1)$ case shows the necessity of using RLC extraction, the delay being twofold greater than the RC extraction. In addition, it can be seen that for line widths greater than 1.0 μm, the RLC delay is greater.

When considering more than three wires switching at one time, the analysis becomes more complex. The case with five switching lines shown in Figure 4.23 shows that the worst-case delay for the center wire for the 0.8-μm spaced wires is for $(1\ -1\ 1\ -1\ 1)$ at 115 ps for the RLC case. Since the increase in delay with RLC versus RC is small, in this case capacitive coupling dominates. If the wires are spaced wider at 2.0 μm, the worst-case delay is for $(1\ 1\ 1\ 1\ 1)$ at 105 ps, indicating a small improvement at a large cost in area. For the $(1\ 1\ 1\ 1\ 1)$ case for both spacings, the large difference in RLC versus RC analysis shows the necessity of including inductive extraction and that the inductive coupling is dominant in this case. The $(-1\ -1\ 1\ -1\ -1)$ case shows that RC analysis alone can also give a higher and unnecessarily pessimistic delay. Also note that in comparison with the three-wire case, the worst-case

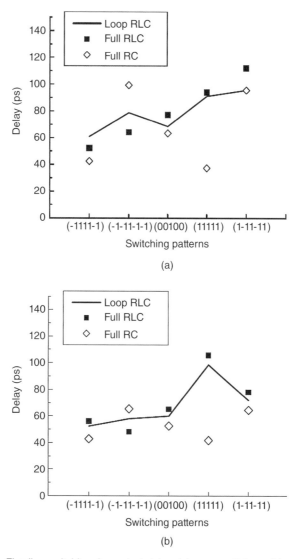

Figure 4.23   Five-line switching-dependent delay: (a) space = 0.8 μm; (b) space = 2.0 μm.

delays increased. Note also that in the three-wire case, the worst-case delay was for (1 1 1), whereas the corresponding pattern (1 1 1 1 1) for the five-wire (0.8 μm spacing) case was not the worst-case delay. This illustrates the differing contributions to delay by inductive and capacitive coupling. Therefore, for a large bus it is necessary to determine how large a window it is necessary to include in the inductive analysis. As larger bus widths are considered, going up to 32 or 64 bits wide, the worst-case delay needs to be determined by detailed extraction and simulation for each wire for all possible switching

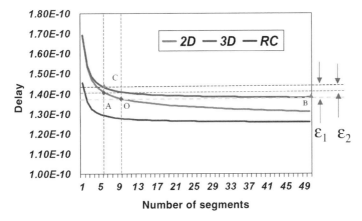

Figure 4.24   Error analysis as a function of number of segments per line.

conditions. Since this is clearly infeasible, an optimal design methodology needs to consider practical extraction and simulation approaches.

4.3.6   Optimized Bus Design

The design approach can simplify the modeling. To facilitate this, we consider the different modeling and error minimizations that are possible. Following Pileggi [6], the complete three-dimensional distributed model includes both parallel and forward coupling, described earlier (see Figure 4.13). For a lumped three-dimensional model, the line needs to be broken into a large number of segments. Reducing the number of segments simplifies the modeling computation, but the lumping error thus incurred increases with the decreasing number of segments. To go from a lumped three-dimensional model to a two-dimensional model involves truncating the far-away mutual terms in order to sparsify the L-matrix [16]. The method of current return shells is used to determine how far away the mutual terms can be truncated by assuming that the partial segment currents return at a finite radius determined by ground-line spacing rather than infinity, which is valid for good power and ground returns. The resulting equivalent circuit still needs model order reduction to be tractable. Error analysis [6] of the modeling in terms of signal delay is shown in Figure 4.24. As shown, the lumping error ε_1 for the three-dimensional analysis going from 49 segments to nine tends to cancel the truncating error ε_2 between three and two dimensional, so that using two-dimensional analysis is a good approximation. The assumption of good return paths requires good decoupling of power and ground. The spacing to the return paths should be much less than the segment length. If the analysis is using a minimum of nine segments per wire length, the spacing to the return paths should be on the order of 1 to 3% of the wire length. For a wire length of 500 μm, this would imply return paths 5 to 15 μm away. This is a significantly larger distance than the return shielding spacing to be shown later for optimum bus delay design, so the assumptions for tractable extraction and simulation should be possible to meet.

Optimization depends on the definition of the cost function to be optimized. The objective function chosen is throughput, defined by Pileggi [6] as

$$P_{\text{norm}} = \frac{\text{throughput} \times \text{transferring distance}}{\text{bus width}} = \frac{nL_s}{\text{maximum delay}[(n+1)s + nw_{\text{si}} + w_{\text{sh}}]} \quad (4.13)$$

where n is the number of wires between two shields, L_s the segment length, s the wire spacing, w_{si} the width of signal wires, and w_{sh} the width of the shield wires. The delay is a function of these and other variables. The algorithm described begins by optimizing the wire and shield sizing and spacing via sequential quadratic programming. Shield insertion is determined by downhill traversal, starting from the most shielded case. Repeater insertion is determined by wire, and repeater tapering by dynamic programming.

The use of repeaters to reduce inductive forward coupling was described in Section 4.3.4 (see Figure 4.22). Using staggered repeaters [10] (Figure 4.25) allows inductive cancellation between neighboring lines in an analogous but opposite sense to the same use of staggered repeaters for canceling capacitive coupling. This fortuitous circumstance allows the use of the same staggered repeaters for capacitive coupling for the $(-1\ 1\ -1)$ signal case and for inductive cancelation for the $(1\ 1\ 1)$ case. One or the other, based on RC or RLC, will determine the critical path.

Using the optimization definition above, bus designs were optimized [6] for two metal technologies (Table 4.6). Using repeaters and optimizing with RC versus RLC, the results are very similar. A similar analysis shows that 0.15- versus 0.36-µm metal technology yields almost the same optimized throughput. Also note that the RC-only optimization has almost the same result, showing that optimal throughput bus designs are dominated by RC effects. The RLC optimization shows that to get the same performance, the addition

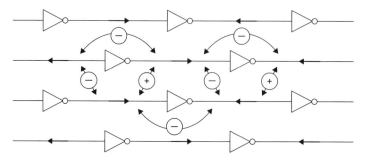

Figure 4.25 Staggered repeaters, showing inductive cancellation for both parallel and forward coupling.

Table 4.6 Optimized bus design for 0.15-μm metal width

Solution	Data Width (μm)	Shield Width (μm)	Wire Spacing (μm)	Buffer Strength	Repeating Distance (μm)	Number of Shielding Periods	Normalized Throughput (Tbps)
Arbitrary	0.4	0.4	0.4	10	1000	1	13.7
Max. number of wires	0.15	0.15	0.15	10	1000	12	17.0
Optimized	0.15	0.15	0.15	6.5	385	6	23.9
RC optimized	0.15	0.15	0.15	6.4	404	—	23.6

of shields or return paths is necessary to compensate for inductive coupling effects.

Another approach to bus optimization would be to treat the maximum delay as a constraint since in most microprocessors, the critical paths are determined elsewhere. Wider lines would allow longer repeater spacing to satisfy repeater placement constraints for the same throughput. Minimizing repeaters and total switched capacitance as a further optimization objective would result in lower power dissipation.

4.3.7 Clock Line Design

4.3.7.1 Estimating Upper and Lower Bounds of Clock Line Inductance

The design of clock networks is driven by a different cost function than for signals. The primary objective is to reduce skew and variation, not delay. Since clock lines drive large loads, they conduct high currents and are particularly susceptible to inductive effects. Several electromagnetic field solvers are available, but these are impractical to use in the early to middle phases of design. This is due to the fact that it is not possible to obtain the exact current-return paths in the early stages, and somewhat later, when the return paths have been defined, the computational task of computing the inductance values is too time consuming. What is needed is a geometry-independent model that can be used to estimate the upper and lower bounds of inductance values of clock lines [17]. For calculating lower and upper bound inductance values, the following assumptions are made:

- A three-wire ground–clock–ground system is described (Figure 4.26).
- Wires in adjacent layers which are orthogonal have no effect.
- Clock wires on the same layer are far enough away not to serve as return paths.
- Skin effect is not considered, valid up to several gigahertz, being 0.93 μm at 5 GHz.

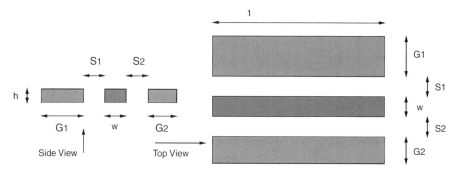

Figure 4.26   Three-wire ground–clock–ground system for lower and upper inductance-bound calculation.

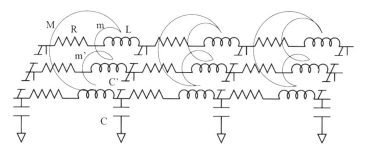

Figure 4.27   Equivalent *RLC* circuit using calculated *L* and *M* values.

$$L = \frac{\mu_0 l}{2\pi}\left[\ln\left(\frac{2l}{w+t}\right) + \frac{0.2235(w+t)}{l} + \frac{1}{2}\right] \qquad (4.11)$$

$$M = \frac{\mu_0 l}{2\pi}\left[\ln\left(\frac{2l}{s}\right) - 1 + \frac{s}{l}\right] \qquad (4.12)$$

Using equations 4.11 and 4.12, the self- and mutual inductance values can be calculated to construct an *RLC* equivalent circuit (Figure 4.27). However, this level of analysis will be too detailed and complicated for practical computation for all segments of the clock system. A further simplification using the loop inductance method is used. The impedances of the two neighboring ground wires are combined into an effective value, and the equivalent circuit is further simplified to that shown in Figure 4.28. The loop inductance is given by

Figure 4.28   Loop Inductance model.

$$L_{\text{loop}} = L_{ss} + 2\alpha_l M_{sl} + 2\alpha_r M_{sr} + 2\alpha_l \alpha_r M_{rl} + \alpha_l^2 L_{ll} + \alpha_r^2 L_{rr} \qquad (4.14)$$

$$\alpha_i = \frac{-Z_{\text{gnd}}}{Z_i} \qquad (4.15)$$

$$Z_{\text{gnd}}^{-1} = \sum_{i=1}^{n} Z_i^{-1} \qquad (4.16)$$

where L_{ss}, L_{ll}, and L_{rr} are the self partial inductances of the clock and the two ground wires, respectively, and M_{sl}, M_{sr}, and M_{rl} are the clock–ground1, clock–ground2, and ground1–ground2 mutual inductances. Z_i is the impedance of the return path i, and Z_{gnd} is the total impedance of all the parallel return paths. L_{loop} can be expressed as a function of the widths and spaces G1, G2, S1, and S2. Using the Lagrangian method, the lower bound will occur at

$$S = s_{\min} \qquad (4.17)$$

$$G = \frac{(-a_1) - \sqrt{a_1^2 - 4a_2}}{2} \qquad (4.18)$$

$$a_1 = h + w + 2S - \frac{2l}{0.7765} \qquad (4.19)$$

$$a_2 = h(w + 2S) + \frac{l(w + 2S - 3h)}{0.7765} \qquad (4.20)$$

The absolute upper bound will be when there are no return paths and the current returns at infinity, giving $L_{\text{loop}} = L_{ss}$. In practical terms, the return paths will be in the power and ground supply system spaced D from each other so that the one farthest is $D/2$ from the clock line. Solving for this upper inductance bound gives

$$\alpha_1 = -1 \qquad (4.21)$$

$$d = \frac{D - G - w}{2} \tag{4.22}$$

The values calculated are compared to FastHenry in Table 4.7.

The delay bounds for the 1-mm segment are shown in Table 4.8. The output waveforms for the upper and lower inductance bounds are shown in Figure 4.29.

4.3.7.2 Circuit and Layout Implementations of Clock Systems

The objective of minimizing across-chip and local skew and jitter in clock systems requires dealing with problems with variation in device and interconnect parameters, power supply fluctuation, and interconnect inductance at high frequencies.

A digital de-skewing circuit is used by Geannopoulos and Dai [18] to equalize two clock distribution spines by means of a digital delay line in each spine under the control of a phase detector circuit and a controller, as shown in Figure 4.30. The phase detector determines the phase relationship between the two spines and generates an output based on the phase relationship, which is then used to adjust one of the delay lines.

A method of correcting systematic skew by means of adjustable domain delay buffers is presented by Kurd et al. [19]. A differential, low-swing

Table 4.7   Inductance values (nH)

Wire config.:	$l = 1\,\text{mm}$ $h = 0.58\,\mu\text{m}$ $w = 4\,\mu\text{m}$ $G = 1.64\,\mu\text{m}$ $s = 0.5\,\mu\text{m}$	$l = 1\,\text{mm}$ $h = 0.58\,\mu\text{m}$ $w = 4\,\mu\text{m}$ $G = 1.64\,\mu\text{m}$ $s = 1\,\mu\text{m}$	$l = 1\,\text{mm}$ $h = 0.58\,\mu\text{m}$ $w = 4\,\mu\text{m}$ $G = 4\,\mu\text{m}$ $s = 4\,\mu\text{m}$	$l = 1\,\text{mm}$ $h = 0.58\,\mu\text{m}$ $w = 4\,\mu\text{m}$ $G = 4\,\mu\text{m}$ $s = 20\,\mu\text{m}$
Analytical method	0.35[a]	0.40	0.55	1.25[b]
FastHenry	0.32	0.37	0.54	1.25

[a]Lower bound.
[b]Upper bound.

Table 4.8   Delay bounds

L_{loop}[a] (nH/mm)	0	0.35[b]	0.55	1.25[c]	1.61
μ_2	2.02×10^{-4}	1.09×10^{-4}	5.55×10^{-5}	-1.31×10^{-4}	-2.26×10^{-4}
μ_3	5.29×10^{-6}	4.55×10^{-7}	-2.31×10^{-6}	-1.20×10^{-5}	-1.69×10^{-5}
Delay (ns)	0.0132	0.0141	0.0152	0.0196	0.0217

[a]μ2 and μ3 are the second and third central moments, where the negative values for the larger inductance values indicate over- and undershoot.
[b]Lower bound.
[c]Upper bound.

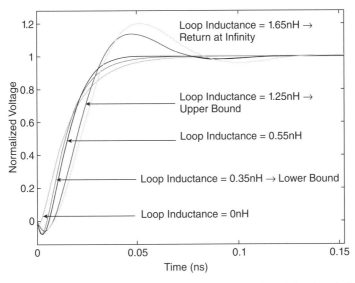

Figure 4.29 Output waveforms for upper and lower bounds of clock-line loop inductance.

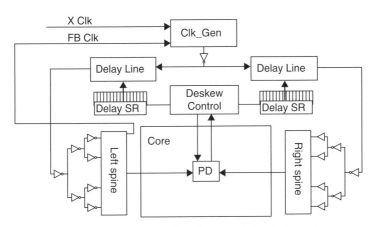

Figure 4.30 Digital deskewing of two clock distribution spines.

external reference clock is amplified by a two-stage double-differential receiver to provide single-ended CMOS levels to the core PLL. A programmable divide-by-N feedback enables the core frequency to be a multiple of N times the external system frequency. Jitter reduction is achieved by massive shielding of the clock lines and by filtering the power supply for the clock buffers using a pMOS implemented RC filter giving a 70-mV IR drop and 2.5-ns RC time constant.

A clock distribution system divided into three spines, each containing a binary distribution tree, drives a total of 47 independent domain clocks

(Figure 4.31). The programmable delay buffers allow skew optimization to correct systematic skew as well as intentional skewing of clocks to improve performance. The programmable delay buffer shown in Figure 4.32 is controlled by register values from a test access port (TAP). A three-level hierarchical clock distribution tree of programmable domain buffers and phase detectors is read out through a scan chain controlled by the TAP (Figure 4.33). The adjusted skew of the clock distribution network is ±8 ps, limited mainly by the resolution of the adjustable delay elements.

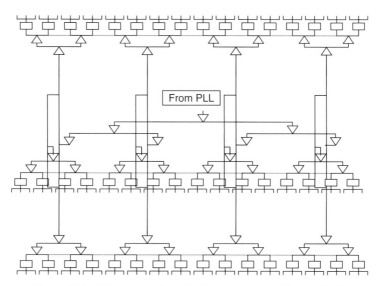

Figure 4.31   Triple-spine clock distribution network. (From ref. 16.)

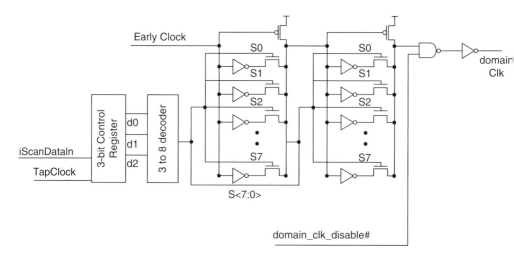

Figure 4.32   Programmable domain delay buffer.

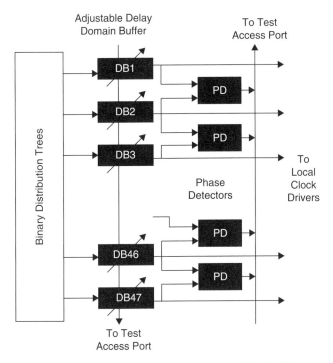

Figure 4.33 Clock tree with programmable delay buffers.

The local clock drivers drive the clock loads with a maximum of 38 ps of skew. There are fast, medium, and slow, pulsed and nonpulsed, clock drivers, enabling clock gating, debug, and calibration. An adjustable delay buffer using self-reset pull-ups allows clock width stretching.

Test data collected from on-die measurements using back-side laser probing, simulation, TAP skew-reduction scan chain, and physical picoprobing show that total jitter was reduced from 65 ps to 35 ps and total skew from 64 ps to 16 ps.

A tunable clock distribution network used on several microprocessors is described by Restle et al. [20]. An ideal global clock distribution achieves low local skew everywhere, thus avoiding timing penalties for signals crossing block boundaries. With process scaling, the total delay of the global clock distribution has not scaled with cycle time and is now dominated by interconnect delay. The majority of jitter is now due to buffer delay variations in the global clock distribution network, caused primarily by power supply noise. The clock distribution strategy described here consists of buffered symmetric H-trees driving a single clock grid for the entire chip (Figure 4.34). Well-designed tree networks can have low latency, low power, minimal wiring track usage, and low skew. In the presence of inductance, a 10 to 90% transition-time criterion is not very useful because of the wide variety of waveforms and

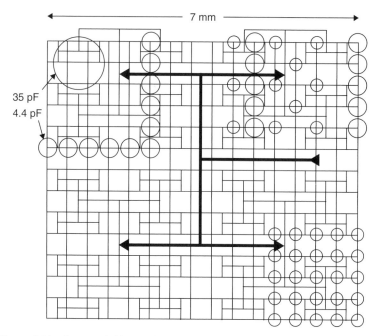

Figure 4.34 Symmetric H-tree with circles representing nonuniform clock loading.

ringing encountered. A more useful measure is the 30 to 70% transition time, targeting 10% of the cycle time.

The choice of a single grid works well only if all the trees have a very similar delay. See Figure 4.35, which shows an X–Y time–space rendering of a clock tree driving a uniform grid. Trees from the central chip buffer to the sector buffers are length matched using wire folding, but trees driven by the sector buffers are not length matched because this would greatly increase the wire tracks and power used, but are tuned by the wire width instead. Critical interconnects are split into as many as eight parallel wires, each surrounded by V_{dd} and ground return paths.

To tune a clock network for nonuniform clock loading and nonideal buffer placement, the grid was cut into small networks, which could then be tuned rapidly in parallel. To approximate the smoothing effect of the grid that is lost, the capacitive loads were smoothed instead. As a result, the tuned metal width variation was as much as a factor of 10. Figure 4.36 shows measured results showing a skew of 70 ps.

A balanced clock tree using active de-skewing driving a grid structure is described by Tam et al. [21]. The clock topology is partitioned into three segments as shown in Figure 4.37. A PLL driven by differential clock inputs running at the front-side bus frequency generates a clock signal at twice the core clock frequency which is then divided by 2 to generate the core clock and

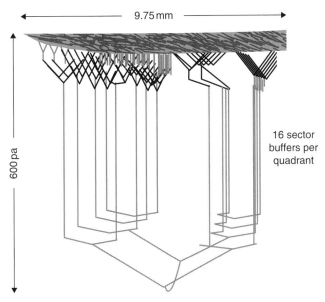

Figure 4.35 *X–Y* time–space rendering of an H-tree clock distribution network with a single grid.

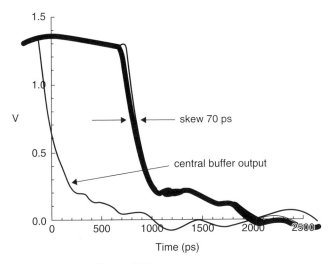

Figure 4.36 Measured skew.

a reference clock which is significantly smaller than the gobal clock in extent and is used to de-skew the local clocks. These two clocks are routed by two identical and balanced H-trees to eight de-skew clusters, each consisting of four distinct de-skew buffers (DSK), each of which is used for active de-skewing.

The global clock tree is implemented exclusively in the top two metal layers and is fully shielded laterally with V_{cc}/V_{ss}. Inductive reflections at the branch points are minimized by properly sizing the metal widths for impedance matching. RLC analysis showed that the inclusion of return path resistance increased line delay by 27% and that the addition of line inductance increased line delay by a further 8% for a total of 35% more than the RC analysis.

The regional clock distribution consists 30 separate clock regions, each consisting of a de-skew buffer, a regional clock driver, and a regional clock grid. The DSK is a digitally controlled delay-locked loop to eliminate skew between its inputs, which permits independent skew reduction due to within-die process variations. The regional clock also contains full lateral shielding, using up to 3.5% of the available M5 and 4.1% of the M4.

Local clock buffers are driven by the regional clock grid and are selected based on the clock load to be driven. Figure 4.38 shows the introduction of extra buffers for intentional clock skew to implement time borrowing. The de-skew buffer architecture in Figure 4.39 shows de-skewing of the regional

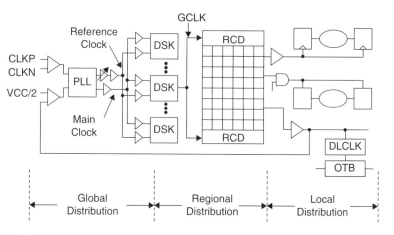

Figure 4.37 Actively de-skewed balanced clock tree driving a clock grid.

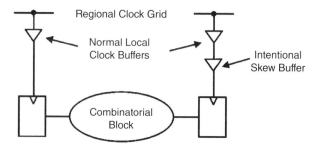

Figure 4.38 Local clock buffers showing the introduction of intentional skew.

clock grid using a local controller consisting of a phase detector and a digitally controlled analog delay line, shown in Figure 4.40. The delay control register can be manually adjusted by a test access port (TAP) for post-silicon in situ timing optimizations to discover and correct critical timing paths. The de-skew operation is performed in parallel at reset for the 30 clock regions, and the register settings are fixed. The push–pull style output stage consists of 12 parallel drivers of which any can be enabled individually via mask options to match the extracted loading to each region. The measured delay range of the DSK is 170 ps, with a step size of 8.5 ps. The total measured clock skew was reduced from 110 ps to 28 ps.

By manipulating the core clock edges, it is possible to shrink the high phase or low phase of any individual clock by use of the on-die clock shrink architecture (Figure 4.41). The total range for any edge shrink manipulation is 200 ps in 14 discrete linear steps.

Departing from a single chip-wide clock, Xanthopoulos et al. [22] in a 1.2-GHz Alpha microprocessor uses one reference and three derived clocks to clock four major clock domains (Figure 4.42). GCLK drives the processor core, and the other three clocks, NCLK, L2LCLK, and L2RCLK, are

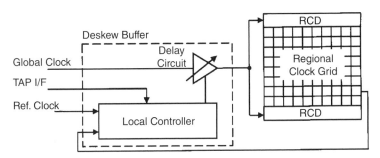

Figure 4.39   De-skew buffer architecture.

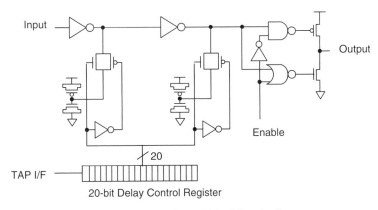

Figure 4.40   DSK variable delay circuit.

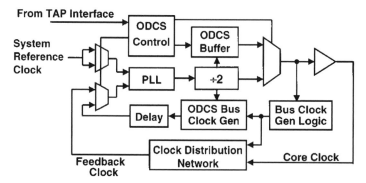

Figure 4.41   On-die clock shrink architecture.

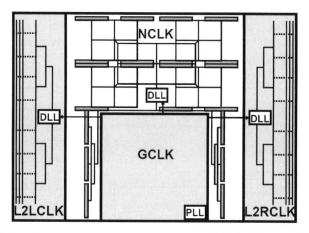

Figure 4.42   1.2-GHz Alpha microprocessor clock domains. L2LCLK, NCLK, and L2RCLK are phase-locked to GCLK.

phase-locked to GCLK by three delay-locked loops (DLLs), which use a dual delay-line architecture (Figure 4.43). The dual-delay line exhibits large dynamic range at the same time as fine-grain delay settings (<10 ps). Due to the large supply noise, the entire clock signal path within the DLLs is powered by a 2.5- to 1.5-V regulated supply, as shown in Figure 4.44. The frequency response attenuates supply noise frequencies in excess of 1 MHz by more than 15 dB. The clock lines are formed by upper-layer low-impedance metal layers with coplanar current-return paths. The clock driver sizing is such that the equivalent output impedance is sufficiently smaller than the grid characteristic impedance to guarantee a high-amplitude incident wave, yet the clock rise and fall times at the driver are sufficiently slow to ensure that the voltage overshoot and undershoot at the far ends are not excessive. A maximum cycle compression of 70 ps was observed.

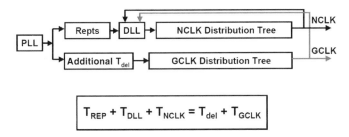

$$T_{REP} + T_{DLL} + T_{NCLK} = T_{del} + T_{GCLK}$$

Figure 4.43 Dual-delay-line Delay-Locked Loop.

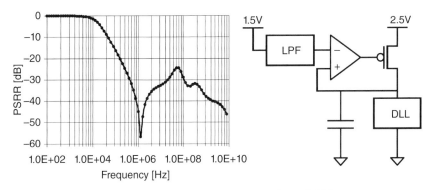

Figure 4.44 Voltage regulator used in each DLL.

A multilevel, balanced H-tree clock distribution system without a grid in a 1.7-GHz Itanium processor code named McKinley is described by Anderson et al. [23] (Figure 4.45). The level 1 route (L1R) consists of a differential pair separated by a center noise shield and two outside noise shielding wires (Figure 4.46). Additionally, there is a parallel $N-2$ layer for shielding noise from lower layers and to provide ground current returns close to the clock signal wires. The L1R delivers the clock to all corners of the die within 52 ps of measured skew, including PVT. In contrast, the level 2 route, L2R, is a width- and length-balanced, side-shielded route in a binary tree. The wire width is modified iteratively until a mismatch of <10 ps is achieved giving 25 ps maximum 1σ skew for the L2R system. The clock system dissipates 30% of the total chip power and achieves 62 ps total measured skew at 1.0 GHz. A test mode allows stop-clock, on-die-clock shrink (ODCS) and locate-critical-path (LCP) debug features. The measured skew is shown in Figure 4.47.

A four-segment clock-distribution system (Figure 4.48) dissipating 25 W and delivering a variable-frequency clock from 100 MHz to 2.5 GHz is described by Tam et al. [21] on a dual-core, multithreaded Itanium processor code named Montecito. A region-based active de-skew (RAD) system reduces the process, voltage, and temperature sources of skew across the 21.5×27.7 mm

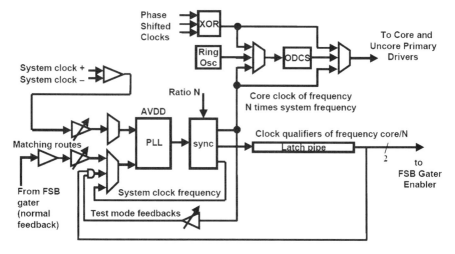

Figure 4.45   Clock generation diagram.

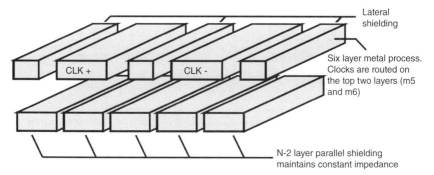

Figure 4.46   Level 1 Clock structure.

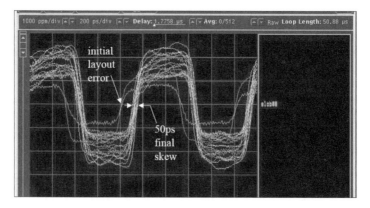

Figure 4.47   Measured on-chip clock skew.

die. Clock vernier devices inserted at each local clock buffer allow 70 ps of adjustment via scan.

The L0 route from the PLL to the digital frequency divider consists of four 5-mm segments that are 400-mV low-voltage swing differential routes shielded on each side and using power and ground wires on level $N-2$, as shown in Figure 4.49. Each segment is terminated resistively at the receiver and is tapered to optimize RLC flight time and reduce power consumption. A two-stage self-biased differential amplifier (Figure 4.50) is used to restore the 400-mV output differential to full swing at each of the three repeaters along the route. At the DFD the route is received by a self-biased amplifier and buffered into the DLL.

The second segment of the route, the L1 route, connects the DFD to the second-level clock buffer (SLCB). The DFD output varies in frequency and operates on a varying core supply voltage. A half-frequency distribution using

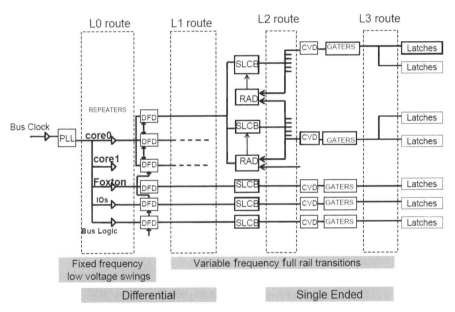

Figure 4.48 Four-segment clock system overview.

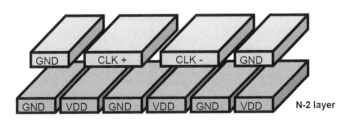

Figure 4.49 L0 route differential signals with shielding.

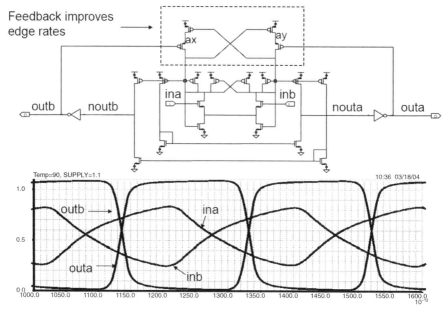

Figure 4.50 Two-stage self-biased differential amplifier used in the L0 route.

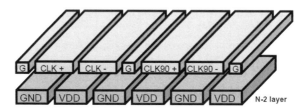

Figure 4.51 L1 route using half-frequency differential clocks.

differential $0°$ and $90°$ clocks is used (Figure 4.51) because the width-limited wire does not support full swings required for good duty cycle in a changing frequency environment.

The third segment, L2, connects the SLCB to the local clock buffers using a skew-matched tree network based on width and space to match RLC delays. A regional active de-skew (RAD) consisting of a hierarchical collection of phase comparators between the ends of the different L2 routes is used to null out the effects of process, temperature, and voltage variation.

The fourth and final segment is the post-gater route. CVDs are inserted into each gater cluster and are controlled via scan and can add 70 ps of delay to any clock in 8 ps increments. The schematic of the CVD circuit is shown in Figure 4.52.

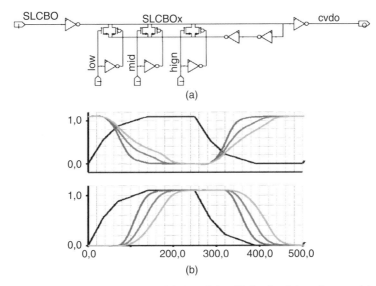

Figure 4.52   CVD circuit and operation: (a) drive fight with feedback (see the arrow) is attenuated with pass gate settings change the delay as desired; (b) SPICE simulations showing low, middle, and high delay settings for SLCBOx (top) and CVDO (bottom).

4.3.8 Power Grid Design

As a result of scaling to the ever-finer dimensions of 45- and 32-nm technologies, combined with the increasing power level and frequencies of emerging high-performance processors, the integrity of the power grid is a serious challenge. Power dissipation of over 250W for some processors, combined with requirements for minimum supply droop in the presence of large di/dt, results in an extremely low impedance requirement for the power supply, on the order of $1\,m\Omega$ for current processors.

Although RC analysis will underpredict the actual total supply voltage drop compared to a complete RLC simulation, satisfying the IR constraints is a minimal and necessary constraint. A simple analysis to determine the static IR drop can be done based on gate size, capacitive load, and vectorless switching activity, which takes into account the nonuniform distribution of current demand. The resulting distribution of current sources driving a complex multilevel grid can be easily solved to give the IR drop values and their location in the grid. Dynamic IR drop caused by programming load or clock gating is usually solved by inserting decoupling capacitors.

A more important problem caused by clock gating or stop clock activity is voltage drop caused by $L\,di/dt$ noise. Good design to reduce loop inductance in the power grid system is required, particularly by providing close, low-resistance return paths, as described earlier.

4.3.8.1   *Measurement and Modeling of Power Supply Noise*

To address power dissipation and provide stable power supply voltage for correct circuit operation, it is necessary to measure supply noise and its effect on circuits. Naffziger et al. [24] describe a two-core Itanium processor code named Montecito. Large changes are shown in Figure 4.53, the time-resolved relative current draw for a particular program, int.256.bzip2. A subsampling method of measuring repetitive waveforms using only two samplers with simple VCO-based ADCs is described. The block diagram is shown in Figure 4.54 and the VCO-based ADC is shown in Figure 4.55. To measure repetitive waveforms, time-dependent distributions, or cyclo-stationary noise spectra, the sampler timing pulses need to be placed with relatively fine resolution within a time basis that spans the longest events of interest. In this coarse–fine

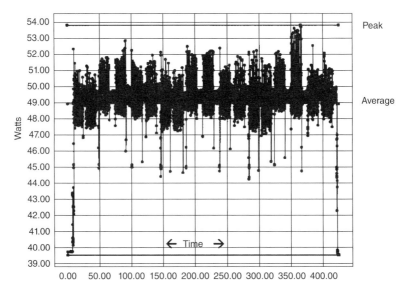

Figure 4.53   Time-resolved relative current draw for int.256.bzip2.

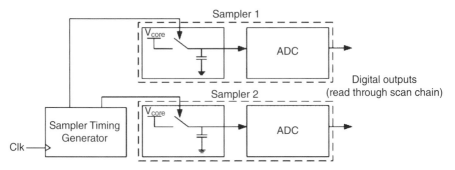

Figure 4.54   Supply noise measurement system.

architecture, the timing signals are placed with clock cycle granularity by a scan-chain-controlled state machine (a counter and a set of comparators), then adjusted with finer steps by a low-fanout inverter-based delay line. For small delay settings, the smp signals from the state machine were retimed by the rising edge of dclk.

Figure 4.56 shows the measured 70-mV peak-to-peak noise with both cores operating in fixed-frequency mode at 1.4 GHz with a nominal supply of 1.05 V; the cores were set to toggle between the power virus program and low activity. As shown by the sharp dips in the distribution, which repeat every clock cycle, and the corresponding pulse train in the noise spectrum, a significant amount of noise is generated by clock-related activity that repeats every cycle. The

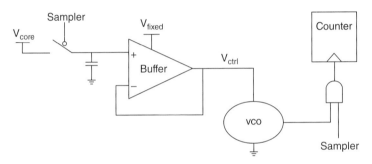

Figure 4.55   VCO-based ADC.

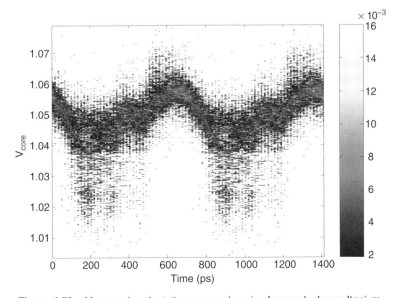

Figure 4.56   Measured cyclostationary supply noise from a dual-core Itanium.

noise spectrum (Figure 4.57) shows the resonance between the package inductance and the on-chip capacitance at about 50 MHz.

Two on-chip noise-probing techniques are described by Nagata et al. [25]. The first consists of a source follower and a latch comparator, as shown in Figure 4.58. Noise picked up by a V_{noise} probe is sensed by the source follower; then output voltage is sampled and digitized by the comparator through successive comparisons with a step reference voltage V_{step} along with shifted sample time, φ. Voltage and time resolutions of 100 μV and 100 ps are achieved over ranges of 1 V and 1 μs. As shown in Figure 4.58, the noise detectors can be added to an existing layout and noise probe lines routed into interesting points for measurement, enabling chip-level substrate crosstalk analysis, well

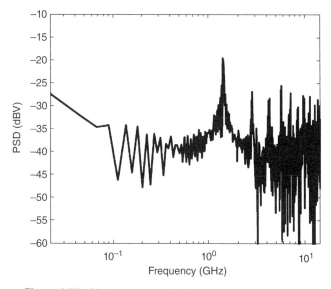

Figure 4.57   Noise spectrum showing resonance at 50 MHz.

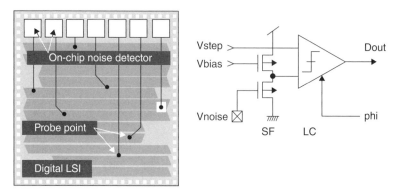

Figure 4.58   Noise sampler using a source follower and latch comparator.

noise measurements in body-biased circuits, and measurements of power and ground noise.

To measure the strong positional dependence of power supply variation, a smaller probe is used (Figure 4.59). In this noise detector, a source follower is followed by a gain stage with a current mirror that drives a time-varying current onto a shared line with 1-GHz bandwidth. It is then read out through a current mirror to an oscilloscope with a termination resistor. The small size of this detector, about that of a flip-flop, enables it to be placed into standard cell rows, providing the capability of mapping noise distributions. An implementation of this circuit for measuring V_{dd} and V_{ss} noise is shown in Figure 4.60. Measured noise waveforms of V_{dd} noise and V_{ss} noise (Figure 4.61) show a drop in power supply voltage and a rise in ground voltage right after the clock edge. The positional dependence of supply noise is shown in Figure 4.62, the difference with a LFSR activated at position 7 with other circuits quiet. The smaller range of V_{ss} sensitivity than V_{dd} is due to being tied to the substrate. N2 and P1 are in the center of the array, while N1 and P2 are at the ends, showing the effects of supply distribution to the center of the array.

Another useful supply noise detector is the on-die droop detector [26], which indicates when the voltage has fallen below a specified level. Droops occur due to large di/dt and IR events, including droops with four different time constants, each due to a different resonance peak in the impedance of the microprocessor's power delivery spectrum (Figure 4.63). Programmability is provided in polarity, duration, and magnitude. The droop detector architecture (Figure 4.64) interfaces to the external world through the standard JTAG port and implements the following features:

- Negative versus positive overshoot detection
- Noise-level and duration detection settings
- Start/stop time for the droop search window
- Process–voltage–temperature (PVT) calibration for analog circuits

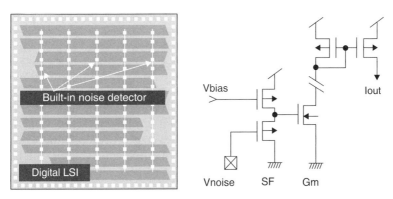

Figure 4.59 Smaller noise detection probes to measure positional dependence.

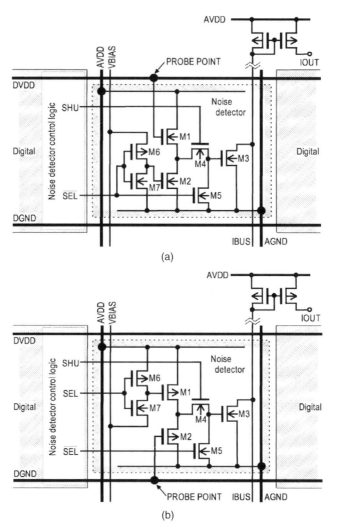

Figure 4.60 V_{dd} and V_{ss} continuous-time noise measurement circuit embedded in a standard cell row.

A shielded, differential analog current pair is supplied by the reference unit (Figure 4.65), which provides better immunity to line resistance and ground offsets as well as minimizing common-mode voltage noise coupling. The detector module is shown in Figure 4.66. In the high-resolution mode, the calibration and threshold settings were 8 and 7 mV, respectively. Noise pulse widths of 100 ps were detected. V_{cc} versus threshold setting across five test modules values for a 90-nm microprocessor are shown in Figure 4.67.

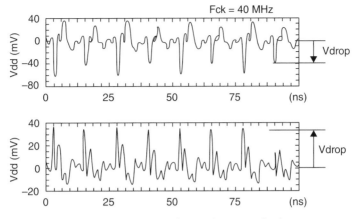

Figure 4.61   Power and ground measured noise.

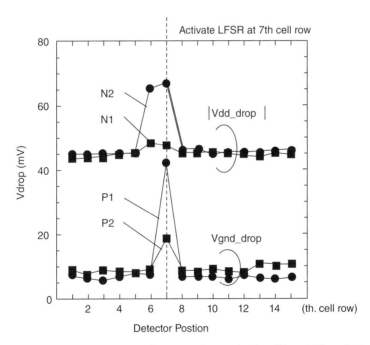

Figure 4.62   Positional dependence of power and ground noise. N2 and P1 are in the center of the array, and N1 and P2 are at the ends.

4.3.8.2 Decoupling Capacitors

Decoupling capacitors are an effective way to reduce the impedance of power delivery systems operating at high frequencies [27]. Large on-chip capacitors are often placed at a considerable distance from the current load they are required to buffer. This problem cannot be alleviated simply by increasing

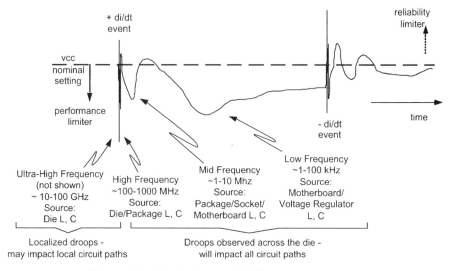

Figure 4.63   Supply droops with different time constants.

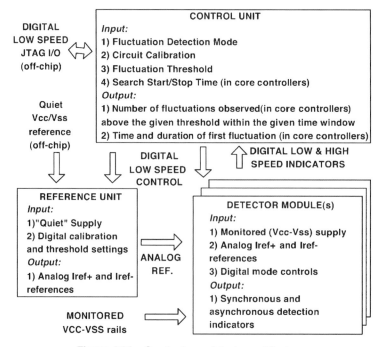

Figure 4.64   On-die droop detector architecture.

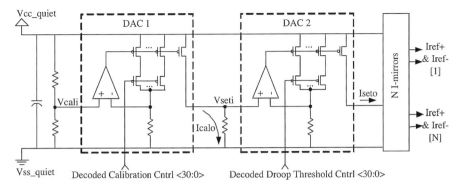

Figure 4.65 Droop detector reference unit.

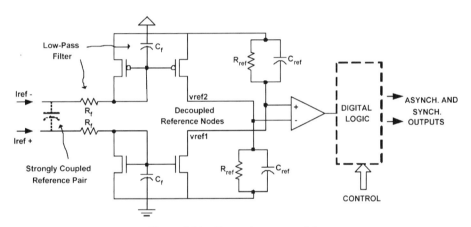

Figure 4.66 Droop detector module.

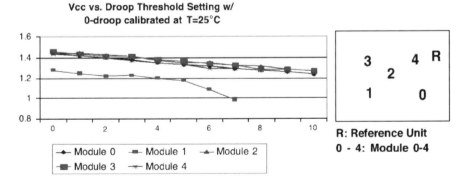

Figure 4.67 Microprocessor V_{cc} versus threshold setting for various module locations.

the size of the capacitors, but they need to be physically close to the current load. A maximum parasitic impedance between the decoupling capacitor and the current load exists at which the capacitor is effective. The schematic of the application of the decoupling capacitance is shown in Figure 4.68. The solution is a distributed system of decoupling capacitors. The circuit model of the distributed capacitors is shown in Figure 4.69. Assuming that the load current I_{max} is 0.01 mA, $V_{dd} = 1.0$ V, $C_1 = C_2$, and the rise time $t_r = 100$ ps, the voltage across C_1 as a function of C_1 in picofarads and parasitic resistance R_2 is shown in Figure 4.70. To reduce the size of the locally placed C_1, the more distant C_2 needs to be larger than C_1 as a function of R_2, as shown in Figure 4.71. The problem of providing large, on-chip decoupling capacitors made from gate oxide is that they take up much space and cause large leakage currents and resulting standby power dissipation. For example, for a 1.5-V 25-W chip and 5% V_{dd} droop, the required decoupling capacitor is 2 μF. The package-die electrical macromodel is shown in Figure 4.72. The time-varying value of core voltage as a function of decoupling capacitance is shown in Figure 4.73.

One solution to the problem of gate leakage for gate oxide decoupling capacitors proposed by Sanchez et al. [28] is to provide a metal–insulator–metal (MIM) capacitor using a high-k dielectric (Figure 4.74). The 8-fF/μm² planar MIM capacitor is formed by physical vapor deposition of alternating layers of HfO₂ and Ta₂O₅ dielectrics between TaN electrodes. The leakage of the MIM capacitor is 2.5 to 3 orders of magnitude less than that of the gate

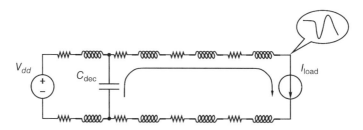

Figure 4.68 Decoupling capacitor with load current and intervening impedance.

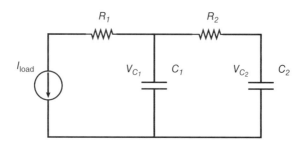

Figure 4.69 Circuit model of distributed decoupling capacitors.

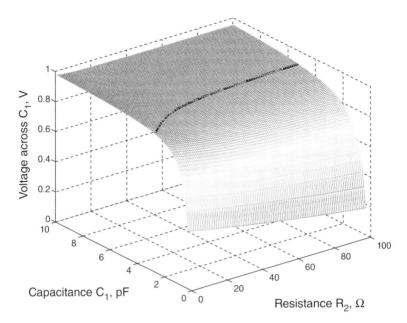

Figure 4.70   Voltage across C_1 assuming $I_{load} = 0.01$ mA, $R_1 = 100$, $C_1 = C_2$, $V_{dd} = 1.0$ V, and $t_r = 100$ ps. For large R_2 and minimum V_{dd}, minimum C_1 is about 3 pF.

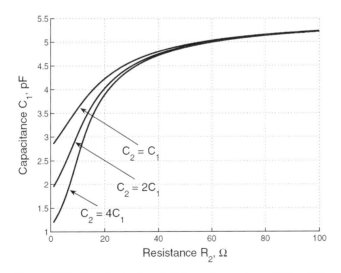

Figure 4.71   Required value of C_1 for various values of C_2 and R_2.

oxide. Figure 4.75 shows the measured speed increase after application of the 250 nf MIM decoupling capacitor. Another example of a MIM capacitor is shown in Figure 4.76. It is implemented by forming a comb structure of narrow fingers in a three-dimensional Damascene process to get a 17-fF/μm^2 capacitor to increase the interelectrode area by a factor of 3.

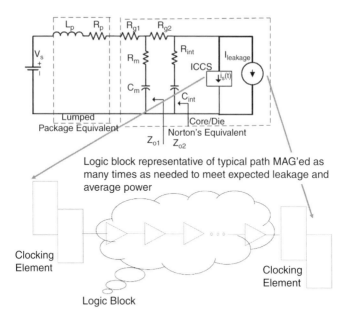

Figure 4.72 Package-die electrical macromodel.

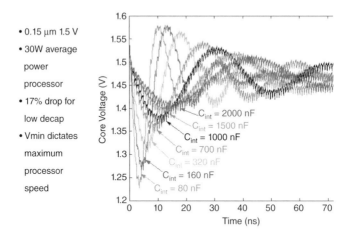

Figure 4.73 Core voltage as a function of decoupling capacitor value.

4.3.8.3 Active Decoupling Capacitors and Circuit Solutions to Power Supply Noise

A low-power, switched decoupling capacitor (decap) circuit is proposed to suppress on-chip resonant supply noise [30]. The resonant noise, typically in the range 40 to 200 MHz, can be excited by a microprocessor loop command or a large current surge during an abrupt startup or termination. The operation of the proposed circuit is shown in Figure 4.77. In the absence of resonant

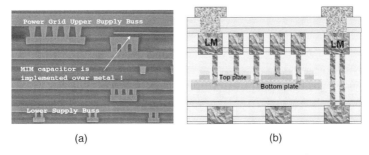

(a)   (b)

Figure 4.74 Metal–Insulator–Metal capacitor using high-k dielectric: (a) die corss section of metallization stack showing integration of MIM into copper/low-k back end; (b) contacts to MIM plates occur from LM down to both top and bottom plates of the capacitor.

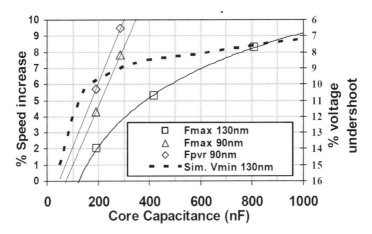

Figure 4.75 Measured speed increase after application of 250-nF MIM decoupling capacitor.

noise, the two capacitors are connected in parallel, storing the maximum amount of charge and serving as conventional decaps. When a supply droop greater than the switching threshold V_{sw} is detected, the decaps switch to a series connection with charge dumped into the supply network to compensate the supply droop. Vice versa, during a supply overshoot, charge is removed from the supply network by switching the capacitors back to parallel. Because the delivered charge of a switched decap ($= 0.5CV_{dd} + C\Delta V_{dd}/2$) is much larger than that of a conventional decap ($= 2C\Delta V_{dd}$), a decap boost is achieved. Figure 4.77 shows the decap boost factor as a function of V_{sw}. A V_{sw} value between 25 and 60 mV leads to a decap boost between 13- and 5-fold, respectively. Figure 4.78 shows the schematic of the switched decap circuit with a digital resonant detection scheme. The noise detection is realized by comparing the delay of a constant delay line (CDL) and a variable delay line (VDL). The supply of the CDL (V_{dd}) is low-pass filtered so that the delay is insensitive to supply droop within the range of regulation frequency (>10 MHz). The supply of VDL is connected directly to the noisy V_{dd}, so that its delay varies

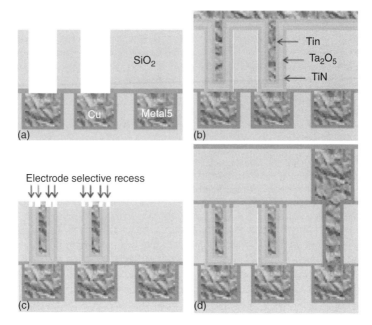

Figure 4.76 TiN/Ta$_2$O$_5$/TiN Damascene MIM capacitor. (From ref. 29.)

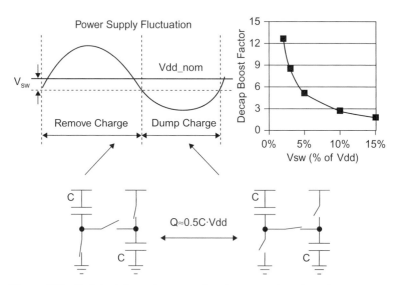

Figure 4.77 Switched decoupling capacitors for supression of supply resonance.

with supply fluctuations. When V_{dd} falls below V_{sw}, the phase comparator switches the decaps. Figure 4.78 shows the switched decap circuit with the digitally controlled resonant detector. The simulated decap performance is shown in Figure 4.79 and the measured supply noise in frequency domain is shown in Figure 4.80.

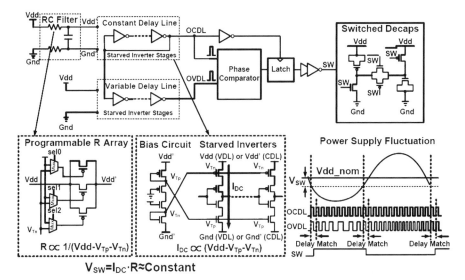

Figure 4.78 Switched decap circuit with digitally controlled resonant detector.

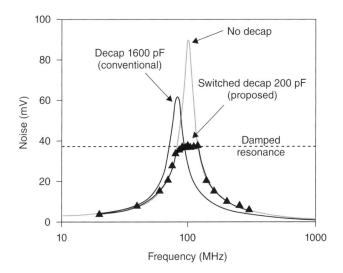

Figure 4.79 Simulated damped resonance using the switched decap circuit.

Another distributed active decoupling capacitor circuit using an operational amplifier was described by Gu et al. [31]. The circuit consists of an operational amplifier and a passive load capacitor C_{load} (Figure 4.81). The feedback loop detects changes in the voltage V_{dd}-ground and drives the load capacitance, effectively amplifying the capacitance seen at the V_{dd} input to $[1 + A(\omega)]C_{load}$ via the Miller effect, where $A(\omega)$ is the gain of the op-amp. Figure 4.82 shows the op-amp design. The noise on the V_{dd} and the ground lines are capacitively coupled to the inputs of the differential stage. A source-follower stage is used to drive the output capacitance load of 1 pF. Simulations showed a dc gain of 9.5, a bandwidth of 535 MHz, a unit gain frequency of 2.7 GHz, and a phase margin of 42°. Figure 4.83 shows measured waveforms of V_{dd} noise and active decap output when turning the active decap circuit on and off. Noise was generated by a single 16-bit linear feedback shift register operating at 120 MHz. The differential noise is measured by the active decap sensor, with a gain of 5 and output biased at 0.32 V. V_{dd} noise is measured at different sensor locations. The wakeup time shown is 200 ns, due to the charging time of the input capacitors. Figure 4.84 shows the measured decoupling

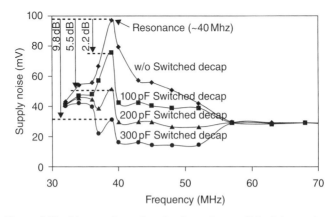

Figure 4.80 Measured supply noise for various switched decap sizes.

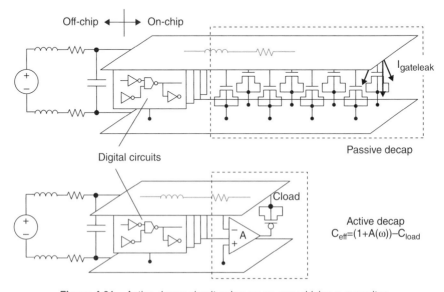

Figure 4.81 Active decap circuit using an op-amp driving a capacitor.

effects of active decaps and passive decaps up to 1 GHz. Measurements show that the 10 pF active decap offers the same decoupling effect as an 80- to 120-pF passive decap up to 500 MHz. The decoupling effect of active decaps decreases at frequencies above 500 MHz, giving a performance similar to that of the 40 pF of passive decap.

An on-chip noise canceler with high-voltage supply lines for the nanosecond-range power supply noise having a fast wake-up time was proposed by Nakamura et al. [32]. A high-voltage supply (V_{ddh}), a switch between V_{ddh} and the normal power supply (V_{dd}), and a level shifter are added to the normal power supply (Figure 4.85) with the level shifter (Figure 4.86). When the logic circuit wakes up, the switch between V_{ddh} and V_{dd} is turned on by the level-shifter and the current from V_{ddh} substitutes the current flowing through

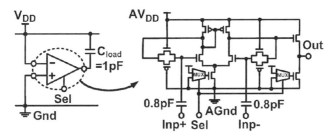

Figure 4.82   Active decap circuit.

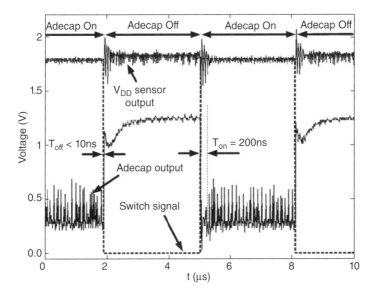

Figure 4.83   Measured waveforms of V_{dd} noise and active decap output when turning the active decap circuit on and off.

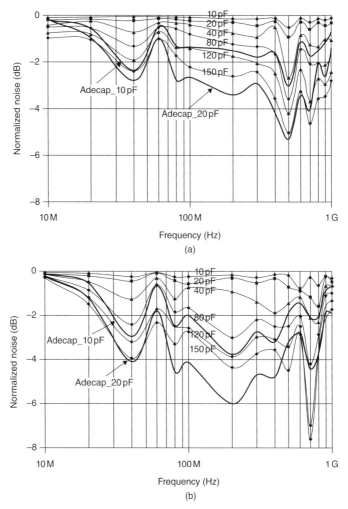

Figure 4.84 Measured decoupling effect for noise generated by (a) LFSR circuits and (b) noise injection circuits. Differential noise V_{dd}-ground is measured.

the bonding wire and the onboard supply lines of V_{dd} (Figure 4.87). Since the noise on V_{ddh} does not influence V_{dd}, the impedance of V_{ddh} supply line can be large compared to the main V_{dd} line as long as V_{ddh} can substitute the current for V_{dd}. The measured performance of the circuit from wake-up is shown in Figure 4.88, showing a 68% reduction in noise.

The several stages of power supply decoupling, die, package, socket, motherboard, and voltage regulator are shown in Figure 4.89. The inductances at every stage need to be neutralized with "next stage" decoupling capacitance [33]. This pole–zero cancellation continues until the caps are very close to the source. It is impractical, however, to place enough decaps on the silicon die

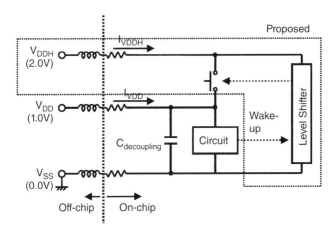

Figure 4.85 High-voltage supply V_{ddh} and switch to nominal supply V_{dd}, controlled by the level shifter.

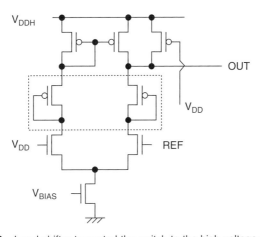

Figure 4.86 Level shifter to control the switch to the high voltage supply V_{ddh}.

to cancel the inductive effects of the previous stage (package and socket), as this requires a prohibitively large silicon area. Because of that, the output impedance of the power network is not flat and exhibits a resonant behavior. In the time domain, the several stages of decoupling translate to different types of global voltage droops, after a clock stop event (Figure 4.90). The first and largest power supply droop is nanoseconds in duration, determined by $L_{package}$ and C_{die} resonance. The simulated first droop with and without die caps is shown in Figure 4.91, showing that the magnitude of the first droop gets significantly attenuated by the decoupling caps. However, the frequency of resonance is lower when decoupling is added. Lower frequency implies higher chances of hitting speed paths. A speed path of several core clocks is

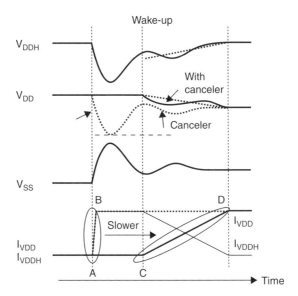

Figure 4.87   Operation of the noise canceler.

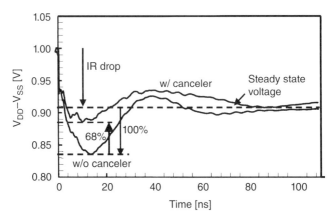

Figure 4.88   Measured 68% reduction in noise when a noise canceler is operating.

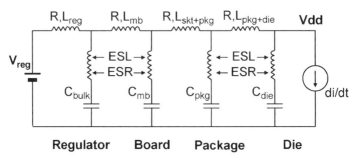

Figure 4.89   Circuit representation of a typical Pentium III and Pentium 4 power delivery network. (From ref. 35.)

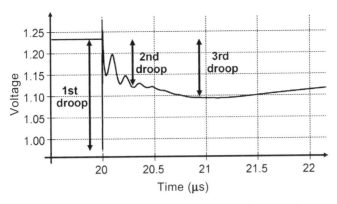

Figure 4.90 Simulated time-domain response to a stop clock event, showing the different stages of inductance and decoupling capacitance responsible for the various droop occurrences. (From ref. 35.)

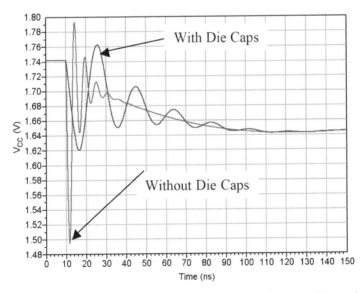

Figure 4.91 Simulated first droop with and without decoupling caps. (From ref. 33.)

insensitive to high-frequency noise due to averaging effects but is very sensitive to lower-frequency noise. It is therefore theoretically possible that on-die decoupling caps slow down the processor speed. To test this, several wafers were built with and without decoupling caps. Figure 4.92 shows the average frequency change between the two lots. The average change is significantly lower than the worst-case simulations would have predicted. Figure 4.93 shows the impact for the top 300 speed paths. As can be seen from this figure, some paths speed up while others slowed down. The worst slowdown is still significantly smaller than predicted initially by the power supply model.

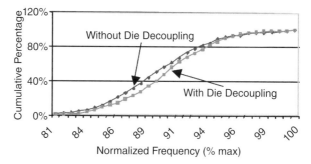

Figure 4.92 Average maximum frequency shift due to decoupling cap removal.

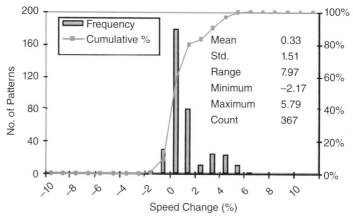

Figure 4.93 Speed change for individual paths of the top 300 speed paths due to die cap removal. Note that some paths sped up, contrary to model prediction.

The unusual conclusion described above, showing that suppressing power supply noise did not improve the performance [33], was analyzed by Rahal-Arabi et al. [34,35]. With a stop clock action, the first droop is typically less than 200 MHz and lasts several clock cycles in a multigigahertz processor. If the critical path is launched by clock edge 1 and latched by clock edge 2, for the case of falling V_{dd}, edge 2 was propagated through the clock distribution at a lower average voltage than edge 1. Edge 2 is therefore delayed with respect to edge 1. This clock stretching partially compensates for the data signal delay induced by the power supply noise, an effect termed *clock/data compensation* (CDC). The dependence of a negative F_{max} margin on clock distribution delay and noise frequency is shown in Figure 4.94. It is possible to increase the CDC by filtering the power supply noise for the clock distribution [34]. The purpose of the filter in the power supply is to shift the noise seen by the clock relative to the noise seen by the data signal and reduce the differential noise (Figure 4.95). A 30-A step change in core current produces

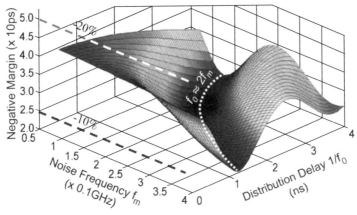

Figure 4.94   Simulated negative F_{max} margin as a function of clock distribution delay and noise frequency. (From ref. 35.)

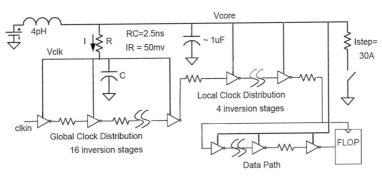

Figure 4.95   Filtered clock power supply. (From ref. 34.)

the core supply noise waveform. The core supply in this example has nearly 1 µF of total capacitance. Figure 4.96 shows the simulated frequency loss due to power supply noise in the presence or absence of the clock filter and power supply capacitance. Two populations of experimental processors with and without the filter were built. The simulated performance with and without on-chip decoupling capacitors are shown in Figure 4.96 and 4.97, and summarized in Table 4.9. The experimental data are shown in Figure 4.98. Focusing on the top few hundred critical paths, with the clock filter removed, all critical paths consistently slowed down anywhere from 0 to 3%.

Adaptive frequency generation is another application of CDC by implementing tight tracking of the clock period to the power supply noise. The sensing of power supply droop must be quickly applied to clock period change. An implementation of this was used by Fischer et al. [36]. Multiple tuned delay lines are deployed to sense the V_{dd} droop, which is then digitized and conveyed

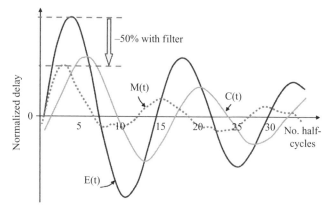

Figure 4.96　Simulated reduction in normalized delay with and without a clock power supply filter with no on-die decoupling capacitance. (From ref. 35.)

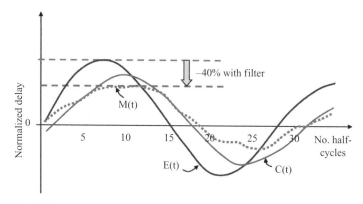

Figure 4.97　Simulated reduction in normalized delay with and without a clock power supply filter with full on-die decoupling capacitance.

Table 4.9　**Simulated normalized F_{max} loss as a function of implementation of the power supply decoupling capacitors and clock filter**

Decap	Filter	Normalized F_{max} Loss (%).	
		Graphical Analysis	Simulated
No	No	100	100
No	Yes	60	53
Yes	No	50	48
Yes	Yes	35	27
Ideal (no noise)		0	0

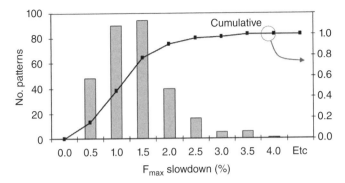

Figure 4.98 Experimental results showing that removing the clock power supply filter slowed down the critical paths from 0 to 3%.

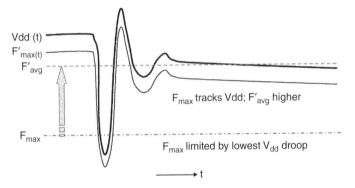

Figure 4.99 Application of an adaptive clock frequency system to track noise in V_{dd}.

to a DLL-based frequency synthesizer, where clock periods are digitally modulated and scheduled on the fly. F_{max} in a nonadaptive system must be set low enough to accommodate the lowest V_{dd} droop. In an adaptive system, F'_{max} tracks V_{dd}, achieving a higher average F'_{avg} (Figure 4.99).

REFERENCES

1. H. J. Barth, Back end of line manufacturing technology, *IEDM 2005 Short Course: Next Generation Semiconductor Manufacturing*, International Electron Device Meeting, Dec. 2005. IEDM 2005, Washington, D.C.

2. B. Arnold, Lithography for the 32 nm technology node, *IEDM 2006 Short Course: 32nm CMOS Technology*, International Electron Device Meeting, Dec. 2006. IEDM 2006, San Francisco, CA.

3. B. Wisnieff, 32 nm BEOL device technology, *IEDM 2006 Short Course: 32 nm CMOS Technology*, International Electron Device Meeting, Dec. 2006. IEDM 2006, San Francisco, CA.

4. J. Belani, Presentation International Microelectronic Assembly and Packaging Society, Sunnyvale, CA, Aug. 2007.

5. R. Ho, Dealing with issues in VLSI interconnect scaling, Tutorial, International Solid State Circuits Conference, San Diego, CA, Feb. 2007.

6. L. Pileggi, Modeling and synthesis of high-speed IC busses and channels, Tutorial, Sun Microsystem, Sunnyvale, CA, Nov. 2001.

7. M. W. Beattie and L. Pileggi, Inductance 101: modeling and extraction, Design Automation Conference, Las Vegas, NV, June 2001.

8. Y. Cao et al., Switch-factor based loop RLC modeling for efficient timing analysis, International Conference on Computer-Aided Design, San Jose, CA, Nov. 2003.

9. Z. Jiang et al., A new RLC buffer insertion algorithm, International Conference on Computer-Aided Design, San Jose, CA, Nov. 2006.

10. L. Pileggi, Modeling of IC electromagnetic couplings including power/ground networks, Research paper, Carnegie Mellon University, Pittsburgh, PA, 2000.

11. X. Huang et al., Loop-based interconnect modeling and optimization approach for multi-GHz clock network design, Custom Integrated Circuits Conference, Orlando, FL, 2002.

12. A. Sinha, Mesh-structured on-chip power/ground: design for minimum inductance and characterization For fast R, L extraction, Custom Integrated Circuits Conference, San Diego, CA, 1999.

13. M. Mondal, Reducing pessimism in RLC delay estimation using an accurate analytical frequency dependent model for inductance, Design Automation Conference, Anaheim, CA, 2005.

14. X. Qi et al., On-chip inductance modeling and RLC extraction of VLSI interconnects for circuit simulation, Custom Integrated Circuits Conference, Orlando, FL, 2000.

15. Simulation and analysis of inductive impact on VLSI interconnects in the presence of process variations, Design Automation Conference, Anaheim, CA, 2005.

16. L. Pileggi, Modeling interconnect parasitics at gigahertz frequencies, Tutorial, Sun Microsystem, Sunnyvale, CA, Nov. 1998.

17. Y. C. Lu et al., A fast analytical technique for estimating the bounds of on-chip clock wire inductance, Design Automation Conference, Las Vegas, Nevada, 2001.

18. G. Geannopoulos and X. Dai, An adaptive digital deskewing circuit for clock distribution networks, International Solid State Circuits Conference, San Francisco, CA, 1998.

19. N. A. Kurd et al., A multigigahertz clocking scheme for the pentium® 4 microprocessor, *J. Solid State Circuits*, Nov. 2001.

20. P. J. Restle et al., A clock distribution network for microprocessors, *J. Solid State Circuits*, May 2001.

21. S. Tam et al., Clock generation and distribution for the first IA-64 microprocessor, *J. Solid State Circuits*, Nov. 2000.

22. T. Xanthopoulos et al., The design and analysis of The clock distribution network for a 1.2 GHz Alpha microprocessor, International Solid State Circuits Conference, San Francisco, CA, 2001.

23. F. Anderson et al., The core clock system on the next generation Itanium™ microprocessor, International Solid State Circuits Conference, ISSCC 2002, San Francisco, CA, 2002.

24. S. Naffziger et al., The implementation of a 2-core, multi-threaded Itanium family processor, *J. Solid State Circuits*, Jan. 2006.

25. M. Nagata et al., A built-in technique for probing power supply and ground noise distribution within large-scale digital integrated circuits, *J. Solid State Circuits*, Apr. 2005.

26. A. Muhtaroglu et al., On-die droop detector for analog sensing of power supply noise, *J. Solid State Circuits*, Apr. 2004.

27. M. Popovich et al., Efficient placement of distributed on-chip decoupling capacitors in nanoscale ICs, Design Automation Conference, San Diego, CA, 2007.

28. H. Sanchez, et al., Increasing microprocessor speed by massive application of on-die high-k MIM decoupling capacitors, International Solid State Circuits Conference, San Francisco, CA, 2006.

29. M. Thomas et al., Reliable 3D Damascene MIM architecture embedded into Cu interconnect for a Ta_2O_5 capacitor record density of $17\,fF/\mu m^2$, *Symposium On VLSI Technology*, Honolulu, Hawaii, 2007.

30. J. Gu et al., A switched decoupling capacitor circuit for on-chip supply Resonance damping, *Symposium on VLSI Circuits*, Honolulu, Hawaii, 2007.

31. J. Gu et al., Distributed active decoupling capacitors for on-chip supply noise cancellation In digital VLSI circuits, *Symposium on VLSI Circuits*, Kyoto, Japan, 2006.

32. Y. Nakamura et al., An on-chip noise canceller with high voltage supply lines for nanosecond-range power supply noise, *Symposium on VLSI Circuits*, Honolulu, Hawaii, 2007.

33. T. Rahal-Arabi et al., Design and validation of the Pentium® III and Pentium® 4 processors power delivery, *Symposium on VLSI Circuits*, Kyoto, Japan, 2002.

34. T. Rahal-Arabi et al., Enhancing microprocessor immunity to power supply noise with clock/data compensation, *Symposium on VLSI Circuits*, Honolulu, Hawaii, 2005.

35. K.L. Wong et al., Enhancing microprocessor immunity to power supply noise with clock-data compensation, *J. Solid State Circuits*, Apr. 2006.

36. T. Fischer et al., A 90-nm variable frequency clock system for a power-managed Itanium architecture processor, *J. Solid State Circuits*, Jan. 2006.

5

ANALOG AND MIXED-SIGNAL CIRCUIT DESIGN FOR YIELD AND MANUFACTURABILITY

5.1 INTRODUCTION

Design of analog and mixed-signal circuits for deep-submicron processes has lead to some interesting challenges to achieve high performance while maintaining a high yield from both a manufacturing and performance standpoint. Within this chapter, specific design and layout methodologies are presented to aid designers developing high-performance analog and mixed-signal circuits on deep-submicron processes. It is critical to ensure a robust implementation between the design and physical layout. A robust design requires a complete verification during the design phase and a thoroughly thought-out layout. In this chapter we provide some generic guidelines for designing successfully on deep submicron processes. Several specific topics are presented and then described in detail.

5.2 DEVICE SELECTION

One of the first decisions that need to be decided during the design process is what devices should be used. On most deep-submicron process, two distinct gate oxides are available: thin and thick. These two distinct oxide thicknesses are meant for different parts of the design. The thin oxide devices are used for the core logic, which needs to operate at the highest speed and may rep-

Nano-CMOS Design for Manufacturability: Robust Circuit and Physical Design for Sub-65 nm Technology Nodes
By Ban Wong, Franz Zach, Victor Moroz, Anurag Mittal, Greg Starr, and Andrew Kahng
Copyright © 2009 John Wiley & Sons, Inc.

resent the highest power dissipation. The thin oxide devices are almost always optimized for the digital logic, so the analog parameters are secondary to the digital performance. The driving advantage of using thin oxide devices is the reduction in the die area and higher fT. The thick oxide devices are included primarily for the input/output (I/O) drivers, which require a higher supply voltage for interfacing to off-chip devices. Use of the thicker oxide devices for the I/O buffers simplifies the design rather than trying to use the core devices at higher voltages. In some cases, a third oxide may be available that provides a median point between the thin and thick oxide I/O devices.

Deciding which device oxide to use is largely dependent on the end application and the supply voltages allowed. If the wrong decision is made, it could impact significantly the yield of the end product. Items that need to be considered include:

1. Operating speed
2. Operating voltage
3. Maximum allowed gate leakage
4. Maximum allowed source-drain leakage
5. Supply voltage available
6. Output impedance
7. Gain requirements
8. Linearity requirements

Additional parameters may be important depending on the specific application.

Figure 5.1 shows the output impedance for a thin and a thick oxide device on a 90-nm process for devices with a minimum channel length and twice the minimum channel length. It is obvious that the output impedance of thin oxide devices is quite a bit lower than that of thick oxide devices. The second parameter to consider is the transconductance, since the transconductance and output impedance determine the overall gain that can be achieved. Figure 5.2 shows the transconductance for the same two oxide thickness and channel lengths. Again, a quick comparison between the two shows that the tin oxide devices possess a significantly higher transconductance. The channel length of the load device may need to be increased if thin oxide devices are used to increase the gain to a value that matches the thick oxide devices.

Analog circuits involve several basic building blocks that are typical for creating all larger subsequent circuits. These circuits include differential amplifiers, current mirrors, and pass gate multiplexers. These basic circuits can be used to evaluate the performance of a process for analog applications. There are several quick metrics that can be made to determine how well a certain device thickness will work for an application. The first metric is the supply voltage/threshold voltage ratio. This metric just takes the ratio of the supply

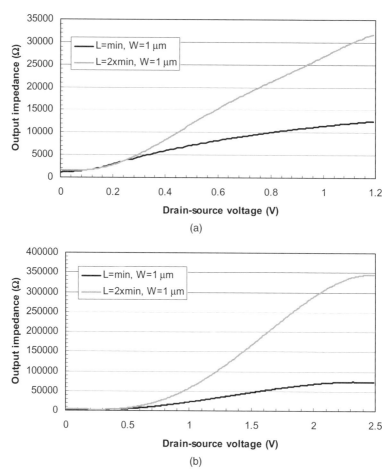

Figure 5.1   Output impedance for (a) a thin oxide and (b) a thick oxide transistor on a 90-nm process for the minimum and twice-minimum-sized channel-length devices with varying drain–source voltage. The gate voltage is set to two-thirds the supply voltage.

voltage over the threshold voltage. Larger values represent a better device from a supply headroom perspective. As a simple example, consider the variation of thin and thick oxide devices on a 90-nm process. The following equation can be written to define the threshold voltage:

$$V_{t\,\text{final}} = V_{t\,\text{nom}} + \Delta V_{t\,\text{TEMP}} + \Delta V_{t\,\text{var}} \tag{5.1}$$

where $V_{t\,\text{final}}$ is the final maximum threshold voltage, $V_{t\,\text{nom}}$ is the nominal threshold voltage, $\Delta V_{t\,\text{TEMP}}$ is the variation from the temperature variation, which is fairly constant regardless of the device oxide or threshold voltage, and $\Delta V_{t\,\text{var}}$ is the variation caused by process variation, which is typically greater for the thinner oxide devices.

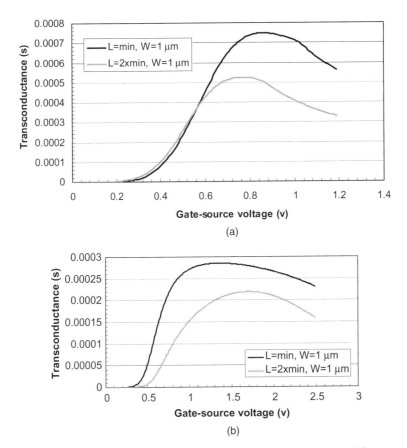

Figure 5.2 Transconductance for (a) a thin and (b) a thick oxide transistor on a 90-nm process for the minimum and twice-minimum-sized channel-length devices with varying gate–source voltage. The drain–source voltage is set to two-thirds the supply voltage.

A simple example is to consider a 90-nm process where the threshold voltage for the thin oxide device is on the order of 270 mV, the process variation will add an additional 100 mV, and the temperature variation (assuming an industrial temperature range) will add another 80 mV. Combining these terms yields 450 mV for the threshold, which give a ratio of 2.67 for a 1.2-V supply. If the thick oxide device is considered, the total threshold is approximately 550 mV, which gives a ratio of 4.5 for a 2.5-V supply, which implies that there will be significantly more supply headroom if thick oxide devices are used.

5.3 "HEARTBEAT" DEVICE SIZE

To design analog and mixed-signal circuits successfully in deep-submicron processes, it is important to follow certain basic rules. One of the primary rules is to use the minimum number of device sizes possible. The reason for following this rule is to minimize dependency on the threshold voltage variation as a function of channel length and width. This means selecting potentially two different device sizes for nMOS and pMOS devices and then building the final device from an array of smaller devices. This succeeds in two things: A smaller set of monitor devices can be used for wafer acceptance, and very accurate models can be developed since they only need to apply to a small well-defined set of devices. Device arraying must be done with some care to ensure that proximity effects do not result in any significant inaccuracies in actual devices versus simulated devices.

Many of the more advanced processes include low, nominal, and high threshold devices, to allow optimization for digital speed while managing power. For analog design it will be advantageous to select only one threshold device to use (typically, the low threshold) to avoid needing to do so as extensive of an analysis of the circuit performance over the entire process corner. The three different threshold devices do not necessarily fully track each other because the various threshold voltages are achieved by using separate implants. Typically, process corner models do not support varying the different thresholds independently. Low threshold devices are typically selected because of the additional supply headroom they typically provide.

5.4   DEVICE MATCHING

Device matching is a major consideration, especially for high-performance analog and mixed signal design. Minimizing the mismatch between devices is crucial for achieving a high manufacturing yield since mismatch can result in devices not meeting performance criteria. Device mismatch can manifest itself as offset voltages, duty cycle distortion, and signal distortion. To minimize mismatch between devices successfully it is critical to understand the sources of device mismatch and what techniques can be used to minimize them. In this section the primary sources of device mismatch are presented along with the techniques that can be used to minimize them.

5.4.1   Shallow Trench Isolation

Device matching represents a critical parameter of many analog circuits. Ensuring matching requires careful physical implementation to minimize the process variation. In general, device matching is driven by the total gate area, but a more accurate representation weights the channel length heavier when the length drops below a certain dimension. Another consideration is the effects of the LOD (length of OD the oxide device), which is the stress-induced change in the drive strength of devices. Modern processes include this effect as part of the process characterization, and they have been part of the extraction flow since the 90-nm technology node. One approach is to do the

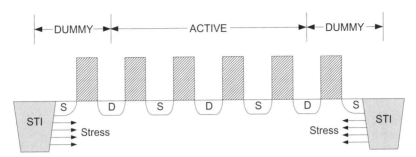

Figure 5.3 Use of a dummy device can significantly reduce the effects of the shallow trench isolation–induced stress effects. A single dummy device is shown for simplicity, but in reality three or more devices are required.

physical implementation, followed by an extraction to run accurate simulations, but a better approach is to do the physical implementation such that the LOD effect is minimized. This will result in less process variation for analog circuits, which it is important to achieve to ensure that analog circuits do not become yield limiters. Figure 5.3 shows a simplified diagram of how a layout can be done to reduce the effects of shallow trench isolation (STI)–induced stress by adding dummy devices on the outer edges of a continuous diffusion. This figure shows a single dummy device on the outer two edges, but in reality more than one device will probably be required to avoid the OD effect. The width of the devices should avoid using the minimum if at all possible since this will also reduce the OD effect. Diffusion sharing should always be used to reduce the stress effect as well. This may require the addition of dummy devices within the array to allow separation of the sources and drains, but it represents a better option rather than separating devices into separate diffusions.

Figure 5.4 shows a simulation from a 65-nm node of how the device IDSAT varies for various dummy devices. To remove the STI effect completely, at least three dummy devices are required. The drastic mismatch between devices is apparent when no dummy devices are used. Exactly the same simulation should be run before any design is done when a new technology node is being evaluated to determine how many dummy devices should be used to reduce the STI effects. A secondary benefit of minimizing the STI stress effect is an overall reduction of the general process variation since the STI stress varies with process as well, although this is not the primary contribution to process variation.

5.4.2   Well Proximity Effects

Well proximity effects (WPEs) also contribute to device mismatch. Figure 5.5 shows how WPEs occur during the manufacturing process. During the well

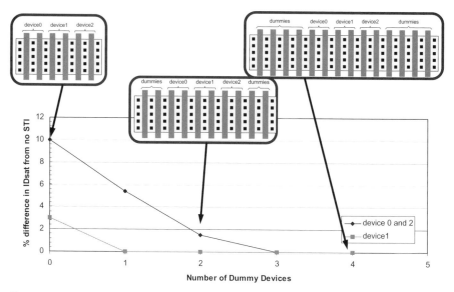

Figure 5.4 Number of dummy devices required to minimize the change in IDSAT for an NMOS device on a 65-nm process compared against a device having no STI influence.

implant processing step, implant scattering within the photoresist can escape and end up implanted in the substrate near the well edge [1]. A doping gradient occurs across the well which makes devices near the well edge have a higher threshold voltage. This will also cause an increase in the variation in the threshold voltage. This effect is not captured in the mismatch models. Typically, a single device layout is used to generate mismatch models, so it is important to know what was used to determine if the mismatch models represent a worst-case mismatch case or a best-case situation. A general rule of thumb is a single device in a single well with the device to well spacing set to the minimum value will represent the worst-case condition. Figure 5.6 shows the change in threshold voltage as the device spacing to the well edge is varied. A slight increase in the device variation is observed as the spacing to the well edge is decreased.

Reducing the well proximity effects is best achieved by increasing the spacing to the well edge. This can be achieved by using the technique shown in Figure 5.7, where the size of the shallow trench isolation is increased to expand the size of the well opening. This technique can be used for cases where sharing diffusion among the various devices is not possible. The drawback to this technique is that devices will still suffer from the STI effect, which is not desirable. A better approach is to find a way to share the diffusions and then add dummy devices to the outer edge, as shown in Figure 5.8. Only one dummy device is used on either end of the device grouping, but in actuality, the number will be similar to that shown for the STI effect since the extent of

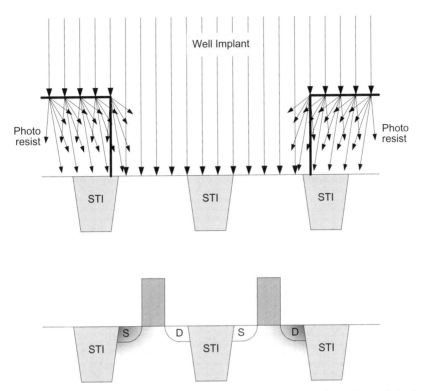

Figure 5.5   Well proximity effect from stray ion scattering caused during the well implant process step. The additional dopant atoms appear near the well edges. If the transistors are arranged as shown above, an asymmetry will be created, resulting in a threshold difference.

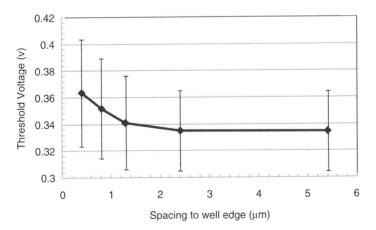

Figure 5.6   Device threshold average value and variation as a function of the spacing to the well edge for a small thin oxide device on a 90-nm process. The variation in the threshold voltage increases as the device gets closer to the well edge.

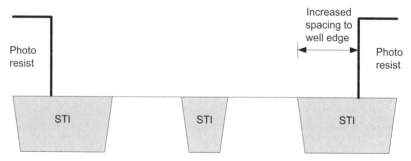

Figure 5.7 Method used to reduce the effects of well proximity by increasing the spacing to the well edge.

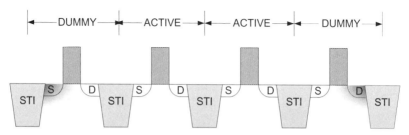

Figure 5.8 Second method for reducing well proximity effects by using dummy devices on the well edges to act as a buffer.

the WPE is on the order of 1 to 2 μm. This ultimately determines the number of dummy devices that should be used.

Modeling of the well proximity effects has improved drastically. Current extraction flows can account for extremely complex layouts, such as the one shown in Figure 5.9. Development of models requires considerable test structures to determine dependencies on all the parameters. The accuracy of a particular manufacturer can vary drastically. The best approach to take for doing a layout regardless of whether it is an analog or a digital circuit is to keep the well structures as simple as possible. See Chapter 6 for further discussion on designing for WPE-related variability.

5.4.3 Poly Gate Variation

Variation of the poly gate can result in significant mismatch between devices. The poly CD control can largely drive the global process variation, although it can manifest itself regionally depending on the uniformity of the poly in the surrounding region. Dramatic changes in the poly density can result in greater variation in the poly CD control. An additional affect is poly edge roughness, which can lead to average poly length variation. Figure 5.10 shows a simplified diagram of a transistor with the edge roughness shown. If the device width is

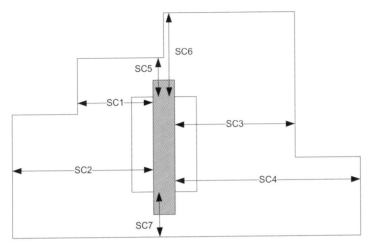

Figure 5.9 Parameter extraction for well proximity effects.

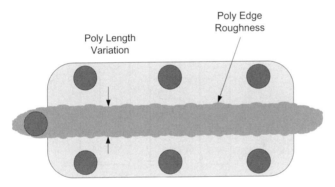

Figure 5.10 Transistor poly gate edge roughness, giving an overall channel length modulation.

too small, poly edge roughness can result in a significant random mismatch between devices.

Another effect that needs o be considered is the actual manufactured device profiles. Figure 5.11 shows a potential layout for an analog circuit. Here, the diffusion is stepped because device sizes were selected that required multiple-device W to be used within a single diffusion strip. Small poly gate extensions were used beyond the diffusion edge. Figure 5.12 shows a section of the same layout post manufacturing. Here the poly and diffusion edges are rounded, resulting in a change in the average channel length and width. Couple the rounding affect with mask misalignments and the variation can become more than a few percent. Figure 5.13 shows an improved layout for the same circuit. The diffusion has been changed to be the same for all the

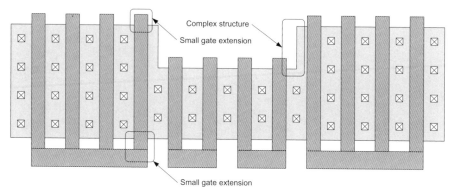

Figure 5.11   Potential layout of an analog block with several issues, such as a small gate extension beyond the diffusion edge and a complex structure which makes for a more complex extraction.

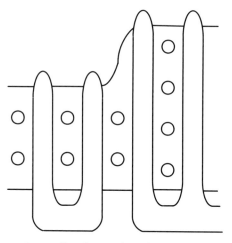

Figure 5.12   Post-processing profile of a section of the layout shown in Figure 5.11. The rounding results in changes in the effect length and width which will vary more between different lots.

devices. If this is not possible, dummy devices should be used at the diffusion step edges to avoid affecting the performance of the functional devices. The poly extension has been increased beyond the minimum-allowed design rule to avoid poly and diffusion rounding issues.

The following guidelines should be used to minimize the poly gate variation effects.

1. Make the device width as large as possible. For analog and mixed-signal circuits, at least twice the minimum allowed width should be used with a fivefold better choice.

Larger gate extension

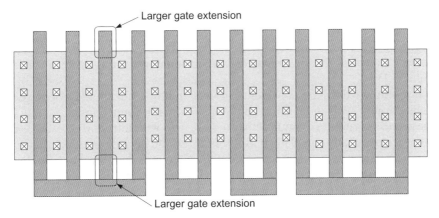

Larger gate extension

Figure 5.13 Improved analog layout based on the diagram shown in Figure 5.11. The device widths have been modified to use a fixed heartbeat size for every device, and the gate extension has been increased to avoid potential poly rounding affects.

2. Poly extensions should be at least 50% greater than the minimum allowed by the design rules.

3. Minimize the use of stepping of the diffusion. If the diffusion must be stepped, use dummy devices at the steps to avoid variation of the devices used in the actual circuit.

4. Maintain a consistent poly density around the actual circuits. This can be achieved by adding additional dummy devices, since this will also help reduce the WPE and STI effect. Figure 5.14 shows a simple example where two layout styles are used with different well structures. Avoid well structures that increase the proximity effect.

5.5 DESIGN GUIDELINES

Numerous guidelines are required and recommended by the semiconductor fabrication facility. These guidelines should be adhered to under most conditions. In addition, it is important to have a clearly defined simulation philosophy up front to ensure that all designers know what they are required to do to verify a design prior to tape-out. In this section we attempt to outline some of the general guidelines that have been applied in the past to ensure that robust designs are created.

5.5.1 Simulation Methodology

Simulation of analog circuits has become more difficult in recent years. A well-defined simulation methodology is crucial for ensuring a robust design, which will equate to a high-yielding product. When developing the simulation methodology, care must be taken not to overdesign the circuit too much such that other parameters, such as the die area or power, suffer. The use of standard digital verification corners may not absolutely cover the worst-case corners. A typical digital circuit focuses on running various process corners (Table 5.1).

If multiple oxides are considered or even multiple threshold devices, additional corners need to be simulated to ensure that a robust design is created. These cross corners are required because devices with different thresholds do not necessarily fully correlate. If different threshold devices are used, they will have the same oxide thickness and poly CD, but the threshold implants will be different, giving rise to different corners. Typically, these cross corners are not available from the foundry and must be created manually. In the case of different oxides, the oxide thickness and threshold implants are different, but the poly length will track. A conservative approach would be to

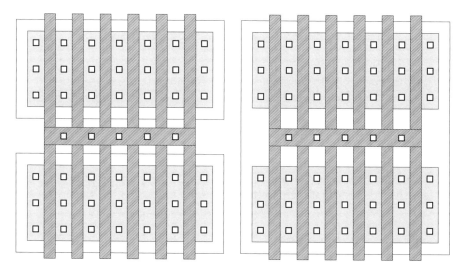

Figure 5.14   Two layouts of two same-sized devices, but one has two wells while the other ones have a single well. The device laid out with a single well will experience less variation because the well proximity effect will apply to fewer edges of the devices.

Table 5.1   Typical process corner simulations for a single oxide device

nMOS	pMOS	Designation
T	T	Typical nMOS, typical pMOS
S	F	Slow nMOS, fast pMOS
F	S	Fast nMOS, slow pMOS
S	S	Slow nMOS, slow pMOS
F	F	Fast nMOS, fast pMOS

consider the different oxide thicknesses to be completely uncorrelated and simulate all the potential corners, which expands the number of corners to 17. If resistors are used in the design, the number of corners increases further. In reality, a minimum of two resistor models (high sheet resistance and low sheet resistance) must be run at each process corner. A similar increase in the simulation corners occurs if capacitors are used in the design, assuming the capacitors are metal. These capacitors could be either a metal finger capacitor to take advantage of a standard CMOS process or metal–insulator–metal (MIM) capacitors if a more specialized mixed signal process is used. If gate oxide capacitors are used, there will be some correlation to the devices, but this must be considered carefully in the context of the models provided.

Another factor that must be considered is the supply voltage. It may be necessary to run all potential supply corners in conjunction with the process corners to achieve a high degree of coverage for a design. This is typically a two- or three-supply value that needs to be simulated, which includes a low supply voltage, a typical supply voltage, and a high supply voltage. Selection of the appropriate voltage values to simulate requires some knowledge of how much expected voltage drop will occur for the real circuit. Simulation should be run by considering the equation

$$V_{supply} = V_{nominal} \pm V_{range} - V_{IR} \pm V_{margin} \tag{5.2}$$

where V_{supply} is the supply voltage to simulate at, $V_{nominal}$ is the nominal supply, V_{range} is the specified allowed variation in the supply voltage, V_{IR} is the expected IR drop, and V_{margin} is the amount of design margin to add to the design. As an example, consider a design with a customer specification that the supply voltage be $1\,V \pm 50\,mV$. For this supply voltage, a margin of $50\,mV$ may be used to ensure a robust design. On top of this, based on a preliminary analysis of the planned power mesh, $20\,mV$ is used as the IR drop. This would suggestion simulating between 1.08 and 0.88V.

A good general approach to use to characterize the general sensitivity of a circuit is to sweep the supply voltage while monitoring the desired parameter. An example of this is shown in Figure 5.15, which shows the delay variation of a clock network as the supply voltage is varied. This simulation illustrates the sensitivity of the clock network to the supply voltage. It also illustrates how the circuit becomes more sensitive to changes in the supply voltage as the supply voltage decreases. Once this type of simulation has been run, it is possible to go back through the circuit and identify what components contribute to most of the variation and then modify the design to decrease the sensitivity. Using this approach will help to improve the overall design and make it more robust.

A common practice is to simulate the performance variation of a circuit as a function of the supply voltage. Figure 5.16 shows the simulation results of a clock network delay as a function of the supply voltage for a 90 nm technology node. Ideally, the delay would not change at all with the supply voltage but for this particular implementation, there is a relatively large dependency on the supply voltage. It is especially important to note how the sensitivity changes at the lower supply voltages. Knowing where the cliff occurs can be extremely important, especially for low power applications where the supply voltage may be reasonably low. If the delay variation becomes too large, yield fallout may occur: A more extensive analysis may be required that uses Monte Carlo simulations to verify the design meets performance requirements.

The last parameter simulated is the temperature. The temperature range that needs to be simulated is largely driven by the application the device will be used in (Table 5.2).

The first question to ask is what the customer specification will be for the temperature since it can be expressed as the junction temperature or the ambient temperature. If the junction temperature is specified, there should be some means for monitoring the junction temperature to ensure that it is not exceeded. If the specification is for the ambient temperature, the thermal conductivity of the package must be included in the analysis.

Figure 5.17 shows a small subset of the simulations that should be run for a circuit. Here, the three primary corners (FF, TT, and SS) are simulated while

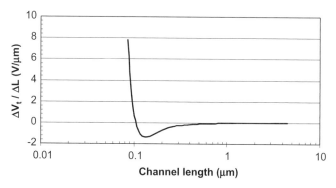

Figure 5.15   Calculated threshold voltage sensitivity to channel-length modulation for a 90-nm process. For the best matching, a channel length of at least 0.3 μm should be selected to minimize poly CD sensitivity.

Table 5.2   Temperature ranges for typical applications

	Temperature (°C)	
Application	Minimum	Maximum
Commercial	0	70
Industrial	−40	85
Military	−55	125

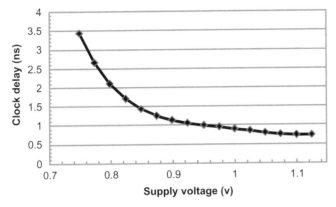

Figure 5.16 Clock network delay as a function of the supply voltage. As the supply voltage is decreased, the rate of change in the clock network accelerates.

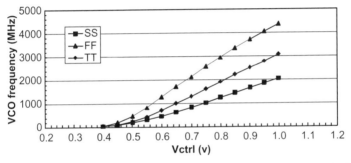

Figure 5.17 Simulation of a VCO for the FF corner at –40 °C and V_{cc} + 10%, the TT corner at 25 °C and nominal V_{cc}, and the SS corner at 125 °C and V_{cc} – 10%. The VCO frequency range changes by a factor of 2.

varying the supply voltage and temperature according to the assumed conditions that will make the performance change the most extreme. It is evident how significant the process conditions are to VCO performance. This set of simulations allows design verification, but it does not enable an understanding of how sensitive the circuit is to individual conditions such as temperature. To understand this effect, it makes sense to run simulations for the circuit against the individual conditions. Ideally, the array of simulations shown in Table 5.3 would be run to obtain a complete understanding of the circuit sensitivity to all possible conditions. This list would need to be repeated for other process corners to allow the process sensitivity to be extracted. The total number of simulations grows quickly but can be reduced by using standard design of experiment (DOE) techniques, which will allow the cross terms to be extracted as well. Three points are used to allow curvature to be extracted since the assumption that the performance variation is linear may not be true, especially for the temperature.

Table 5.3 TT process corner simulation for understanding the sensitivity of a circuit to individual conditions

Temperature (°C)	Voltage (V)
−40	$V_{cc} - 10\%$
	V_{cc}
	$V_{cc} + 10\%$
25	$V_{cc} - 10\%$
	V_{cc}
	$V_{cc} + 10\%$
125	$V_{cc} - 10\%$
	V_{cc}
	$V_{cc} + 10\%$

The same approach as that used for testing the process and supply sensitivity can be used for the temperature. In many cases it is important to understand how a circuit will perform against various environmental conditions for comparison against characterization results. The weakness of the design may also be exposed, which can be corrected before tapeout as opposed to post tapeout when yield issues arise.

5.5.2 Monte Carlo Analysis

The use of Monte Carlo analysis cannot be overstressed as a critical tool in evaluating a circuit performance in the presence of process variation as well as in understanding the expected yield of critical circuits. This verification and optimization technique can prove one of the most useful tools for ensuring a high manufacturability of the resulting design. For very simple circuits, use of Monte Carlo may not be required since it is possible to analyze the design directly to determine the effects of threshold offsets on the circuit, but this may not be possible for larger circuits, an entire system, or a large subsystem. In these cases it may be necessary to use Monte Carlo to understand the interaction between various devices within the system. As an example, consider a simple digital counter which is used to divide a voltage-controlled oscillator (VCO) output to produce a divided clock. Two outputs are generated, but the phase relationship between the two counter outputs is critical for timing. Monte Carlo analysis can be used to look at the distribution of the phase error between the two counter outputs to ensure that timing issues are not observed. If the variation is too large, circuit modifications can be used to reduce the variation. Figure 5.18 shows simulation results for a configurable counter used as part of a phase-locked loop with multiple outputs. These simulations show how much error there can be between two counter outputs. This information can be especially important if the phase relationship between the

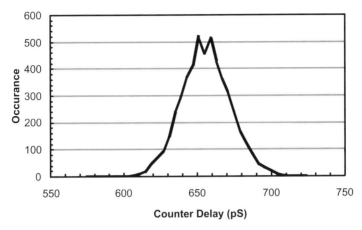

Figure 5.18 Counter delay variation from a 5000-run Monte Carlo simulation. The counter has a maximum delay variation of 112 pS with a standard deviation of 16 pS.

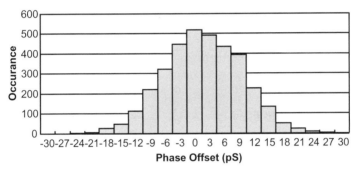

Figure 5.19 Monte Carlo simulation of a VCO phase error at a fixed process corner, supply voltage, and temperature. The phase error is measured against the ideal phase value for a four-stage differential VCO.

counters is important, such as a case where the counters are used for parallel-to-serial or serial-to-parallel applications.

A second example of how Monte Carlo can be applied is with the simulation of a VCO specifically looking at the phase error between multiple outputs. A four-stage differential VCO is simulated using Monte Carlo analysis to determine how significant the phase error is between different stages because the VCO will be used in a clock data recovery (CDR) circuit. In this case, phase interpolation techniques will be used, so the phase error between the various stages is important for determining the clock jitter recovered. Figure 5.19 shows the results from this series of Monte Carlo simulations. The measured phase error is approximately 8 pS, which is too high for the end application. Based on the Monte Carlo analysis, further circuit optimization can be done to reduce the variation and meet the end targets for the product. Without

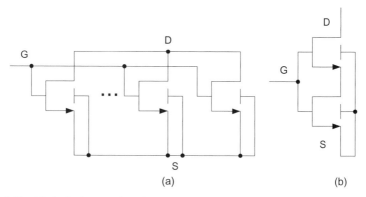

Figure 5.20 Methods for creating devices with (a) larger widths and (b) longer channel lengths. The method used to create larger channel widths works quite well, but the method used to created the longer channel length should be avoided if possible, especially for high-speed switching applications since the body bias effect can be extreme and is typically not well modeled, especially for the transient cases.

the use of Monte Carlo analysis, implementation of the original circuit would have resulted in yield fallout, potentially significant. Using Monte Carlo analysis allows a better understanding of how a circuit will perform in the presence of process variation.

5.5.3   Device Configuration

The use of heartbeat device sizes was discussed in Section 5.3, but using this method does require some care. Figure 5.20 shows typical ways in which devices with larger total widths and longer channel length can be created. The method used to create the larger width works quite well and has been used extensively, but the method used to create longer channel lengths may not work as well because of inaccuracies in the body bias effect and the transient behavior of the circuit. Longer channel lengths are typically required for current mirrors, where the output impendence needs to be reasonably high to ensure good matching. A better approach to take is to use two different device heartbeat sizes, one with a smaller channel length to be used for devices that require high-speed operation, such as the differential pair in a VCO and then a second channel length for current mirrors, which is longer, to provide the high output impedance.

5.6   LAYOUT GUIDELINES

The physical implementation of analog and mixed-signal circuits is important for the overall performance and yield of a design. In this section we provide a summary of the various issues that designers should consider when developing the physical implementation of a design.

5.6.1 Analog Layout Design Rules

Typical deep-submicron process design rules are generated around optimization for digital design. This results in minimizing the size of the design rules to maximize the logic density, which will not necessarily be the best thing for analog circuits. General design rules are as follows:

1. Make the contact-to-gate spacing larger than the minimum rule by 40%. The contact-to-gate spacing will modify the stress effect on the device, resulting in a varying performance. A nonminimum spacing will make the poly-to-poly spacing larger, which makes the poly CD effect have less of an effect on the device performance.
2. The poly end should be longer than minimum to avoid poly rounding effects.
3. The poly-to-diffusion edge should have longer spacing than the rule recommends.
4. Use a uniform poly density at and around critical circuits to avoid etching effects at poly gate array boundaries and chemical–mechanical polishing affects.
5. Use a single poly direction to minimize poly orientation variation. Typically, foundries provide greater accuracy for poly gates oriented in a single direction.
6. The channel length should be longer than minimum to reduce poly CD effects.
7. Diffusion-to-well edges should have larger spacing, to minimize well proximity effects.
8. Use dummy devices to add enough at the ends of critical diffusions to minimize the shallow trench isolation and well proximity effects.
9. Use a consistent surrounding area to minimize systematic offsets caused by varying STI sizes.

As the process continues to shrink, a new host of issues become significant enough that they must be included in the modeling and considered during the layout phase of the project. Several of the secondary effects that should be considered include the shallow trench isolation and well proximity effects. It is important to follow a set of design rules for analog circuits that minimize the secondary effects since these will contribute to the variation in analog circuits.

A second general guideline to follow is always to avoid complex shapes in the layout. Complex shapes in the layout result in more complicated extraction parameters, which typically results in less accurate simulations. Although the most recent SPICE models support modeling of many of the secondary affects, the general accuracy is still dependent on having the appropriate test structures in order to generate the models. It is not possible to generate all the possible cases that could arise in real layouts in test structures, but the most common cases are supported.

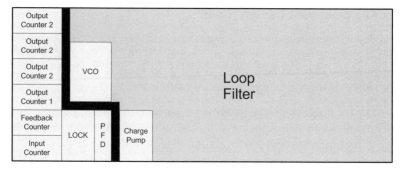

Figure 5.21 Potential floor plan for a phase-locked loop with a clearly defined boundary between the digital and analog supplies. The dark gray shaded area represents the isolation zone between the various blocks to reduce noise coupling. These regions could consist of guard rings or other isolation structures.

5.6.2 Floor Planning

A carefully thought-out floor plan can greatly improve the yield of a circuit. Floor plans should consider the power and signal routing. In general, it is important to group blocks on the same power domain together to allow a robust power mesh to be generated, especially given that it is typical not to extract the power mesh when running post-extraction simulations. It is also important to consider the signal flow since long routes can result in slower edge rates, making a circuit more sensitive to process variation and supply noise. Figure 5.21 shows a simplified floor plan for a typical phase-locked loop. This basic floor plan allows for a reasonable power busing and signal flow.

While floor-planning an analog or mixed-signal block, keep the following items in mind:

1. Group blocks that share common supply voltages to allow a more robust power mesh to be constructed.
2. Plan for a power mesh up front. This means running simulations to estimate IR drops and the expected amount of supply noise to determine if the mesh is adequate to meet the performance targets.
3. Plan for supply decoupling capacitors. Do not add decoupling as an afterthought to fill unused space. The required decoupling must be calculated up front and be planned into the layout.
4. Avoid using minimum width and spacing metal routes, especially for the routing that connects blocks together. This reduces the RC component of the signals, which helps keep edge rates fast and reduces the critical area of the circuit.
5. Keep blocks that interconnect close to each other to reduce the amount of routing required.

6. Avoid allowing dummy file layers to be added to a circuit. A dummy file is used to provide a consistent density, which is important for manufacturability, but the addition can result in undesirable parasitic. If the extraction simulations do not include the dummy file layer, the added loading may not be detected. If a dummy file is used, make sure that extractions are performed with it included in the database.

5.6.3 Power Busing

The voltage drop has already been discussed in Section 5.5.1. To minimize the voltage drop requires careful planning of the power bus. Circuits with high IR drops are typically more process sensitive, which can lead to greater yield fallout if sufficient design margin is not included. The only way to know if enough design margin has been included is to run simulations to verify that the power grid is sufficient. Figure 5.22 shows the two power domains for the PLL floor plan shown in Figure 5.21. Partitioning the floor plan in this way allows a more robust power mesh to be created, which should reduce the IR drop to the underlying circuits. In many cases, analog circuits will draw static current. This is especially true for differential voltage-controlled oscillators or current mode logic circuits. Often, these circuits are used because they draw static current and very little dynamic current, since the current is steered rather than switched, but high-speed operation requires considerable current. Ensuring a robust power bus is paramount in the overall performance, and therefore yield, of a circuit. A typical approach, even on analog and mixed-signal circuits, is to use a fixed mesh that matches the analog cell layout. The mesh can be created before the layout has begun by using schematic-based simulations to determine the current draw for the underlying blocks and then creating a mesh using resistors to represent the power mesh. Inclusion of the

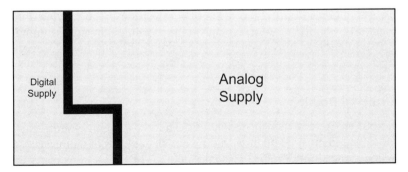

Figure 5.22 Power domain mapping for the global phase-locked loop circuit. By properly floor planning the design, it is possible to optimize the overall power busing, which will ultimately lead to improved performance and yield by reducing the dependency on the process.

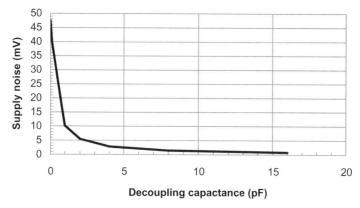

Figure 5.23 Simulated effects of adding decoupling on a small mixed-signal block on a 90-nm process with a 1.2-V supply. Once the decoupling becomes greater than approximately 2 pF, the reduction in supply noise decreases significantly less.

via resistance may be important as well to ensure that accurate *IR* drop information is obtained. Creation of the mesh can be done using an actual metal mesh layout and then running an extraction to obtain the mesh resistance. This represents an easier way to generate a realistic power mesh, but it may not be easily scaled for investigating the effects of varying wire widths and spacing. It may be important to be able to generate a scalable mesh to allow the wire width and spacing to be investigated. This is easiest to handle through some form of script. Current sources can be attached to the approximate location of various blocks and simulated using both transient and dc techniques. Use of dc current sources to represent the underlying blocks will allow a static *IR* drop to be determined. Using transient current sources will allow an peak *IR* drop number to be estimated and on-die decoupling capacitance to be determined. Figure 5.23 shows a simulation for a small mixed-signal block consisting of an analog VCO with a conversion stage to rail-to-rail signals. Simulations are run with the actual power mesh and varying amounts of decoupling while monitoring the supply voltage noise. Running this type of simulation is important for determining the amount of decoupling required.

5.6.4 Signal Routing

Signal routing is very design dependent. The exact requirements can vary dramatically and must be dealt with on a per signal basis in many cases. Critical analog signals may require special shielding or isolation, while a critical clock signal may require nonminimum width and spacing to minimize the *RC* component. In general, the following items should be considered when planning the signal routing.

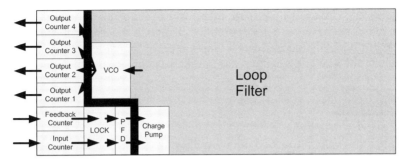

Figure 5.24   Signal flow for the phase-locked-loop floor plan shown in Figure 5.21. All signal routes are relatively short, which helps maintain fast edge rates, making the design less sensitive to process variation.

1. Minimize the length of the routes as much as possible.
2. Extremely long routes should consider the possibility that the photoresist could collapse. Consult with the foundry to find the maximum routing length possible.
3. Consider the need to include antenna diodes during the design phase. Adding diodes later can become difficult and suboptimum.
4. Use nonminimum wire widths and spacing if possible [2].
5. Maintain a uniform metal density if at all possible.
6. Avoid leaving open areas on the metal layers since these will be filled in by the dummy file routines. This can result in undesirable parasitics that are not included in extracted simulations.

Figure 5.24 shows the signal flow diagram for the PLL floor plan shown in Figure 5.21. The floor plan has been developed specifically to minimize the signal routing length, which will reduce the wiring required for connecting the various blocks and should reduce the power required since smaller buffers can be used.

5.6.5   Dummy Diffusion–Poly–Metal

Dummy layers are required on deep-submicron processes to ensure uniformity of the various layers. The dummy layers are used to fill in open space to maintain a constant density factor to avoid uneven chemical–mechanical polishing. Most foundries have special tools for adding these dummy layers into a layout. In general, these fill programs work reasonable well for maintaining a constant density, but they do not take into account the potential impact on the associated circuits. Dummy fill can add additional parasitics that are unwanted. The best approach to take is to fill in the layers manually for critical circuits to avoid the undesirable effects of the dummy layers. If the fill programs must be used, make sure that post extraction is run with the dummy fill included in the layout. The potential change in net capacitance can vary greatly but typically increases as a function of the signal length, so the longer the signal line, the greater the potential increase in additional capacitance.

5.7 TESTING

One area that will be touched upon briefly is testing, since the ability to verify quickly that a circuit is performing correctly and meets datasheet parameters is critical for production release of a product. How a circuit will be tested must be included as part of the design effort. Critical parameters must be easy to measure in a reasonable amount of time. The items that should be considered are:

1. *What tester will be used for the product.* Knowing what the tester capabilities are will be important for determining how to partition on-chip test features with tester-implemented test features.
2. *Margining requirements.* How much tester margin will be used for determining whether a part is good or bad. For example, if a PLL must run at 1.6 GHz, will it be tested to 1.8 GHz at a particular temperature and supply to guarantee that it meets the 1.6-GHz value across a broad temperature and supply range?
3. *Planned test list.* A list of what tests will be performed will allow an estimate of the test time. If this time is too long, a modified design may be required to speed up the test time.
4. *Tester limitations.* If the tester has certain measurements that it has a limited capability to make, additional consideration will need to be given to the methods that can be used to make the measurements.

REFERENCES

1. T. Hook et al., Lateral ion implant straggle and mask proximity effect, *IEEE Trans. Electron Devices*, vol. 50, pp. 1946–1951, Sept. 2003.
2. G. A. Allen and A. J. Walton, Critical area extraction for soft fault estimation, *IEEE J. Solid State Circuits*, vol. 11, no. 1, pp. 146–154, Feb. 1998.

DESIGN FOR VARIABILITY, PERFORMANCE, AND YIELD

6.1 INTRODUCTION

The rapid growth of the semiconductor industry in the past three decades has largely been due to a drive toward scaling and miniaturization and has been predicted to continue for the next two decades. But as devices approach the scale of the silicon lattice, the precise atomic configuration will critically affect the macroscopic properties. Furthermore, this scaling exacerbates the variability since the granularity of control is now limited to monoatomic dimensions. Especially important is the fact that the interaction between the design and the process fabrication steps can affect the manufactured components' parametrics. This will introduce a new set of systematic variability that will affect not only the process control but also modeling, simulation, timing, and chip integration. Our goal in this chapter is to outline the key technology issues that exacerbate variability, discuss their impacts on circuits, and suggest potential solutions at the device, circuit, and system levels. We also go over some design techniques that can minimize the impact of random variability. Even more important are the techniques used to harness systematic variability to improve circuit performance and reduce or eliminate the need for design margin by making systematic variability deterministic.

Nano-CMOS Design for Manufacturability: Robust Circuit and Physical Design
for Sub-65 nm Technology Nodes
By Ban Wong, Franz Zach, Victor Moroz, Anurag Mittal, Greg Starr, and Andrew Kahng
Copyright © 2009 John Wiley & Sons, Inc.

6.2 IMPACT OF VARIATIONS ON DESIGN

In nano-scaled technology, device feature sizes approach atomic dimensions. The gate oxide is on the order of a few atomic layers until a new higher-dielectric-constant (high-k) material is available to replace nitrided silicon dioxide. High-k gate insulator will not be available for the 65-nm nodes and most foundry 45-nm technologies. With such thin gate oxide we can expect higher gate leakage as well as larger variability, causing a large spread in I_{off} and I_{on} as well as gate leakage. The low operating voltage as a result of the thin gate insulation decreases the signal-to-noise ratio for analog as well as digital circuits. Signal integrity for digital circuits is also going to be challenging, and attention must be paid to power distribution networks as well as crosstalk. In Chapter 4 we go into detail as to signal and power integrity issues, as well as possible design solutions. Furthermore, the device channel area is getting to be so small that random dopant fluctuation will be significant for such devices.

Reactive ion etch (RIE) is now the preferred etching technique, but with a serious side effect, as it can induce threshold voltage shifts. If the antenna length connected to the matched transistor is significantly different, the resulting threshold voltage shift can be different enough to cause mismatch. Identical layout at every layer that is connected to the gates of the matched transistor is a must to avoid RIE-induced V_t shifts. An antenna diode can help mitigate but may not completely eliminate the V_t shift.

The increase in variation and parametric spread will require extra design effort from memory, analog, and even digital designers [1]. The two most critical components of the memory design are the sense amplifiers and bitcells. Both components rely on pairs of transistors that have to be identical for the circuit to work with a reasonable margin. Transistor mismatch in a sense amplifier design results in amplifier offset, which manifests itself as pattern sensitivity in the memory. An amplifier with an offset will tend to favor one logic level. If the complement of the favored logic level is being sensed, the sense margin as well as the speed will be affected adversely, while the favored level will have a large margin, resulting in systematic pattern sensitivity.

In this chapter we provide more design examples to deal with process variability. We also provide some examples pertaining to how we could harness systematic variability to our advantage to improve performance and reduce leakage power. See Chapter 11 in the book by Wong et al. [1] for further discussion of variability issues and design strategies for variability.

6.2.1 Impact of Variation on Bitcell Design

For a well-behaving six-transistor bitcell as shown in Figure 6.1, the transistor pairs M1/M4, M2/M5, and M3/M6 must be matched, where M1 is matched to M4, M2 to M5, and M3 to M6. In addition, the ratio of the drive strength of M1/M2 and M4/M5 must be held at the design point over process corners. All this is required to achieve a usable bitcell with a margin between the read and write levels so that it is stable during reading, yet can be written when the

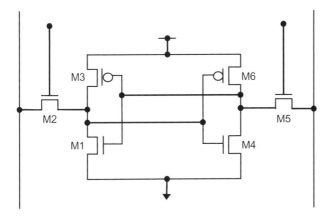

Figure 6.1   Schematic of a six-transistor bitcell.

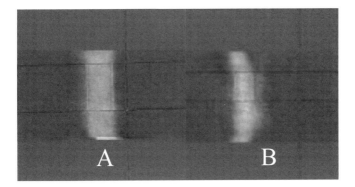

Figure 6.2   Scanning capacitance image of charge loss.

bitlines are driven to the writing levels. The voltage scaling, a consequence of the dimensional scaling, reduces the voltage headroom, resulting in a reduced read–write margin. The dimensional scaling also introduces increased transistor parametric variations, which along with the lithographic distortion changing the device characteristic is yet another perturbation that can cause failure of the bitcell and the memory system built with such a bitcell design.

Random dopant fluctuation can cause bitcell failure due to transistor mismatch. The scanning capacitance image of M1 and M4 (Figure 6.1) transistors is shown in Figure 6.2, showing charge loss in transistor B. This difference in dopants of the transistors causes the bitcell to fail at the low end of the operating voltage of the system. We have seen from Chapter 1 that dopant location is also critical for matched transistor designs (see Figure 1.2). Such failures due to random dopant fluctuation cannot be determined using any other imaging tool. This failure was at first classified as a nonvisual defect (NVD), which was not very helpful to product and fabrication engineers, as it provides no information for corrective action. At the insistence of one of the authors, further analysis was performed on the hunch that this is a case of random dopant fluctuation. The transistor characteristics were measured, which indicated V_t mismatches, which was then confirmed by scanning capacitance imaging (Figure 6.2).

Systematic and random variability is getting worse at each subsequent technology node, and designers will need to learn to design for variability. In this chapter we offer perspectives to deal with variability other than those discussed by us elsewhere [1]. Systematic variability is a result in part of the new mobility enhancement techniques: process proximity effects due to well, lithographical, and shallow trench isolation (STI). The random component is contributed mainly by process control of film thicknesses, critical dimensions line-edge roughness (LER), and the implant step. Ion implantation is a statistical event and the dimensions of the devices are getting so small that this effect will begin to grow in importance [2]. Random dopant fluctuation can cause large mismatches of adjacent transistors, as shown in Figure 6.2.

Random device mismatch has a significant impact on yield of SRAM arrays as well as in analog design, where transistor matching is very important. Analog design is covered in much greater detail in the next section and in Chapter 5. Random device mismatches cause many failure mechanisms, such as read disturbance, write failure, performance failure, random single-bit failures at the low end of the supply, and data retention failures in SRAM arrays. Nonintrusive read bitcells are beginning to show up for designs in the nano-CMOS regime to address the ever-increasing variability issue. The nonintrusive read bitcells will have no read current fed into the storage node: hence the term *nonintrusive*. Therefore, the nonintrusive read bitcells will no longer require a trade-off between the write and read margin but require two or more transistors, hence a larger cell size [3,6,7]. Other cells, such as the 7T bitcell, have also surfaced to address the shrinking static noise margin by cutting off the regeneration path of the cell [7].

Patterning-induced variability, and design techniques to minimize its effects, are covered in detail in our earlier volume [1], and readers should refer to that volume for the effects of lithographic distortions on the SRAM bitcell. We will show the impact of the OPC recipe on the lithographic distortions, to equip designers with sufficient knowledge to work with their foundry suppliers to mitigate its effect on the design, as well as to design with OPC in mind.

Figure 6.3 shows one design of a six-transistor bitcell with two different OPC recipes applied, with the corresponding poly gate patterning contours labeled A and B. The OPC recipe for contour A resulted in a "light bulb" (bulging) effect at the line end. Most lithographers do not comprehend the effect of their work on the electrical performance of the circuits they will be patterning. One of their main concerns is line-end shortening. To prevent line-end shortening, many lithographers would overcorrect for line ends, resulting in the light bulb effect, most of which occurs in bitcells simply because most bitcell design pushes line-end overlap of the active rule to mini-

Figure 6.3 OPC affects the performance of critical circuits.

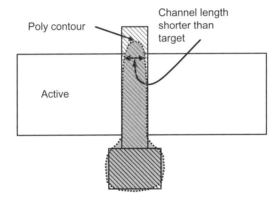

Figure 6.4 Shorter channel L resulting from line-end shortening.

mize bitcell size. This makes it more challenging for lithographers to make sure that line-end shortening will not result in high leakage when the line-end overlap of the active region is insufficient where the channel length of the device close to the line end is shorter than target, as shown in Figure 6.4.

One of the goals of the bitcell design is to maintain the beta ratio over processing corners and patterning overlay (misalignment) errors. Table 6.1 shows how well each of two OPC recipes, which are applied to the bitcell in Figure 6.3, maintain the beta ratio of the bitcell over a ±40-nm overlay. It is clear that the light bulb effect is detrimental to the ability to hold the beta ratio over misalignment between active and poly. This study did not include the nonlinear behavior between V_t and channel L, as shown in Figure 6.5. Minimum channel L is positioned after the V_t roll-off point for nano-CMOS processes. If included, the difference would be more pronounced. It has also been observed that even in the absence of misalignment, there is still a difference in the behavior of the two cells, which we believe is caused by the obstruction or scattering of the halo implants by the bulging tip.

Table 6.1   Impact of misalignment with different OPC recipes

Misalignment	Beta Difference (%)		Sigma CD	
	A	B	A	B
0	2.26	1.44	1.68	0.51
40	−3.68	−0.59	3.55	1.66
−40	7.11	5.04	4.12	0.59

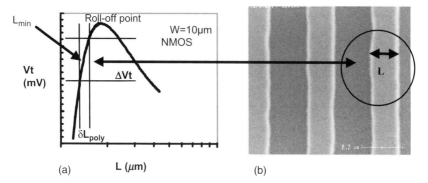

(a)   L (μm)   (b)

Figure 6.5   Minimum L follows V_t roll-off; LER has a major impact on I_{off}.

6.2.2   Impact of Variation on Analog Circuits

Local transistor mismatch due to random dopant fluctuation causes amplifier and comparator offset as well as current mirror accuracy. Random dopant location in a device channel depletion region is now getting to be significant as junction depth decreases with technology scaling (see Figure 1.2). Dopant location in the channel depletion region is determined primarily by dopant implant energy differences. There will always be an energy tail for all dopant implanters, which is difficult to get rid of. Therefore, dopant location differences between devices are here to stay, affecting device matching. Dopant channeling is another mechanism by which dopant location can vary between devices.

The stress proximity effect is a newly exacerbated effect, the result of differences in device layout in terms of contact placement and pitch, device proximity to STI, and poly pitch that will skew the device drive current of the various devices, resulting in device mismatch. Care must be taken in layout to ensure that the matched devices see the same stress proximity, to avoid mismatch due to this newly exacerbated effect. Chapter 3 covers the physics of this effect in detail.

Bit accuracy of analog-to-digital and digital-to-analog converters is dependent on how well the designer implemented the physical design of the circuit.

The key to a successful design is to understand each of the effects mentioned earlier so that implementation strategies can be worked out to eliminate or minimize their effects on the physical implementation of the circuit. We provide some sample implementation strategies later in the chapter.

Power supply noise for a mixed-signal chip is inherently noisier due to digital circuits, especially synchronous digital circuits on the same piece of silicon. The current state-of-the-art isolation scheme includes the use of a triple well with analog circuits in deep NWELL. Some manufacturers' processes for the deep NWELL can result in characteristic differences in the devices within the deep NWELL versus devices outside. Since this is also another high-energy implanted step, it will be another step that results in a well proximity effect at the well edge, due to dopant scattering at the edge of the photoresist.

Mixed-signal design will require better power supply rejection (PSR) and a better common-mode rejection ratio (CMRR) in differential circuits. Device choice in the nano-CMOS technology and supply headroom will be limiting factors in achieving better PSR and CMRR.

Passive component matching will be affected by lithographic distortion and is best combated by using virtually identical physical layout and proximity context. Active devices that need to be matched should also be implemented with identical physical layout and proximity context. The use of a common centroid layout will be required.

It has been observed that varying well voltage under the matched transistors translate to varying V_t. It is important to place the well contacts at the right location between the transistors to equalize the well voltage under the transistors. The source and direction of well noise must be well understood to determine the placement of well contacts around the matched transistors.

6.2.3 Impact of Variation on Digital Circuits

Local as well as global transistor mismatch results in clock skew in a synchronous digital design, consuming the design timing margin. Worse is the hold-time failure as a result of unaccounted skews and failed timing races, which will result in a nonfunctioning chip irrespective of the clock frequency.

Wire trench etch (causes wire bottom variability), as well as CMP-induced top-of-metal variability, can lead to RC mismatch that causes clock skew (Figure 6.6). Other causes of wire variations include copper plating defects and impurities that will affect the resistivity of the wire. Wire width variations also result in RC variation and can be due to variability of the etching rate as well as lithographic distortion (Figure 6.7), which may be process or design density related. Lithographic distortions can change the drawn design shape significantly, resulting in resistance and capacitance change compared to the design as drawn, which will not be captured by a simple parasitic extraction tool. The use of a calibrated contour-based extraction will be needed to capture the distortions of drawn shapes as seen in the examples in Figure 6.7.

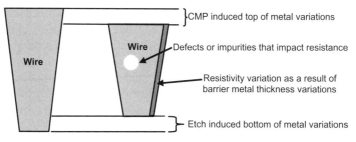

Figure 6.6 Wire variations.

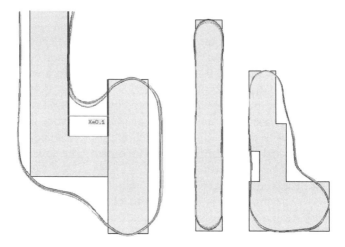

Figure 6.7 Lithographical distortion of design in the nano-scaled regime.

The more intricate is the drawn design; the worse is the distortion. Tools are currently available for contour-based extraction, which takes into consideration the shape distortion and provide an extraction that is much closer to the silicon result. More details on contour-based extraction and the corresponding tool are provided in Section 6.7.3.

The temperature and supply voltage gradient around the chip can vary during circuit operation, depending on the circuit activity, and can be a source of variation that affects the timing. Today, there are many tools that can assist designers to avoid the variation in timing that results from varying temperature as well as supply voltage.

High-performance digital circuits will feel the effects of stress proximity and must be designed with those effects in mind, especially race and self-timed circuits, where mismatched timing will result in catastrophic failure. Tools are available to aid in the design for these effects. However, such tools require substantial calibration, which must be repeated if the carrier mobility enhancement process changes. Once calibrated, the tool would be capable of extracting changes in device drive current based on its proximity as well as the layout construct of the device. Without such an extraction, one can have errors as high as 10 to 15% compared to actual silicon behavior. An error this large can cause even digital circuits to fail, and such extraction is mandatory for analog design unless the designer is very careful to make sure that matched devices are identical in every aspect. In the cell-based digital designs, where it is nearly impossible to guarantee the consistent proximity of every device in a chip, such a tool will provide the means to design for systematic device variation as a result of a different proximity and layout construct.

Figure 1.13 shows an example of the limitations of BSIM4 LOD modeling, where M1 and M2 can have a different drive current as a result of the different context but are otherwise identical transistors. There is no way to describe the difference in context using the BSIM4 LOD model with SA, SB, and SD parameters alone. It will also be difficult to predict the transistor context of every transistor in a large design. Therefore, designers need an extraction tool to sieve through the layout to create a representation of the stress proximity with different layout construct of the transistors to enable designing for the systematic variation due to stress proximity and mobility enhancement modulation as a result of different layout context. In Section 6.7.2 we discuss how we turn this variability to our advantage by making it deterministic, which then does away with the need for a design margin to cover for the unknowns. We also show how we can harness this effect for performance and power improvement, and to some extent chip size reduction, without sacrificing performance.

6.3   SOME PARAMETRIC FLUCTUATIONS WITH NEW IMPLICATIONS FOR DESIGN

6.3.1   Random Dopant Fluctuations

Analog designers had been contending with random dopant fluctuation, which is now showing up in the digital world in the nano-CMOS regime as severe performance and yield limiters [5]. As the active area of transistors scales to ever smaller dimensions, we can expect the number of dopants to decrease as well. This is worse especially in lower-V_t transistors, where the number of doping atoms is even lower [6].

If X is the number of dopants in the channel depletion region, V_t is proportional to X, while $\sigma \Delta V_t$ is proportional to the square root of X. The smaller the value of X, the worse $\sigma \Delta V_t$ is. For example, if $X = 5000$, $V_t \propto 5000$ and $\sigma \Delta V_t \propto \sqrt{5000}$ or $1.4\% V_t$. For the nano-scaled transistor we can expect that $X = 150$ and $\sigma \Delta V_t \propto \sqrt{150}$ or $8\% V_t$, which is getting to be significant. Figure 6.8 shows the scaling effect on the number of dopants in the channel depletion area as well the random dopant variation for a device with a W/L ratio of 4:1. The effect on V_t and the σ of V_t due to each of the intrinsic parameter

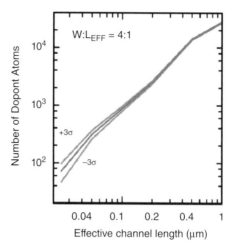

Figure 6.8 Scaling effect on the number of dopants in the channel. (From D. J. Frank et al., *VLSI Technol.*, pp. 169, 1999.)

Table 6.2 Effect of different physical fluctuations on V_t and σV_t for a 35-nm MOSFET

Fluctuation[a]	$\bar{V}_T$ (mV)	σV_T (mV)	Calc. σV_T (mV)
RD	133	33.2	
LER	126	19.0	
OTF	122	1.8	
RD and LER	126	38.7	38.2
RD and OTF	123	33.9	33.3
LER and OTF	113	22.8	19.1

Source: Ref. 24. Data courtesy of Professor Asenov of Glasgow University, UK.
[a]RD, random dopant fluctuation; LER, line-edge roughness; OTF, oxide thickness fluctuation.

fluctuations, random dopants (RDs), line-edge roughness (LER), and oxide thickness fluctuation (OTF) individually and combined is presented in Table 6.2.

The mean variation can be mitigated by good design, but $\sigma \Delta V_t$ is more difficult for designers to mitigate, as it is inversely proportional to the square root of the device area $(W \times L)$ [7]:

$$\sigma \Delta V_t = \frac{A \Delta V_t}{\sqrt{WXL}} \qquad \text{(after Pelgrom [3–5])} \qquad (6.1)$$

where $A_{\Delta Vt}$ is the slope of the plot of $\sigma \Delta V_t$ versus $1/\sqrt{WL}$.

Device area reduction comes with technology scaling, and if a designer decides to increase the device size to reduce $\sigma \Delta V_t$, it defeats the choice of

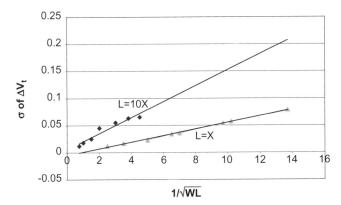

Figure 6.9 Anomalous behavior of $\sigma\Delta V_t$ when $L \geq 10\times$ minimum.

going to a more advanced technology node. This is one reason why analog designs do not gain much from an advanced technology node.

Another deviation from the well-understood behavior of $\sigma\Delta V_t$, which is inversely proportional to the square root of $W \times L$ for non-minimum L, occurs when the value of L increases to 10 times the minimum L and beyond. For devices with such long L we find that the $\sigma\Delta V_t$ becomes worse and follows a totally different trend [the slope $A\Delta V_t$ in equation (6.1) changes compared to those with L close to the minimum. For example, a device with the same area $(W \times L)$ but longer L (10 times the minimum L) and smaller W would have a worse $\sigma\Delta V_t$ than another device with the same area but shorter L. This effect is the result of the use of "halo" or "pocket implants around the device drain and source active regions to improve device short-channel-effect performance. As the channel length increases, the halo implants begin to separate and there is a region in the channel between the halo implants with very low doping, which results in this new behavior of $\sigma\Delta V_t$ or the changing $A_{\Delta Vt}$. It is therefore important for designers to have this phenomenon well characterized to determine at which channel length the trend deviates from normal, and design their circuits accordingly. Figure 6.9 shows the anomalous behavior of the $\sigma\Delta V_t$.

6.3.2 Line-Edge Roughness Impact on Leakage

In an attempt to push device performance, most processes has positioned the device minimum channel length after the V_t roll-off point, as shown in Figure 6.5. The result of this choice is that channel-length variation as well as line-edge roughness (LER) would result in very large V_t variation. For critical designs it is advisable to increase channel length past the roll-off point to minimize the sensitivity of the device V_t to channel-length variation.

LER is the result of the use of chemically amplified resists in the patterning of device structures, including the poly gates [6]. During the lithographic process, in which the mask or design pattern is exposed onto the resist, standing waves can result from a reflection from the bottom of the resist interacting with the incoming wave. Every attempt to minimize this effect has been exercised by the process engineers using bottom antireflective coating (BARC), which is not perfect. The result of the standing waves is that we will see steps along the edge of the developed resist. To minimize this effect, process engineers use post-exposure baking before developing. The baking process reduces the standing-wave effects significantly but introduces a post-baking thermal and chemical image blur, resulting in an irregular edge profile after development. The other causes of LER are fluctuation of the total exposure dose as well as photon absorption positions due to nonuniform resist composition [25].

The LER in the resist is transferred onto the underlying material during etch. This affects the discrete locations of dopant atoms during the implant process as well as the varying channel length, which leads to fluctuations in device V_t, thus affecting the device characteristics and increasing leakage [10]. LER also contributes to CD variability and becomes part of the CD control budget.

6.3.3 Poly Critical Dimension Uniformity and Oxide Thickness Variation

As the poly CD scales, it is getting more difficult to meet the poly CDU requirements for some critical circuits. For example, the typical channel length of a 45-nm device is below 40 nm. Even at 40 nm, a 4-nm variation represents a 10% poly CD variation. It is doubtful that the absolute CD variation will scale at the proportion of the channel-length scaling. The CDU budget would include:

- OPC accuracy and reproducibility (Figure 1.10)
- Mask CDU
- Mask error enhancement factor (MEEF)
- Lithography process (focus and exposure dose consistency) (Figure 6.10)
- Thermal–chemical blurring of exposed image
- LER
- Etch variations

Even if we assign 1 nm to each factor listed, we would still blow the 4-nm CDU budget. MEEF is getting to be significant in low-$k1$ imaging, and it is the multiplier of the mask error. The higher the MEEF, the worse would be the mask error on the printed image. $k1$ is the process constant in Raleigh's equation. See Chapter 2 for further discussion of MEEF and Raleigh's equation (see also ref. 1, Chap. 3).

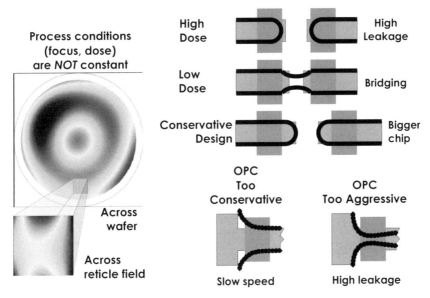

Figure 6.10   Lithography conditions vary across the wafer.

To step up to the challenge, process control engineers have developed feedforward and feedback process control for poly CDU. The feedforward part will look at the after-etch poly CD and control the implant dose to correct for V_t. If the CD is narrower than the target, the halo implant dose will be increased, and vice versa (narrower poly CD results in lower V_t). In the feedback loop, the *resist CD*, also known as *develop inspect CD* (DICD), is used to control the etch time. If the DICD is lower than the target, the etch time is decreased to compensate for the final poly CD, and vice versa (a shorter etch results in higher poly CD). With this closed-loop control, the mean CD, as well as the mean V_t, can be held very close to target. However, the σ or spread of the CD and V_t are still dependent on the tools and conditions. This means that there are still opportunities for designers in the nano-CMOS regime to make a difference.

A short list of opportunities is as follows [11,12]:

- Unidirection poly achieves about a 4% improvement in poly CDU.
- Critical circuits on three predetermined pitches or less achieves about a 4% improvement in CDU; minimizes ID_{sat} variations as well (more later).
- Wider poly CD for hold-time cells, as well as large slack paths for analog circuits, but bear in mind the anomalous behavior of the device matching, as shown in Figure 6.9.

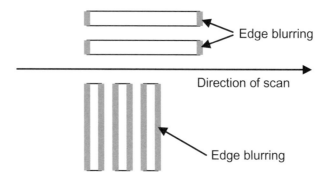

Figure 6.11 Edge blurring with respect to the direction of scan.

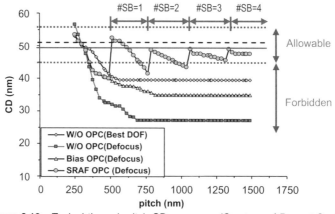

Figure 6.12 Typical through pitch CD response. (Courtesy of Puneet Gupta.)

The patterning of the design in nano-CMOS technology nodes is accomplished with a step-and-scan scanner. It has been found that the CDU of poly lines drawn vertically is different from the CDU of poly lines drawn horizontally. This has been attributed to the synchronization between the stage and the mask motion during the scan, blurring the image. In well-synchronized scanners the CDU contribution is very small but still eats into the meager CDU budget. This is worse along the scan direction. The poly lines that are drawn perpendicular to the scan direction result in worse CDU than that of lines drawn in the direction of the scan. Lines drawn in the direction of the scan will have blurring of line ends, which can tolerate a larger variation in the edge placement than can the gate edge of a device when drawn perpendicular to the scan direction (Figure 6.11).

A chip designed with a fixed poly pitch will have better CDU performance, as the context of the critical poly lines would be constant. Even a design with more than one poly pitch can have very tight CDU distribution, provided that they are set at pitches that coincide with the placement pitch of the subresolution assist features (SRAF or scatterbars), as shown in Figure 6.12. See Chapter

2 for a full discussion of the purpose and principle of operation of the SRAF (see also ref. 1, Chap. 3).

CDU variation translates to V_t variation as well, resulting in a large spread in the device leakage. A chip with large CDU variation would have a difficult time meeting both the performance and leakage specifications. If one would like to have a high-performance part, the leakage would be very high as well, making the resulting chip unattractive to customers. Keeping in mind the rules of thumb as listed above can help alleviate some of these issues in the design.

Another technique used to reduce leakage as a result of poly CD variation is to bias the channel length longer for circuits not in the critical path. A detailed discussion of this technique is presented in Chapter 7. Other designers have selected simply to use longer channel length for the noncritical paths and reserve minimum channel length for the critical paths. Intel has selected this approach to minimize chip leakage and leakage variability as a result of V_t variation and band-to-band tunneling leakage. Our analysis shows that they are using 35-nm channel length for critical paths and 50 nm in noncritical paths in the 45-nm design. The $\sigma \Delta V_t$ is about 12 mV for a transistor with a 50-nm channel length, while it is about 30 mV for a transistor with a 35-nm channel length.

Gate oxide thickness (EOT) in an advanced technology (with SiON) is about 12 Å. This translates to about four or five atomic layers. It is unlikely that the gate oxide will scale below this thickness until high-k gate oxide is deployed, at which time the EOT may reach 8 Å while the physical thickness will be about 2.5 times thicker and gate leakage reduces two orders of magnitude. In the SiON process, losing one atomic layer will represent a 20 to 25% thickness variation, which is very significant. For most merchant processes the next node with a high-k dielectric would be the 32-nm node, and possibly in the later part of the 45-nm technology.

Many designers use MOS devices as on-chip decoupling capacitors, and since a large area is required and the leakage through the gate oxide can be significant, many tens of amperes. To mitigate the high leakage, thicker oxide I/O transistors have been used to serve as on-chip decoupling capacitors. The rule of thumb for the decoupling capacitance required can be calculated from the power consumption of the chip:

$$P = CV^2 f \quad \text{and} \quad C = \frac{P}{V^2 f} \tag{6.2}$$

where C is the equivalent switching capacitance, and the decoupling capacitance is usually set at 10 times the values of C where space permits. This is usually a large enough area to worry about gate oxide integrity (GOI) as well as leakage current, which is why it would be prudent to use the thicker gate oxide device as the decoupling capacitor at a corresponding loss in area efficiency.

In some designs, such as the loop filter of a phase-locked-loop circuit, where leakage of the loop filter can cause unrecoverable phase offset, one will again resort to a thicker gate oxide device to prevent this functionality problem. The trade-off is a larger area.

6.3.4 Stress Variation of Straining Film and eSiGe with Different Contexts

All advanced technology nodes have some form of applied strain for device performance enhancement to keep pace with Moore's law and to meet *International Technology Roadmap for Semiconductors* (ITRS) targets for sub-90-nm CMOS devices [16]. In some technologies, embedded silicon–germanium (eSiGe) is also used to further improve the pMOS performance besides the strain nitride films. There will be some inherent variation of the stress applied by the film due to processing temperature fluctuation as well as the recess depth of the eSiGe, causing some random variation. The largest variation, however, is systematic, caused by the layout of the design [16]. This is the part that cannot only be designed for, but can be made deterministic and on the high end of the drive current distribution to improve performance, and at the same time, not needing design margins for this systematic variability.

The force applied by the strain nitride film is proportional to the volume and proximity of the film to the channel (Figure 6.13). The film volume and proximity are modulated by:

- Poly-to-poly distance
- Contact CD and contact-to-poly gate distance
- Contact pitch

A similar circuit, when laid out differently, can have a major impact on the circuit's performance as well as the drive current variability. The current

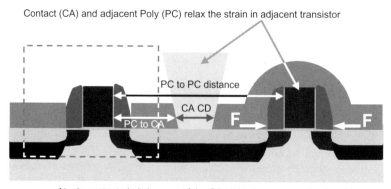

Contact (CA) and adjacent Poly (PC) relax the strain in adjacent transistor

PC to PC distance

CA CD

PC to CA

F

F

At min contacted pitch, most of the CA nitride is consumed by the CA

Figure 6.13 Contact perforation relaxes straining film.

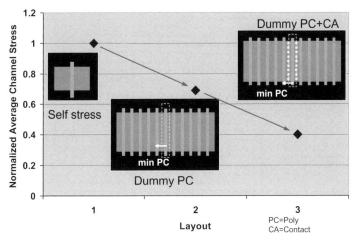

Figure 6.14   Effect of contacts and minimum pitch poly on channel stress.

BSIM4 LOD model uses parameters SA, SB, and SD. As shown in Figure 1.13b, the two devices are similar in size, but SA and SB have very different stress proximity. The current BSIM4 LOD model cannot differentiate the two devices, yet in reality they can have quite different drive currents, due to the different stress context. Design aids have surfaced to differentiate the device differences, so that the correct drive modulation is applied to the respective devices on the chip. The importance of being able to extract the correct drive modulation of the devices on the chip grows as the gain of the device drive improves as a result of stress engineering. Refer to the Section 6.7.2 for further details on the extraction technique for stress proximity that works around the weakness of the current BSIM4 LOD model.

Minimum pitch poly reduces the film volume in proximity to the channel and therefore reduces the mobility-enhancing strain in the channel, therefore reducing the device drive current. Minimum contact pitch perforates the strain nitride film, thus relaxing the film strain and reducing the device drive (Figures 6.13 and 6.14).

For technologies with eSiGe for pMOS drive current improvement, the polypitch affects the eSiGe recess depth. The device channel strain, as well as the channel strain variation, increases with increased recess depth (Figure 6.15). Also note the steep slope of the plot of channel strain with respect to $L_{p\text{-}p}$ at the minimum $L_{p\text{-}p}$ value in the 45-nm node. $L_{p\text{-}p}$ is the space between the poly lines and is modulated by the poly CD. Poly CD variation in such a device would have the double whammy of V_t variation as well as channel strain variation, which is additive, making for a larger spread in the drive current variation. Narrower poly CD would increase drive as well as leakage, as it reduces V_t. At the same time, the $L_{p\text{-}p}$ increases, increasing the channel strain, which further increases the drive. For some critical circuits one may choose

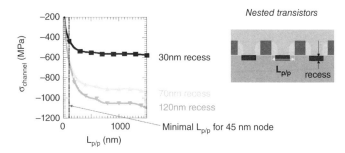

Figure 6.15   Poly-to-poly space impact on recess depth and channel strain.

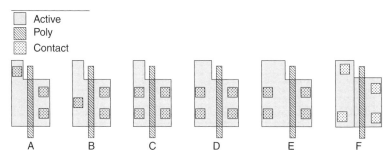

Figure 6.16   Contact placement effect on transistor drive.

to design their circuits in the plateau part of the channel strain versus $L_{p\text{-}p}$ plot to minimize the effect of poly CD variation triggering another strong response in drive current variability that is additive to the reduction in poly CD. Also, circuits designed in this region of the curve would result in the higher drive devices.

Figure 6.16 shows different layouts of a transistor with the same device size. Device A will have the maximum stress as a result of the reclusive single V_{ss} contact. However, it suffers from higher source resistance to the V_{ss} node. This device will have at least four times the source resistance of device C, which will have the lowest source resistance of all the devices. Also, device A may experience higher device failure due to the un-landed contact, not recommended as a performance boosting technique.

In Figure 6.16, device B is no better for an un-landed contact than device A but has better source resistance and somewhat reduced device drive. Device B, which has a single contact but minimum spacing to the channel, and devices C, D, and E were simulated for the change in channel stress as a result of progressively moving the contacts away from the edge of the channel. It has been observed that a device with contacts that are 60 nm away from the channel edge would resemble a self-stress device, being the highest for performance. The results are presented graphically in Figures 6.17 and 6.18, which show slightly different configurations where the contacts at both the source

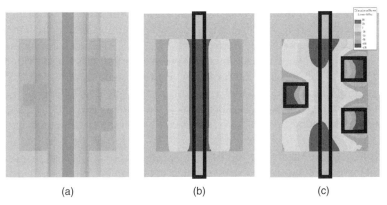

(a)   (b)   (c)

Figure 6.17   Stress distribution around contacts: (a) top view of an sCESL MOSFET with three contacts; (b) stress distribution before contact etch; (c) contact-induced stress relaxation.

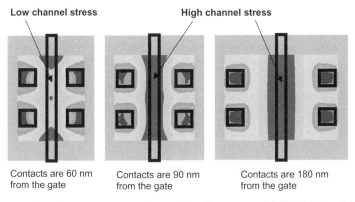

Low channel stress   High channel stress

Contacts are 60 nm   Contacts are 90 nm   Contacts are 180 nm
from the gate   from the gate   from the gate

Figure 6.18   Effect of contact-to-poly edge spacing. Contacts are (a) 60, (b) 90, and (c) 180 nm from the gate.

and drain were moved away from the channel in a progression of 30 nm. From all the simulation results we believe that device F would be the highest performance device without sacrificing yield due to another yield loss mechanism as well as excessive area or source resistance and higher drain capacitance.

6.4   PROCESS VARIATIONS IN INTERCONNECTS

Figure 6.6 shows some of the processing variability that can affect interconnects. First, the thickness variation can be induced by CMP or etch, as well as both together. CMP results in copper loss variation, which affects the top part of the interconnect, whereas the bottom variation is the result of trench etch variation in the trench where the interconnect is encased. Both of these variations are density dependent, which in turn is design dependent.

The next class of interconnect resistance variation is due to defects in the interconnect as well as plating impurities embedded in the copper. These are totally process dependent and there is nothing the designer can do to minimize them. The third class of interconnect resistance variation is the barrier thickness variation. This is again independent of the design.

The final class of interconnect variation is subwavelength distortion of the design pattern (see Figure 6.7). This class of variation is expected to get worse in future technology nodes. Tools are available to assist designers to check for areas of design that can result in catastrophic failures due to necking (electromigration hazard on top of increased resistance) as well as potential shorts. However, long before you see such catastrophic failures, you will find that the interconnect pattern distortion has changed the designed characteristic sufficiently to cause timing failures for long-haul routes [8]. The extraction tools that most designers are used to only provide extraction of as-drawn design, not after the subwavelength distortion of the drawn design. This will create an accuracy gap that will widen in future technology nodes beyond 65 nm. Contour-based extraction will be needed to close this gap. More on contour-based extraction later.

The use of a nonminimum interconnect width for a longer routing wire improves delay as well as reducing variability, as ΔCD will be a smaller percentage of a wider wire. It will be clear when we look at the equation for a given cross section of the wire X:

$$\sigma \Delta X = \frac{A_X}{\sqrt{WH}} \tag{6.3}$$

where the larger the width (W) of the wire, the better $\sigma \Delta X$ is. Although H cannot be changed by the designer for a given layer, it can be influenced by the design. For example, the trench depth (H) is influenced by the line width, where the minimum end of the allowed line or wire width the trench depth (H) decreases (Figure 6.19). Due to CMP dishing, using too wide a wire will result in reduction of wire thickness (H) (Figure 6.20). For normal signal line width, CMP dishing would not have set in yet, but the designer is advised to note this effect, depending on the fabricator of the chip. However, power supply lines are normally wide enough to be a problem. Therefore, it would be advisable to use multiple power lines instead of a single very wide wire, which may end up with high resistivity as well as poorer electromigration performance.

The pattern-dependent interconnect variability due to dishing, erosion, and trench etch can be improved to some extent by means of metal fill [13]. However, this results in additional capacitance and coupling if the fill structures are not tied to V_{dd} or V_{ss}. Figure 6.21 shows that the impact of dummy fill can be as much as 10% of the RC of the interconnect. This also increases database size and mask data preparation time, which in turn increases design-

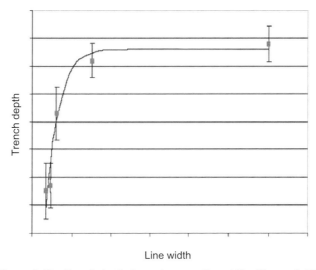

Figure 6.19   Trench depth dependence on line width. (From ref. 27.)

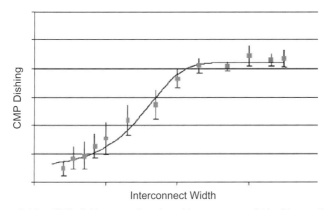

Figure 6.20   CMP dishing as a function of interconnect width. (From ref. 27.)

to-mask building turnaround time (TAT) as well as design to prototype TAT, delaying product to market. There exist tools that can place fills that are aware of the parasitic capacitance effect on the design performance and will satisfy both the density requirement for improving interconnect processing variability as well as the timing requirement.

The future of scaled interconnects will require a change in design methodology to deal with the increasing interconnect resistance at each subsequent node, as shown in Figure 6.22. Figure 6.23 shows that the short interconnect lines are insensitive to interconnect resistance, hence to resistance variability as well. Therefore, the new design methodology will require careful placement of cells that need to be interconnected to minimize routing distance. The

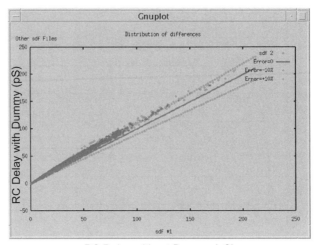

RC Delay without Dummy (pS)

Figure 6.21 *RC* delay difference with a dummy can be as much as 10%. (From ref. 27.)

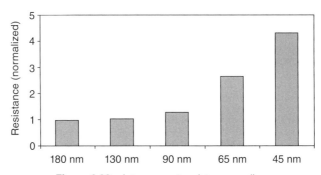

Figure 6.22 Interconnect resistance scaling.

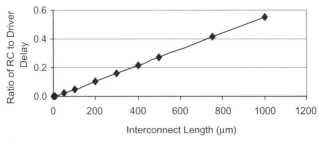

Source: Nagaraj

Figure 6.23 Short routes of about 20 μm are insensitive to interconnect resistance. (From ref. 27.)

University of Bonn in Germany has been developing a new concept in chip design whereby block boundaries are no longer along straight lines or bounded by square or rectangular blocks to minimize route length to address this problem. There may be a need for repeaters to reduce total route length but at the cost of increased leakage and dynamic power. The long-haul interconnects will require wider lines to improve their resistivity as well as to minimize variability.

6.5 IMPACT OF DEEP-SUBMICRON INTEGRATION IN SRAMs

It is clear that process variability will get worse in future technology nodes and is not expected to get better, which makes a lot of the design goals unrealizable: for example, the drive to ever-smaller bitcells, together with the need for lower minimum operating V_{dd}, a stable and yet writable cell cannot be met all at once. To achieve these goals requires trade-offs and creative circuits to address the problems: for example, the need for write-assist circuitry to write a cell designed to be very stable for better static noise margin (SNM). However, this increases total array size and reduces array efficiency.

As we push cell size, the number of weak bits will increase. The traditional trade-off is to increase cell size, but that negates the use of advanced technology nodes. Another option is to use redundancy or error correction rather than make every cell perfect, to reduce the yield impact on the RAM array due a to weak bits. This may be a good trade-off for a large array, but a smaller array may be better off with larger bitcells. SRAM-layout-specific SPICE models will be needed for design closure [14]. All foundries currently offer SRAM-specific transistor models just for this purpose.

To reduce the power of operation, many have selected to operate their circuits at a low voltage to take advantage of the square term for V_{dd} in the power equation. Even some high-performance applications such as graphics processors (GPUs) are also resorting to reduced V_{dd} to control dynamic power due to the power constraints of the GPU card in the computer system. Although additional power can be brought into the GPU card using additional power dongle rather than relying on the power supplied from the card interface to the main board, the power density would be too high to cool effectively. There is also the limitation of form factor for the GPU card to a card slot for competitive reasons thus, limiting the size of the fan sink that can be mounted on the GPU. The reduction of V_{dd} of any chip with embedded SRAM is limited by the cell stability [15]. In the best case, this limitation is brought on by the inability to scale gate dielectric thickness as well as V_t mismatch due to dopant fluctuation is constant with scaling. Therefore, it is critical for the designer to contribute to the reduction of all other components of mismatch due to other systematic mismatch resulting from the cell layout: for example, stress proximity as well subwavelength distortion and misalignment-induced mismatch. Table 6.1 shows the performance of the cell in Figure 6.3 over ±40nm of

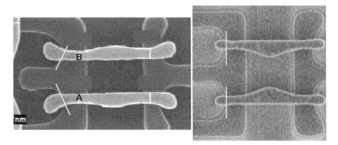

Figure 6.24   Slight layout difference affects the yield of the SRAM significantly.

overlay misalignment between the poly and the diffusion mask. The goal is to keep the beta ratio constant over the process perturbation. This analysis is done with the distortion of the poly only and would be worse if the distortion of the diffusion is included. Figure 6.24 shows two different bitcell layouts that will have significantly different process windows. The cell where the poly lines have the same tangent over the rounded diffusion will be much better than the other cell where the angle of tangent is different. See Section 6.6.4 for a detailed explanation of the differences and how they affect the process window (see also ref. 1).

Some of the design options proposed in the recent literature to improve the low V_{dd} performance of bitcells are as follows [15]:

1. Larger cells
2. Separate array *Vdd*
3. Alternative cells
    a. 7T SRAM [7]
    b. 8T SRAM [6]
    c. 10T SRAM [17]

6.6 IMPACT OF LAYOUT STYLES ON MANUFACTURABILITY, YIELD, AND SCALABILITY

6.6.1 Insufficiency of Meeting Design Rules Alone

Figures 1.3 and 1.4 show catastrophic failures under certain process conditions, thereby limiting yield, which clearly illustrates the deficiency of the Boolean-based design rules. Figure 1.3a shows a failure that can be tested, and Figure 1.3b shows an insidious defect as it represents a reliability hazard. This failure was detected only after burn-in stress, which finally caused the short. All the designs with failures met the design rules. These are only a small sample of the failures that were seen in a typical nano-CMOS design and articulate the fact that design rules cannot cover all possible hazards of unstructured layout, even with the many additional rules [25]. The solution is not more rules that add many more layers of complexity to the design process, but a new design methodology to improve the design process interaction.

These failures are due to the difficulty of a subwavelength patterning system to deal with two-dimensional patterns that are random. In the case of Figure 1.4, the failure is due to a contact pad being next to a line above it, also with a contact pad. That combination of patterns creates a bright region next to the contact pad B, thereby changing the process window of this area compared to other parts of this design. To pattern the area around contact pad B without causing a short would require a lower exposure dose (reduced exposure intensity). However, this may affect the CD of the other areas in this layout, which is also undesirable. The best solution to this problem is to normalize the pattern density of the entire design so that the image intensity is about the same everywhere. This will allow the lithographer to optimize the dose for best CD and pattern fidelity using a single dose setting for the entire design. It is not possible to adjust the dose setting for different areas of a design even though the preceding text may suggest that as a possibility. Lithographers use subresolution assist features (SRAF) to help normalize pattern "brightness" in all areas of the design, but it is impossible to use SRAF in a dense random layout. See Chapter 2 for further details of the principles behind lithography and for a more detailed discussion of the lithography and resolution enhancement techniques.

In the future, the degrees of freedom for layout will be restricted, as patterning problems in the 45-nm nodes and below will be even more difficult. The lithography process is lagging the dimension scaling, as shown in Figure 1.11, where the $k1$ value has gone way below 0.33, the currently accepted robust manufacturing condition, for nodes below 65 nm.

The circle of influence of the layout context will be increased in subsequent nodes as a result of dipping $k1$ value. The circle of influence is tied to the $k1$ value as well as the wavelength λ of the illumination used in patterning the design. The current λ is stuck at 193 nm, as shown in Table 6.3. There are not many choices of usable illumination source for lithography, and the limitation is the physical properties of the materials used to generate the illumination. The circle of influence of layout context is between 4 and 6λ [12]. With λ stuck

Table 6.3   History of lithographic light sources

Wavelength (nm)	Light Source	Year Introduced
436	g-line	1980
365	i-line	1989
248	KrF laser	1995
193	ArF laser	2002
193 i	ArF laser	2006
157	F2 laser	Failed
13.4	EUV	?

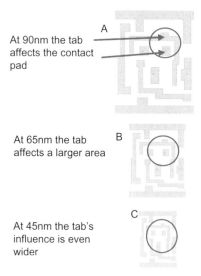

At 90nm the tab affects the contact pad

At 65nm the tab affects a larger area

At 45nm the tab's influence is even wider

Figure 6.25   Circle of influence of context increasing.

at 193 nm, the circle of influence will be fixed at between 772 and 1158 nm, which is substantial in the 45-nm nodes and beyond. Using the cell from Figure 1.4, the circle of influence would look as in Figure 6.25 for the 90-, 65-, and 45-nm technologies. With influence this far reaching, it would be really difficult to describe in a rule; the designer will need a model-based tool to cover the bases in the 45-nm node and beyond to deal with the proximity effects.

Figure 6.10 illustrates the problems faced by the design in the nano-CMOS regime. The lithographic process conditions (focus and dose) vary across the wafer. In an area where the exposure dose is high, we have worse line-end shortening, resulting in higher device leakage. In the area where the dose is lower, we risk having shorts. One way to design for this is to use conservative design rules, but that results in a larger, less cost-effective design. It is important to know where the design can use conservative design rules for free: that is, without growing the die size. With the use of model-based DFM tools, it is possible to flag regions that have problems and require attention and fix only those layouts, rather than indiscriminately increasing the spacing of all polygons in the design. This keeps the die size in check and at the same time eliminates regions that will result in catastrophic failures or parametric shift (e.g., higher leakage) that can alter the designed characteristics of the circuit.

Figure 6.26 is a scanning electron microscope (SEM) of a failure. Using an available tool, the simulated contour predicted a failure. If the designer ran the tool that predicted the failure before committing to silicon, this failure and the re-work required to fix this problem would have been avoided. It is very expensive and time consuming to have to re-work a design in the nano-CMOS regime, due to the mask cost and long fabrication throughput time (TAT).

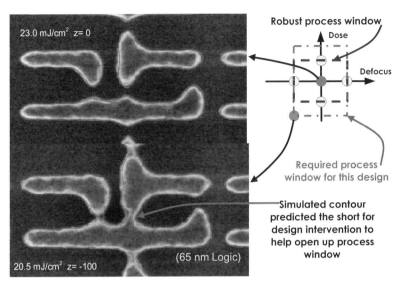

Figure 6.26 Early detection of failures is key.

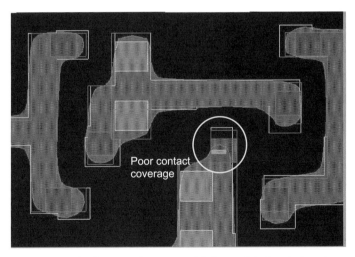

Figure 6.27 Problem can be corrected if detected during design stage.

Figure 6.27 shows a design with a contact coverage problem, even though it meets all design rules. The problem can be fixed when detected early in the design cycle. It is getting important that designers invest in new model-based design tools to help find problem areas in the design so that they can be fixed, thus avoiding costly design re-working. Rule-based tools like the classic design rule checker will require supplementary checks using model-based design tools to find these potential problems before the masks are burned.

6.6.2   Wire Uniformity and Density; High Diffusion Density

The best interconnect design for manufacturing is one with one pitch and width in only one direction. Such a design restriction will be very difficult to comply with, especially when an auto place-and-route tool is used to improve design productivity. Wire uniformity and density can be corrected to some extent if the designer obeys some simple rules of thumb, which include:

1. Set a preferred route direction for each metal layer and abide by it as closely as possible.
2. Minimize the number of pitches allowed and bear in mind the density correction rules using either metal fills (for CMP and lithography) or SRAF (for lithography). A design that cannot be corrected by either can limit the manufacturing window and yield. An example is shown in Figure 1.16, where the spaces between the lines in the first four metal layers is large enough to cause CMP problems but not wide enough for fills. Since the same pattern is repeated through all four layers, the cumulative dip causes too large a depression on the fifth metal layer, making it impossible for CMP to clear the copper, resulting in shorts.
3. For interconnects for which timing is critical, it may be important to note the wire spacing which results in the lowest coupling capacitance between lines. As the lines are spaced apart, the line-to-line capacitance drops, but the fringing capacitance to the top and bottom layer metal increases. At a certain space between lines, a minimum total line capacitance can be achieved (please see ref. 1 for further details). At this point the coupling capacitance is also at its minimum, which depends on the metal stackup and dielectric.
4. Nonpreferred direction route wires need to be wider and spaced further apart than minimum. The most difficult routes for manufacturing are minimum width and spaced routes with short jogs in the nonpreferred direction.

During shallow trench isolation (STI) etching, the diffusion is masked by a nitride layer. This layout is stripped after STI etching to expose the silicon for silicide formation on the source and drain diffusion. When the density of diffusion is too high in some areas, the nitride strip may not be complete for the dense areas, resulting in poor silicidation of the source and drain diffusion, causing poor contact, affecting device performance and reliability.

6.6.3   Minimum-Spaced Wire over Wide Metals

When minimum-spaced interconnects are routed over wide metals below them, the dishing of the wide metals below will cause a dip in the dielectric on which the wires are laid. The dip will make it difficult for the CMP process to clear the copper, resulting in shorts. This is especially true if minimum spaced interconnects are routed orthogonally over a series of minimum-spaced wide wires below. Therefore, it would be prudent to space out the routes that must pass over orthogonally wide wires below.

6.6.4 Misaligment Issues Coupled with Poly Corner Rounding and Diffusion Flaring

The first example is shown in Figure 6.28, where the device size increased by about 25%, due to diffusion flaring. Such a layout would also be subjected to device size variation, due to poly and diffusion overlay misalignment. This example was actually a part of the sense amplifier in a high-performance microprocessor cache design. The device was used as an equilibrating device for the sense amplifier. Due to the subwavelength distortion that increased the size of the device from the designed size, the driving gate of this device was not sized to anticipate the 25% increase in device size, resulting in bad equilibration timing, causing the cache to fail self-time. The reason for the failure is that the driver was designed to drive 256 such devices in the cache (for the 256 columns). The 25% increase in capacitive load and the *RC* of the long line connecting all the devices overwhelmed the driver. The designer SPICED the circuit with the layout-extracted netlist. However, the current layout extractor does not include the patterning distortion in the parasitic extraction, which is why this problem was not detected before fabrication of the chip on silicon.

   This problem would have been made apparent with contour-based extraction (more later). The correct-by-construction design methodology would have reduced the potential of such a problem as well. If the device size were chosen such that the diffusion was drawn as a straight rectangle without the "dog bone," as in this layout, the device size would remain very close to the

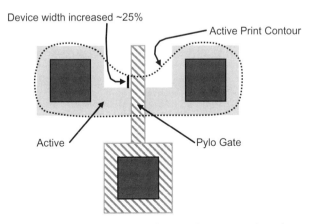

Figure 6.28 Typical layout of a sense amplifier equilibration transistor in a cache memory design.

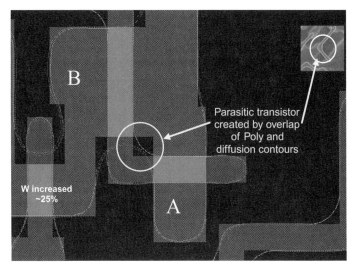

Figure 6.29 Patterning distortion of unstructured layout can result in circuit failure.

designed intent rather than growing 25% as in this case. The key to surviving patterning distortion is to reduce the number of vertices in the device layout if the designer does not have the ability to perform contour-based extraction.

The second example is shown in Figure 6.29, where there is an L-shaped poly forming two transistors. Due to the diffusion and poly flaring, a parasitic transistor was created, resulting in a leakage path between nodes A and B, which was not anticipated.

Figure 6.30 is another example of poly and diffusion flaring that changes the design intent of the circuit. In this design, NM1 and NM9 are designed to be identical, and the circuit operation relied on matching the two devices. Due to the difference at which the devices were laid out, the flaring of the diffusion is different, resulting in device width mismatch. Figure 6.31 shows the difference in timing between normal extraction and contour-based extraction. Contour-based extraction was able to uncover the device mismatch and showed that the designer needed a 240% margin for this circuit to ensure that it works. No design can afford such a hefty margin; therefore, it would be advisable to re-layout this circuit, making sure that the flaring effects will not cause a device size mismatch. The key is to make sure that the transistors are laid out with the same context, and better yet, use precision or analog layout rules for the matched transistor pair. See reference 1 for precision and analog layout guidelines.

Figure 6.24 shows two slightly different bitcell designs with one designed with the poly gates intersecting the rounded diffusion edge at different tangents, while the other has the same tangent. The cell on the left with the

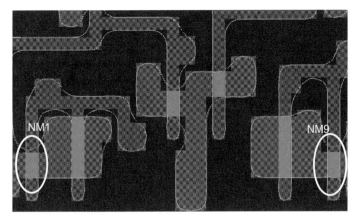

Figure 6.30   Mismatch due to a poor layout of differential transistors.

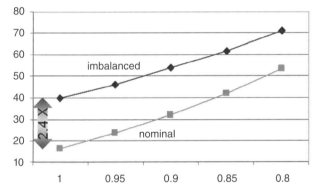

Figure 6.31   Timing difference with and without patterning distortion.

different tangent is sensitive to misalignment [1]. If the poly misaligns upward with respect to the diffusion, transistor A will increase in size as it rides up the tangent, while transistor B decreases in size as it rides down the tangent. This size change as a result of misalignment is yet another variability that is systematic as a result of poor layout that will affect the static noise margin of the cell. In this cell the matched transistor sizes are altered by poly-to-diffusion overlay misalignment. The cell on the right does not experience this problem and is not sensitive to misalignment. The light bulb effect at the end of the line also causes the cell to be sensitive to misalignment as well, where the cell on the right is again a better design (see also Section 6.2.1).

Figure 6.32 shows a design with a narrow bridge to V_{ss}. The cross section of the narrow bridge shows that there is a very narrow ledge of diffusion for silicide formation, due to the encroachment of the gate spacer, and if there is poly-to-diffusion misalignment, the ledge area may be too narrow to form

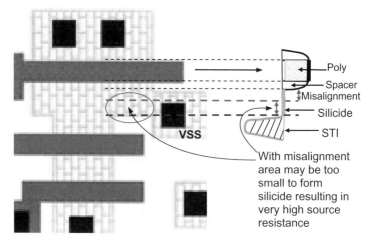

Poly

Spacer

Misalignment

Silicide

VSS

STI

With misalignment area may be too small to form silicide resulting in very high source resistance

Figure 6.32 Narrow diffusion bridge to V_{ss}.

silicide, and even if silicide is able to form, it is likely to crack, resulting in high source resistance to V_{ss}.

6.6.5 Well Corner Rounding Issues

The problem caused by the well corner rounding issue has only surfaced in the 65-nm node, as a result of the use of different resolution enhancement (RET) precision for corner rounding correction of the wells versus the RET precision used for diffusion. Well rounding correction has been deemed to be unimportant, and for the most part this is true. Motivation for the use of less precision in the well rounding correction is mask cost (more on mask cost and complexity later).

The result of the lower correction precision results in more severe corner rounding of the wells as well as line end shortening. Although for most cases this would be fine, as the well structures are large with few corners; there are cases that this would be a pitfall. The NWELL in an SRAM array would be one of them, as the NWELL is relatively long, about 100 µm, with a width of 0.4 µm. The problem occurs when a designer places a diffusion minimum-spaced to the NWELL edge and at the NWELL corner. The NWELL corner rounding can cause poor NWELL coverage of such a diffusion, resulting in leakage (Figure 6.33).

6.6.6 OPC and OPC Reproducibility

Figure 1.10 shows two poly gates that are identical as drawn. However, after processing they are slightly different, which can be attributed in part to the reproducibility of the optical proximity correction (OPC) applied to the gates.

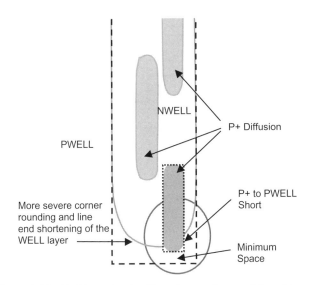

Figure 6.33   Problems due to different correction precision of layers.

This has implications for matched transistors and has to be taken into consideration in the design. In time, this will begin to affect digital circuits. There will come a time when restrictive design rules (RDRs) will be an important part of the design methodology. Design symmetry and regular context will be needed to minimize this problem and the first to enforce RDR are the integrated device manufacturers such as Intel and the IBM microprocessor divisions. In the case of Intel, where "dry" lithography is used at the 45-nm node, this is mandatory. Many layout constructs that are difficult to pattern with dry lithography are banned.

6.6.7   Timing Implications of Cell Mirroring

Analog designers have been designing circuits without mirroring to avoid device-matching issues. We have reached a stage where such device matching is affecting digital circuits. Today, cell mirroring will result in different copper thickness profiles of the interconnect, transistor parasitic resistance difference, difference in random dopant fluctuation, and so on. All these effects cumulatively are sufficient to cause timing differences for the mirrored cell. Circuits that rely on timing matching will need to step and repeat the cells rather than mirroring.

6.6.8   Mask Cost, Complexity, and Building Throughput Time

Figure 6.34 is the mask cost trend since 1992, projected to 2009. It is clear from this graph that mask cost keeps increasing at an alarming rate. At some point

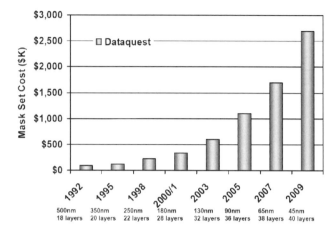

Figure 6.34 Mask cost trend.

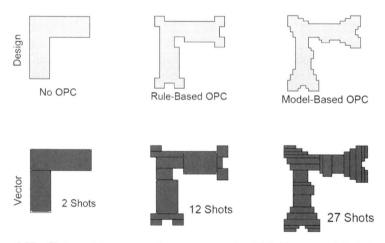

Figure 6.35 Shot count increases with more aggressive OPC. (Courtesy of Chris Progler.)

in time, this may itself be a limit to the ability of the industry to scale the technology. The cost cross-over point would be too far out in the future for advanced technology to be adopted right after it has been qualified.

With some basic understanding of how the design influences mask cost, we will be able to implement steps to keep this escalating cost in check. Figure 6.35 shows that the driver of database size as well as shot count is the aggressiveness of the OPC used. Shot count is one of the drivers for mask cost. The shot count increases the higher the figure count, resulting in a larger database as well as a longer writing time. This has created a bottleneck in the mask-making process, which includes mask data preparation [19]. Longer writing

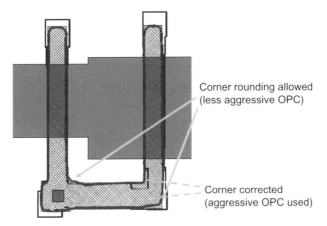

Figure 6.36   Examples of opportunities to reduce shot count. (Courtesy of Chris Progler.)

time and higher shot count also increase the inspection time, all of which increases mask building throughput time and cost.

The shot count increases with the aggressiveness of the OPC. The design shown in Figure 6.35 without OPC would require only two shots to write. As the OPC increases in aggressiveness, the short count gradually increases. Twelve shots for the rule-based OPC and 27 shots for a model-based OPC. In an ongoing effort to control shot counts and the cost of the mask, many manufacturers are resorting to the use of the less aggressive OPC where possible, without causing problems in the design. For this reason, some of the larger polygons, such as NWELL layers, use the least aggressive OPC. Designers must be aware of such a trade-off at their manufacturers and design accordingly to avoid the problems described in Section 6.6.5.

Figure 6.36 shows an example of where less aggressive OPC can be used to reduce shot count even on a layer that normally requires aggressive OPC. It would be necessary to understand the principles behind the choice of the aggressiveness of the OPC used and design accordingly to assist the manufacturer to maximize the use of less aggressive OPC. Less aggressive OPC will be applied to areas where corner rounding will not be a problem. In the example in Figure 6.36, the corner of the poly that is far away from the edge of the diffusion, more pronounced corner rounding is allowed, hence requires less aggressive OPC. The outer corner without a contact can also use less aggressive OPC. The inner corner close to the diffusion edge as well as the outer corners covering a contact will require more aggressive OPC.

Figure 6.37 is an example of the simple changes in a design that can convert the design to a fracture-friendly design with an opportunity to reduce database size as well as shot count. The motivation to reduce database size is to speed up transmission of the design to the mask shop. See Chapter 2 for further details.

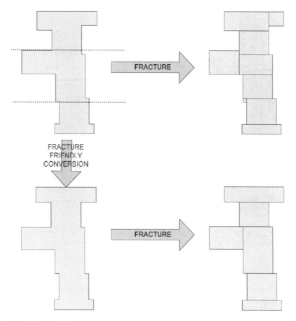

Figure 6.37 Fracture-friendly design. (Courtesy of Chris Progler.)

6.7 DESIGN FOR VARIATIONS

6.7.1 Limiting the Degrees of Freedom to Improve Design Uniformity

6.7.1.1 Restrictive Design Rules

Restrictive design rules (RDRs) have existed in some form or another under different names to deal with mainly subwavelength lithography issues. In the future, RDRs will cover all other manufacturing artifacts, such as chemical–mechanical polishing (CMP) dishing and erosion effects. Halo implant effects penetrate more deeply under the mask, due to the angled implantation affecting the V_t of the devices at the edge of the mask. Device V_t drift is a result of the source–drain implants of the SRAMs, where the nMOS and pMOS are very close to each other with a common poly line. During anneal, after implantation, diffusion of the n-type implant species to the pMOS channel regions, and vice versa, along the common poly line can change the work function of the device, resulting in altered V_t.

Integrated device manufacturers (IDMs) can be in a position to edict the adoption of RDRs to their designers. Therefore, RDRs are used mainly at IDMs such as Intel, IBM, and some Japanese IDMs. RDRs are practically nonexistent at foundries (only as recommended rules) simply because they are a competitive disadvantage. Foundry customers will choose the foundry that does not require the use of RDRs in their design because that requires a lot more work from the designers with very little gain if any for them and in fact may come with an area penalty. Designers view RDRs as a push more in the favor of the manufacturing side, in terms of manufacturability and performance trade-off [19]. Although electronic design automation (EDA) tools that support RDRs are emerging [9], they are still not very widely adopted in the nonfabrication design community. Most IDMs have some form of internal EDA tools that support RDRs to maintain design productivity with RDRs.

Figure 6.11 shows an example of RDRs in which the poly gates are limited to one direction to improve across-chip line-width variation (ACLV) as well as enabling the use of special illumination for patterning. The use of single poly orientation can also lend itself to the use of two direction halo implants rather than four direction halo implants. This will improve the halo implant–induced variation of the device characteristics.

The most common RDRs that are being employed are shown in Figure 6.38, where A is the poly pitch, B is the line-end pass diffusion edge, C is the contact-to-poly space, D is the poly bend-to-diffusion spacing, and E is the diffusion bend-to-poly spacing. In time, E will be banned altogether, due to the fact that the corner-rounding radius will not be improving when the spacing needs to be reduced in future nodes, and this restriction will have too much impact on the area. The other restriction, which is currently banned by all nano-CMOS technologies, is the use of bend gates on diffusion. Figure 6.32 is another example of a layout that will not be allowed in future nodes. Figure 6.39 shows a layout with the contact that is too close to the poly corner and will be fixed if rule D in Figure 6.38 is obeyed.

Layout using gridded glyph geometry objects (L3GO) is the IBM version of the restricted rules design methodology with the aid of their internal tool, L3GOlizer. L3GO restricts drawn objects to straight lines and rectangles which are placed on a grid. This guarantees that the poly lines will be on a

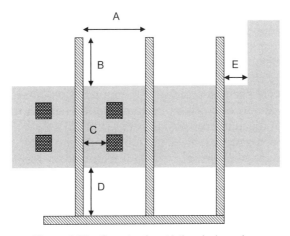

Figure 6.38 Sample of restrictive design rules.

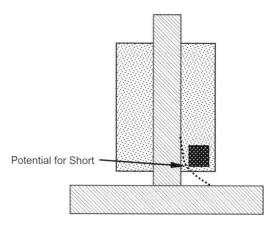

Potential for Short

Figure 6.39   Potential for short due to corner rounding plus misalignment.

uniform pitch, which improves ACLV (see Figure 6.12). This restriction allows IBM to reduce the number of design rules needed for this methodology. The trade-off is some increase in area, which comes with longer interconnect and potentially higher dynamic power as well as reduced performance.

Pileggi has come up with a similar RDR methodology at Carnegie Mellon University, named the "bricks" [18]. Library cells with regular pitches were created for poly and they also implement IBM's L3GO restrictions without area penalty. The drawback is that the cells need to be handcrafted by a master layout designer. We still do not have the technology to continue this regularity using state-of-the-art routers without creating route congestions or increasing routing time to an unacceptable level. Fortunately, interconnect CD is not as critical as poly CD. However, down the road of dimensional scaling, interconnect CD will become critical enough to require the same restrictions as poly lines.

Figure 6.24 shows the print contours of two-bitcell designs in which yield can be as different as 20%, with the one on the right being the better-yielding design, as explained in Section 6.6.4. Figure 6.40 shows another two-bitcell design in which cell A's V_{dd} line is snaking back and forth to pick up the V_{dd} contact on the cell. The first problem with this cell is that there will always be a minimum space between either the BL or the BLB lines, with V_{dd} along the entire column of the memory array. In contrast, cell B only sends a tab to pick up the cell V_{dd} contact. As long as the tab is wide enough, it will not be susceptible to line-end shortening, creating a high contact resistance when the metal line does not cover the contact completely as a result of line-end shortening. The result is that cell B will yield better, due to having a lower critical area than cell A. On top of that, the lower BL and BLB capacitance to V_{dd} also improves the CV/I of the cell by reducing the C component. These two examples show the possibility for designers to contribute to a better-yielding

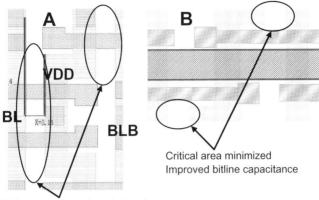

Critical area minimized
Improved bitline capacitance

Critical area along the whole column

Figure 6.40   Simple changes can improve yield as well as performance.

as well as higher-performing design even without RDRs, by understanding some of the principles that we have provided in this book.

6.7.2   Stress Proximity Extraction

As mentioned earlier, in a complex design it would be very difficult and error prone for the designer to hand-edit layout-induced carrier mobility changes for every transistor in the design, even if a look-up table were available to the designer for every situation. It would also be close to impossible for the designer to figure out what correction is needed under each layout construct. Therefore, a model-based approach is called for to provide the designer with the appropriate correction for every situation and layout construct. Synopsys has implemented a model-based approach in the tool Seismos, but extensive model calibration would be required before the tool can be used. Other EDA companies are working on a version of their own but have not disclosed details to the authors.

The Seismos implementation of the stress proximity extraction tool replaces SA/SB and other relevant parameters by the mobility and V_t modifier as shown below:

- Original Spice netlist
  - Netlist without Seismos, SA and SB stress parameters used
  - M1 d g s b l = 50 nm, w = 0.6 μm, SA = 0.18 μm, SB = 0.18 μm
- Seismos annotated netlist
  - Stress parameters SA and SB replaced by mobility multiplier and ΔVt_0
  - M1 d g s b l = 50 nm, w = 0.6 μm, ~~SA = 0.18 μm, SB = 0.18 μm~~, +MULU0 = 1.2, DELVT0 = 0.003

The new parameters MULU0 and DELVT0 are both standard BSIM 4 parameters supported by Hspice, Hsim, and Nanosim, as well as other circuit simulators.

It is now possible to harness this systematic variability with the aid of this tool to improve the performance as well as to reduce the leakage of the design. The strain on the device channel due to the mobility enhancement films can be altered in many ways. The strain effect is proportional to the volume and proximity of the film to the channel. The film volume and proximity are modulated by:

- The contact size and contact-to-device channel edge
- The contact pitch
- The poly pitch

In embedded SiGe (eSiGe) source–drain transistors, eSiGe recess depth modulates the channel strain, while the recess depth is modulated by the poly pitch.

The first is the contact perforation of the film, which relaxes the film, thereby reducing the drive current enhancement. Figure 6.17 shows the effect of contacts on the stress distribution. Other factors that affect stress on the device channel are contact dimension, contact pitch along the device channel, and contact space to the channel edge. Figure 6.18 shows the effect of contact space on the edge of the device channel. As the contact is moved away from the device channel edge, we see the channel stress increase—hence the increase in drive current for better performance without having to push the process.

Figure 6.14 shows the effect of minimum pitched contact and poly lines next to the device. When the device is isolated and has no contacts, the channel stress is at the maximum normalized to 1. However, such a device is unusable, so we have to add contacts to use the device in a circuit. It is here that the right choice and number of contacts in the layout can result in a higher-performance device. The middle icon in Figure 6.14 shows the effect of having minimum pitched poly lines next to the device. The final icon on the right shows the worst-performing device, where we have minimum poly pitch and fully contacted minimum pitched contacts that are minimumly spaced to the device edge.

The improvement that one can realize from judicial contact and poly placement comes for free, and if the designer understands this effect, the designer can take advantage of it for better performance. If the designer desires to reduce leakage, this can be applied to increase the drive of a higher V_t transistor to maintain the required drive while resulting in reduced leakage. Alternatively, the improved drive can be taken advantage of by reducing the device size, thus improving leakage and dynamic power while reducing the chip size. With this tool, another knob becomes available to the designer for circuit optimization. Furthermore, the designer no longer needs to add margin to the design for this systematic variability, as it is now deterministic, further improving the design. For complete details on the physics of the stress engineering techniques and the interaction between layout and device performance, see Chapter 3.

6.7.3 Contour-Based Extraction: Design in the Era of Lithographic Distortion

Figures 6.28 to 6.30 show the extent of the patterning distortion of the design in the nano-CMOS nodes. The failure of the cache described in Section 6.6.4, with a device drawn as shown in Figure 6.28, would have been found in simulations if the models for this distortion existed during the design phase. The timing of the circuit shown in Figure 6.30 with and without contour-based extraction is plotted in Figure 6.31, which shows that the designer needed to assign a 240% margin to the design to capture the degradation in timing due to the subwavelength distortion. Figure 6.7 shows the distortion of the interconnect, and the as-drawn extraction may not capture the change in capacitance as well as the resistance of intricate routes as shown [20–22].

Figure 6.41 shows the results of library variability analysis performed on a 65-nm standard cell library containing more than 700 cells. Using model-based lithography and etch simulation at different process conditions as described previously, contours were generated for the drawn layers of each standard cell, and the resulting transistor-level netlists were simulated for cell delay and leakage. The context-specific discrepancy was calculated by finding the difference in delay and leakage at a particular defocus point among different instances of the same cell in the randomly generated context layout. Under defocus conditions, simulation results show up to a 10% discrepancy in gate length, 15% discrepancy in delay, and a 200% discrepancy in leakage between different instances of the same cell under the same operating conditions, but with different neighbors.

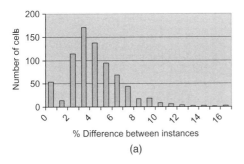
(a)

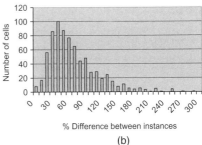
(b)

Figure 6.41 Discrepancies in (a) delay and (b) leakage between instances of the same cell, but with different neighbors. The difference between instances is based on the best- and worse-case instance of the cell. The cell count is the number of cell types (AND2, NAND2, NAND4 etc.) that have the given difference between instances.

6.7.4 Design Compensation for Variations

6.7.4.1 Importance of the Input Slew Rate; Fanout Check

Since the pMOS and nMOS have separate V_t implants, they can vary independently, affecting their relative drive strength, causing the input trip points of gates to change. If the input edge is sharp, this change in trip point will have a small difference in delay. However, a sluggish input signal will amplify the V_t variability [1] and translate it to a large timing variability (Figure 6.42).

Driver fanout check is designed to find heavily loaded nodes which will be sluggish, so that the designer can either repartition the logic or add buffers and repeaters to reduce the load. Sluggish nodes will also increase the probability of V_t shift due to hot carrier injection, as slow output nodes provide a long high-voltage bias on the drain of the device, which is the recipe for inducing hot carrier injection. Slow nodes also increase the overlap current, as the pull-up and pull-down devices are on at the same time for a longer period.

6.7.4.2 Longer L for Matched Transistors and Leakage-Sensitive Circuits

Figure 6.43 shows the effect of poly CD that is ±10% from the target. When the CD is higher than the target, we notice that the V_t variability is a lot lower than even at target CD. However, when the poly CD is lower than the target, the V_t spread is very large, with some devices in the negative V_t range, or extremely high leakage. For leakage-sensitive circuits and circuits that have a lot of timing slack, the poly L can be biased higher to improve the V_t variability as well as leakage.

Figures 1.6 and 1.7 shows the importance of controlling the low end of the poly CD. A 6% decrease in CD causes the part to fail due to leakage. The margin for failure of the high end of the CD is wider and is better controlled, as shown in Figure 6.43, which suggests that it makes sense to bias the CD higher. If the CD of the devices in circuits that have sufficient timing slack are biased higher, the resulting part would have a lot less leakage with minimal, if any, impact on performance. Figure 1.7 shows the importance of CD control,

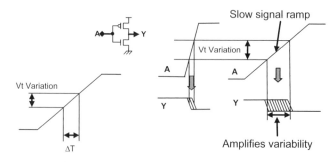

Figure 6.42 Sluggish signals amplify variability.

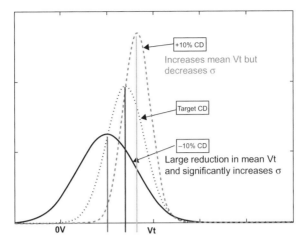

Figure 6.43   Effect of poly CD deviation from the target.

which is getting more and more difficult as the technology scales. The margin and yield drops off very quickly as the CDs decrease from target.

Limits to Poly Biasing for Leakage Variation Reduction   While there is a lot of motivation to bias CD higher, if this is an afterthought on a completed design, using the Blaze tool will be limited. The contacted poly pitch includes:

- The poly width
- The contact width (CD)
- Spacer and overlay tolerances
- Competing requirements for contact resistance, contact–poly capacitance, contact–poly breakdown voltage, and maximum poly bias for leakage

If poly biasing is designed into the cells, these limitations will not be an impediment at a slight increase in area. However, spacing out the contact from the edge of the poly also has performance advantages, as described earlier.

6.7.4.3   Clocked Feedback Safer for Variation

Most flip-flops and latches are designed with a ratioed feedback for a lower transistor count (area) as well as a lower clock load [13]. However, the use of ratioed feedback is severely compromised by trade-off between writability versus noise margin over the process corners. The higher variability in the nano-CMOS nodes significantly reduces the design margin for ratioed feedback design. Clocked feedback eliminates trade-off where the storage element can be written easily without having to fight the feedback driver. Once the storage node is written, stronger feedback can be turned on by the clock to hold the state with excellent noise immunity and improved tolerance to feedback and data path variations due to process. Clocked feedback can also be designed to improve radiation hardness against alpha particles, since the active feedback can be made stronger to improve the critical charge of the storage element without having to worry about writability.

6.7.5   Layout Optimization to Maximize Useful Stress and Reduce Variability

6.7.5.1   Impact of Layout on Stress in the Transistor Channel

Considering that stress propagates into a layout neighborhood about 2 μm around the stress source, it is evident that layout scaling leads to a larger number of neighbors affecting the performance of each transistor. For example, a cell height of about 1.25 μm at the 32-nm node means that a transistor in a library cell will feel stress coming from its own cell, from the first layer of adjacent cells, and even from the second layer of neighbor cells. Therefore, it is affected by about 25 cells.

How do you make sure that a cell works for all possible layout contexts? On the one hand, having as regular a layout as possible helps to suppress layout-induced variations to some extent. On the other hand, accounting for a specific layout context with a tool such as Seismos takes care of the rest by providing the circuit simulator with the appropriate input on transistor performance. Besides, knowing all stress sources that are used in a particular process, the layout can be designed to enhance the overall circuit performance. In the following sections we describe several possible ways for doing that.

6.7.5.2   Isolated Transistors Versus Nested Gates

It is shown in Chapter 3 that isolated transistors have different V_t values as well as different stress and therefore different driving currents than those of nested transistors. This introduces undesirable layout-induced transistor performance variations. One way to deal with this is to account for such variations by modeling all necessary proximity effects. Another way is to modify layouts such that such variations are suppressed.

For example, it is possible to avoid isolated transistors completely by replacing STI with isolating gates (Figures 6.44 and 6.45). The isolating gates that are depicted in Figure 6.45 are connected to ground and the power supply for the nMOS and pMOS, respectively. To further reduce static leakage currents, they can be connected to negative voltages and those higher than the power supply voltages.

Replacing isolated transistors by isolating gates can help to drastically reduce variations of V_t, stress, and driving current. The single diffusion area can be extended beyond each individual library cell and connect diffusion areas of the adjacent cells into one continuous diffusion area that can potentially span the entire chip.

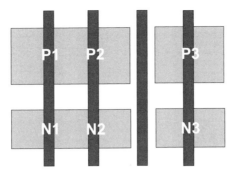

Figure 6.44 Typical layout with nested transistors N1, N2, P1, and P2 to the left of the dummy poly and isolated transistors N3 and P3 to the right of the dummy poly.

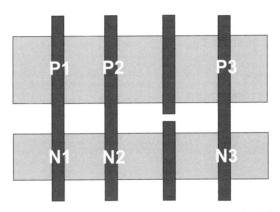

Figure 6.45 Alternative layout with all nested transistors and an isolating gate instead of STI.

6.7.5.3 *Placement of Well and DSL Masks*

It is shown in Chapter 3 that well proximity effects increase V_t and reduce the drive current of transistors located near well boundaries compared to transistors located farther away from the boundary. One way to exploit such an effect is to put the critical path transistors as far from the well boundary as practical to make them stronger. The rest of the transistors should be placed as close to the well mask as practical to suppress their OFF current (Figure 6.46). This improves the speed of the chip and reduces the standby power consumption simultaneously.

The distance from the DSL mask to the nMOS and pMOS determines their stress patterns and therefore modulates the drive currents. For the standard (100) wafer with <110> channel direction, it is beneficial to move the DSL mask as close to pMOS as practical, because it creates beneficial tensile transverse stress on the side of compressive CESL and harmful compressive transverse stress on the side of tensile CESL.

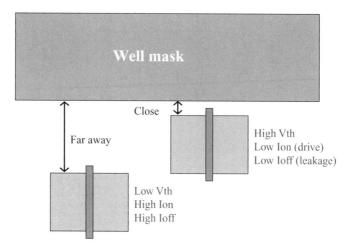

Figure 6.46 Placing transistors to take advantage of WPE such that the delay and the power consumption improve simultaneously.

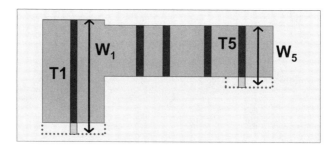

Figure 6.47 Making the weaker transistors wider to balance the circuit.

6.7.5.4 Adjusting Channel Widths

Once the stress-induced performance variations are determined as shown in Section 3.5.6, several approaches can be used to fix or compensate them. One approach would be to increase the channel widths of the unexpectedly weaker transistors and/or decrease the channel widths of the stronger ones. Let's consider a library cell depicted in Figure 3.46. Figure 6.47 illustrates how this can be done for the two weaker transistors T1 and T5 considering the mobility variations shown in Figure 3.47. This will alter the shape of the diffusion area and will therefore alter the stress distribution and related mobility enhancements. Several iterations will be required to find the layout where channel widths and stress effects combine to create transistors of the right strengths.

Besides altering stress distribution, this approach changes the gate capacitances proportionally to W. This changes the load for the adjacent circuit and therefore requires additional iterations, with the circuit simulation to get all

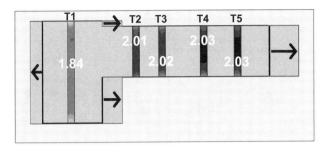

Figure 6.48 Stress-induced mobility variation in five pMOSFETs, with layout modifications as shown by the arrows.

three components: the transistor strengths, stress effects, and load capacitances in balance.

6.7.5.5 Adjusting the Shape of the Diffusion Area

An alternative approach is to modify the shape of the diffusion area in a way that suppresses stress nonuniformities throughout the circuit. If this is done without changing the channel widths, there is no need for iterations because the load capacitances do not change. An application of this approach to our library cell is illustrated in Figure 6.48. The layout is modified specifically to improve stress uniformity throughout the cell.

The four narrow transistors, T2 through T5, exhibit fairly similar enhancements that are higher than in the original layout, due to the larger overall SiGe area. The double-width transistor, T1, is still weaker than its width implies, but this time by only 10%. This is an improvement compared to the original layout, but still not a satisfactory solution. A combination of this approach with channel width adjustment is required to balance the circuit.

6.7.5.6 Adding Dummy Diffusion Features

Instead of changing the shape of the diffusion area that contains transistors, it is possible to introduce dummy diffusion features around it. The main advantage of this approach is that it does not change any capacitances and therefore requires fewer adjustments of the surrounding circuit. Figure 6.49 shows that the introduction of a dummy SiGe feature reduces the transistor performance variation to 7%, which is only half of the variation exhibited by the original design. The dummy diffusion introduced creates additional compressive longitudinal stress in transistor T1 and therefore helps to make it stronger.

6.7.5.7 Using Uniform Layout

The more uniform a layout is, the more uniform is the stress distribution. It is not always possible to make the layout completely uniform, but it is a useful approach to take whenever possible. Fortunately, this is possible in our

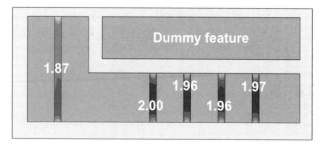

Figure 6.49 Stress-induced hole mobility enhancements in a layout with a dummy diffusion feature.

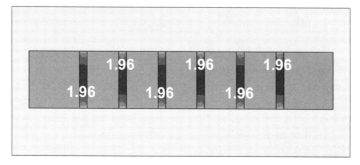

Figure 6.50 Stress-induced hole mobility enhancements in a layout that is made as close to uniform as possible.

example by splitting the double-width transistor into two narrow transistors that are connected in parallel. Figure 6.50 illustrates that this indeed gives a perfectly uniform stress enhancement for all six narrow transistors.

6.7.5.8 Accounting for Stress-Induced Variations in SPICE Simulation

There might be cases where it is not practical to suppress completely all performance variations introduced by the layout-sensitive stress effects. In such cases the variations can be estimated and communicated to a circuit simulation tool. Once the stress effects are accounted for in circuit simulation, the circuit can be modified in a way that enables its proper operation. This would require estimating layout-specific stress effects and creating appropriate instance-specific transistor models. The circuit modification strategy used will determine if the corresponding layout modifications are similar to the approach described above or perhaps completely different. Typically, several iterations will be required to converge to an acceptable solution where the circuit is fine-tuned to work for a specific design implementation.

Stress-induced transistor performance is shown to exhibit considerable variations due to the differences in layout. The silicon stress depends on the

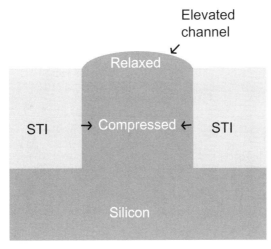

Figure 6.51 Relaxing undesirable STI stress by either elevating the channel or recessing the STI.

shape of the diffusion area and on the density of the adjacent layout. The transistor performance variations are demonstrated to result in noticeable changes in the inverter delay and ring oscillator frequency. Based on accurate stress analysis, the layout can be adjusted to compensate for the stress-induced variations in transistor performance and to improve the parametric yield.

6.7.5.9 *Relaxing Harmful Stress Components*

There are several ways to relax the undesirable stress components that degrade transistor performance. One way is to move the harmful stress source either above or below the channel. For example, if the surface of compressive STI is recessed relative to the channel surface, it introduces much less stress into the channel (Figure 6.51). Another example is removing tensile longitudinal channel stress generated by NiSi that is located on top of source and drain regions. Such stress is beneficial for nMOS; therefore, most nMOSFETs have either a flat or a slightly recessed source–drain surface (see Figures 3.19 to 3.21).

However, such stress is harmful for pMOS, but it can be reduced drastically if NiSi is moved above the channel, on top of elevated SiGe source–drain, as shown in Figures 3.24 and 3.31.

An alternative way to relax a harmful stress component is to directly relax the undesirable stress source. For example, compressive STI introduces compressive transverse stress that is harmful for both nMOS and pMOS. Etching trenches in STI as shown in Figure 6.52 relaxes STI stress and therefore improves the performance of the adjacent transistors.

The opposite is also true: By relaxing desirable stress away from the channel, it might be possible to increase beneficial stress in the channel. For example,

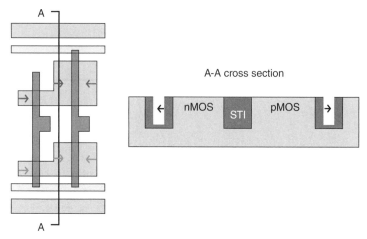

Figure 6.52 Relaxing undesirable STI stress by etching trenches in the transverse direction.

the SiGe source and drain squeeze the silicon channel in between as well as the gate stack. The gate stack helps the channel to balance part of the compressive force in longitudinal direction. Removing the polysilicon gate relaxes stress above the channel and simultaneously redirects the entire SiGe stress to the channel.

The metal gate with metal-last approach used in the structures depicted in Figure 3.31 does exactly that—the polysilicon gate is removed before the metal gate deposition. Numerical stress analysis using Fammos [26] indicates that upon polysilicon gate removal, the channel stress increases by over 50%. Subsequent metal gate deposition is performed in between the already deformed spacers and therefore does not affect channel stress.

6.8   SUMMARY

It is becoming evident that continued CMOS scaling into the nano-CMOS regime is causing some new and unexpected variability that the designer working with such a technology node needs to be well aware of as well as understanding the principles behind it in order to be able to find effective solutions. Although RDRs have been touted by many IDMs as a solution for all these issues, at some cost in area and performance, it is no panacea [23]. Lithographic distortions of drawn objects, especially transistors, cannot be fully described in rules, nor will it disappear altogether with RDR limitations. Certainly, there will be yet other effects that cannot be fully described by rules and will require model-based solutions. For example, there will still be a need for the contour-based extractions to incorporate the parasitics as well as a device parametric shift as a result of the shape distortion, which is predicted to get worse before the deployment of EUV lithography. The stress context would be another especially difficult effect to be described by rules alone. Furthermore, RDRs are out of the question for foundries, as it will be viewed as a competitive disadvantage; therefore, the model-based solution will continue to have its place. Some of the techniques described in this chapter to harness the systematic variability for performance can be used only with the help of a model-based tool, even for custom designs.

Fanout check will be needed to help eradicate slow nodes and reduce circuit sensitivity to V_t variability. It is through the use of model-based tools that the designer can be armed with the necessary insights to turn some systematic variability into deterministic, performance-boosting techniques that also reduce the design margin required. The tools, coupled with an understanding of the principles, will be sufficient for any designer to initiate robust, complex product designs that are at the same time the smallest, highest-yielding, and best performance chip in the nano-CMOS regime.

With the complexity of the product designs these days, there is a clear need for an autocorrection tool that juggles the many trade-offs needed for an optimal layout. This would require a cascading series of slight edge movements throughout the layout. The many edge movements required to arrive at the optimum trade-off would be difficult to reproduce manually, and even if it was possible, this would be the work of a few master designers. Emerging tools that we had a chance to validate are capable of fixing unstructured layout automatically, and the reduced device variability improvement is measurable and reported in a paper at the Society of Photographic Instrumentation Engineers [9]. Chapter 7 is devoted to tools and the technology roadmap, highlighting where the gaps and possible showstoppers are and the future of DFM in filling the gaps and bridging the showstoppers until the technology catches up.

REFERENCES

1. B. P. Wong et al., *Nano-CMOS Circuit and Physical Design*, Wiley, Hoboken, NJ, 2005, Chap. 11.
2. K. Agrawal and S. Nassif, Statistical analysis of SRAM cell stability, Design Automation Conference, San Francisco, CA, July 2006.
3. M. Pelgrom, Nanometer CMOS: an analog challenge! IEEE Distinguished Lectures, Fort Collins, CO, May 11, 2006.
4. M. Pelgrom et al., Matching properties of MOS transistors, *IEEE Solid State Circuits*, vol. 24, no. 5, p. 1433–1439, 1989.
5. H. Tuinhout, Impact of parametric mismatch and fluctuations on performance and yield of deep-submicron CMOS technologies, Tutorial, Philips Research, Edindhoven, The Netherlands, 2002.
6. Tze-Chiang Chen, Where Si-CMOS is going: trendy hype vs. real technology, International Solid State Circuits Conference Plenary, San Francisco, CA, Feb. 2006.

7. K. Takeda et al., A read-static-noise-margin-free SRAM cell for low-VDD and high-speed applications, *IEEE J. Solid State Circuits*, vol. 41, no. 1, pp. 113–121, Jan. 2006.

8. B. P. Wong, Bridging the ROI gap between design and manufacturing, SNUG 2006, Santa Clara, CA.

9. K. K. Lin et al., Silicon-verified automatic DFM layout optimization: a calibration-lite model-based application to standard cells, Society of Photographic Instrumentation Engineers, Monterey, CA, Aug. 2007.

10. B. Arnold, Lithography for the 32 nm technology node, International Electron Device Meeting, San Francisco, CA, Dec. 10, 2006.

11. B. P. Wong, Part 3: Challenges and opportunities for design in nano-CMOS technologies, DFM Tutorial, DATE 2007, Nice, France, Apr. 2007.

12. A. Kahng, Part 4: Manufacturing-aware, model-driven design optimizations for 65 nm, 45 nm, DFM Tutorial, DATE 2007, Nice, France, Apr. 2007.

13. J. Farrell, Incorporating variability into design, Designing Robust Digital Circuits Workshop, UC Berkeley, Berkeley, CA, July 2006.

14. L. Pileggi, Regularity for reduced variability, Designing Robust Digital Circuits Workshop, UC Berkeley, Berkeley, CA, July 2006.

15. S. Crowder, Low power CMOS process technology, International Electron Device Meeting, Washington, DC, 2005.

16. R. Borges et al., Strain engineering and layout context variability at 45 nm, Semiconductor International, Nov. 1, 2007.

17. B. H. Calhoun and A. Chandrakasan, A 256 kb sub-threshold SRAM in 65 nm CMOS, International Solid State Circuits Conference, San Francisco, CA, Feb. 2006.

18. K. Y. Tong et al., Design methodology of regular logic bricks for robust integrated circuits, International Conference on Computer Design, San Jose, CA, 2006.

19. L. Capodieci et al., Toward a methodology for manufacturability-driven design rule exploration, Design Automation Conference, San Diego, CA, June 2004.

20. N. S. Nagaraj, Dealing with interconnect process variations, International Workshop on System-Level Interconnect Prediction, San Francisco, CA, 2005.

21. N. S. Nagaraj et al., BEOL variability and impact on RC extraction, Design Automation Conference, Anaheim, CA, 2005.

22. N. S. Nagaraj. Benchmarks for interconnect parasitic resistance and capacitance, International Symposium on Quality Electronic Design, San Jose, CA, 2003.

23. D. McGrath and R. Goering, Designing without a net: restricted design rules challenge DFM's role, *EETIMES*, July 24, 2006.

24. G. Roy et al., Simulation of combined sources of intrinsic parameter fluctuations in a "real" 35 nm MOSFET, Department of Electronics and Electrical Engineering, University of Glasgow, Glasgow, UK.

25. K. Nowka, Survival of VLSI design: coping with device variability and uncertainty, Lecture, Texas A&M University, College Station, TX, Oct. 23, 2007.

26. Fammos User's Manual, v. 2007. 06, Synopsys, Mountain View, CA, 2007.

27. N. S. Nagaraj et al., Interconnect process variations and *RC* extraction. DAC DFM Tutorial, San Francisco, CA, 2006.

III

THE ROAD TO DFM

7

NANO-CMOS DESIGN TOOLS: BEYOND MODEL-BASED ANALYSIS AND CORRECTION

7.1 INTRODUCTION

Looking forward to ≤45-nm nano-CMOS process generations, it is clear that semiconductor manufacturers face ever-greater business challenges of capital cost and risk, along with ever-greater technical challenges of pitch, mobility, variability, leakage, and reliability [1]. To enable cost-effective continuation of the semiconductor roadmap, there is a greater need for (1) design technology (i.e., EDA tools and methodologies), to provide equivalent scaling, and (2) product-specific design innovation, to provide "more than Moore's law" scaling of product value.

Equivalent scaling, which refers to the introduction of design automation technologies that reduce power or improve density without requiring innovation in process technology, is required to continue Moore's law trajectories of performance, density, and cost. Conservatively, half of a process node of power, a third of a node of area, and one full node of performance can easily be gained, the only question being whether the industry will recognize and invest sufficiently in this opportunity. Design innovation (i.e., multicore architecture, software support, beyond-die integration, etc.) will be the workhorse for more than Moore's law scaling which goes beyond what underlying process and design technologies can achieve.

One can envision a roadmap aligned to the continued delivery of semiconductor value, with that value arising from a combination of manufacturing

Nano-CMOS Design for Manufacturability: Robust Circuit and Physical Design for Sub-65 nm Technology Nodes
By Ban Wong, Franz Zach, Victor Moroz, Anurag Mittal, Greg Starr, and Andrew Kahng
Copyright © 2009 John Wiley & Sons, Inc.

technology, design technology, and design innovation. Finally, the balance of these contributors to scaling must be determined by a new understanding of system scaling. In the future, Dennard's classical scaling theory [17] no longer holds, nor will ITRS-style scaling in terms of such parameters as A factors (the number of square half-pitches occupied by a memory bitcell), FO4 delays, or CV/I metrics (normalized measures of clock period and transistor switching speed). The future of semiconductor technology scaling will instead be dominated by application- and product-specific system constraints on reliability, adaptivity, cost, reusability, software support, and others.

Technical concerns with scaling to ≤45 nm will inevitably include the following:

1. *Variability.* Critical dimension (CD) control and process variations will challenge both manufacturing and design until the end of the CMOS roadmap. In mature 45-nm products, local pattern and pitch dependencies in resist and etch processes will be ameliorated by restricted layout rules and improved dummy structure methodologies, but these issues are still problematic. Back-end-of-the-line (BEOL) *RC* performance will exhibit increased variability even as percentage-wise control of chemical–mechanical polishing (CMP) remains constant or improves. This is due to, for example, nonscaling of barrier thickness that magnifies *RC* impacts of CMP-induced thickness variation, and CD variation induced by etch and lithographic defocus [15].

2. *Leakage currents* (subthreshold leakage, gate leakage, junction leakage, band-to-band tunneling, etc.) remain a dominant concern into the 45-nm node. Leakage power is not only "wasted," but also compromises achievable form factor, integration density, packaging choice, reliability, and other product metrics. Further, variability impact on leakage is substantial: Subthreshold leakage scales exponentially with respect to both operating temperature* and transistor gate CD (i.e., channel length), and total leakage can vary 5- to 20-fold across chips from the same wafer lot. Mitigation of leakage through multi-V_t, MTCMOS, or higher-level design techniques incurs area overhead and design process complexity, along with added variability.†

3. *Stress and reliability.* Today's scaling of ON current and device speed is based on stress and strain engineering (embedded SiGe, stress memory, dual stress liner, etc.). Shallow trench isolation (STI) stress can affect device ON current by up to 40%. FEOL stress changes the mobility and threshold voltage of transistors; BEOL stress affects interconnect integration and reliability. By the late 45-nm node, design tools and methodologies must actively modulate

---

*Each additional °C of ambient temperature (e.g., due to global warming) will increase subthreshold leakage by roughly 5%.

†For example, random dopant fluctuations and reduced supply voltage headroom make triple-V_t strategies less viable in future nodes. In the long term, the former will be ameliorated by metal–gate and vertical (e.g., finFET or dual-gate FET) transistor architectures.

stress to improve timing (mobility change) as well as leakage (threshold voltage change).

Nano-CMOS chip design must holistically comprehend unpredictability from wear-out (NBTI, TDDB, electromigration, etc.), parametric variation (line-edge roughness, random dopant fluctuation), and transient phenomena (particle strikes, supply noise). The challenge will be to minimize the cost of on-chip monitoring, redundancy, and reconfiguration structures. Of course, many other challenges will compete for attention, ranging from next-generation lithography and consensus on radical layout restrictions to software development for the highly concurrent, multicore SOCs, to which many applications have converged. Moreover, a number of design technology gaps (e.g., electrical design for manufacturability, standards for modeling and characterization of process variability) must be addressed if the industry is to achieve a true "design for value" (maximizing profit per wafer) capability.

7.2 ELECTRICAL DESIGN FOR MANUFACTURABILITY

By the ≤45-nm mode, parametric failures—chips that fail to meet power and timing requirements—will become a dominant yield-limiting mechanism. In this context, there are many opportunities for design for manufacturability (DFM) tools that bridge chip design/implementation and process/manufacturing know-how, to deliver high-value equivalent scaling advances. Three precepts must be kept in mind for design technology eventually to enable a true design for value (maximizing profit per wafer) capability:

1. Bring design awareness into manufacturing as parametric yield-driven, feature-specific criticalities, tolerances, or biases.
2. Bring manufacturing awareness into design as optimizable objectives.
3. Maintain adoptability by working within existing design environments without requiring *any* major changes to the design flow, the design sign-off, the handoff to manufacturing, or the process flow.

Electrical DFM focuses on objectives that the chip designer or product engineer cares about: leakage power, dynamic power, timing, timing and power variability, process window, and even reliability. As illustrated in Figure 7.1, the drivers for such optimizations consist of analysis engines that comprehend a full spectrum of physical and electrical implications of manufacturing. The "knobs" or degrees of freedom to achieve the optimization goals include changes to placement, wiring, and vias, and even the dimensions of individual transistors.

Several prototype or near-production techniques exemplify how electrical DFM solutions can take into account design-specific information to improve design analyses and optimizations. These include (1) iso-dense awareness of pitch-dependent through-focus CD variation, to reduce timing guardbands and improve timing robustness [12]; (2) post-layout transistor gate-length

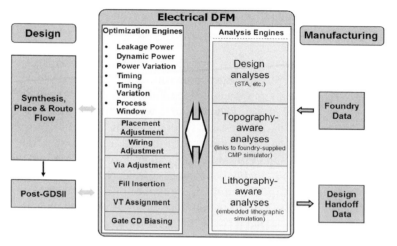

Figure 7.1 Electrical DFM with analysis engines comprehending a full spectrum of manufacturing process effects.

biasing, specified at tapeout but realized in the foundry's OPC flow, to reduce leakage and leakage variability [10]; (3) self-compensating design techniques, which minimize the inherent sensitivity of critical paths to various sources of process variation (dopant density, oxide thickness, L_{eff}, etc.; see, e.g., ref. 3), and (4) timing- and SI-driven CMP fill, which maximizes both timing robustness and post-CMP wafer uniformity [13,14]. These techniques lie along a necessary trajectory for the industry as it addresses manufacturing variability:

1. Address systematic variation ("model–predict–compensate" or "measure–model–mitigate") [12].
2. Make designs robust to variation: for example, by forcing to zero the sum of sensitivities to a given variation source (wire thickness, defocus, V_{th} shift, etc.) [3].
3. Address remaining random variations: for example, through statistical timing and leakage optimization techniques.

In Chapter 6, we have already discussed the concept of process-aware analysis of design performance and power. In the remainder of this section we elaborate on another example of electrical DFM: gate-length biasing for leakage and leakage variability mitigation.

Transistor gate-length biasing [10] takes advantage of the inverse exponential relationship between transistor gate length and subthreshold leakage current. A very small increase (positive bias) in the length of a transistor gate results in a disproportionately large reduction in subthreshold leakage. For a 65-nm process, biases of up to 3 nm per edge (6 nm total) are within the

Transistor on noncritical path: maximum bias

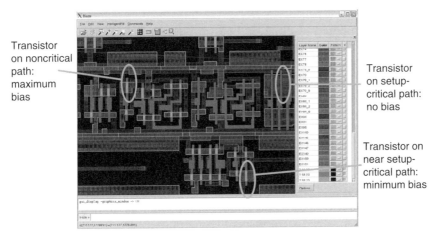

Transistor on setup-critical path: no bias

Transistor on near setup-critical path: minimum bias

Figure 7.2   Transistor gate-length biasing.

tolerances allowed by the process. Even biases of as little as 1 nm per edge can significantly reduce leakage power when applied to millions of transistors. As illustrated in Figure 7.2, because positive biasing makes a device slower, it is important to apply such biases only to transistors whose setup timing is not critical. Gate-length biasing can also reduce leakage variability: key device variabilities (e.g., of threshold voltage) are understood to be typically proportional to $(WL)^{-0.5}$, where W and L are, respectively, the device width and length; hence, increasing length will reduce these variabilities.

A recent software tool, Blaze MO from Blaze DFM, Inc. [13], implements gate-length biasing to reduce subthreshold leakage power and leakage variability. The tool performs transistor gate-length biasing in a design-driven manner by reading timing and power constraints along with golden parasitics and library models, performing internal timing and power analyses, and then selectively biasing transistors whose setup timing is not critical. In a typical design, the vast majority of cells, perhaps 80% or more, are not in any critical path. This presents a rich opportunity for leakage reduction since those non-timing critical cells are all eligible for optimization. Although any combinational and sequential cell-based logic can be biased, typical methodologies usually prohibit biasing of memories, clock trees, analog and mixed signal circuitry, or I/O cells.

It is useful to review the methodology around Blaze MO gate-length biasing in light of the precept of adoptability noted above. The tool does not change the drawn shape of the transistor gate, but rather, creates directives on a separate layer of the GDSII file that guide the optical proximity correction (OPC) tool to implement the biasing. Blaze adds annotation shapes on a separate, standard layer (defined in the foundry's design kit) to communicate the biasing directives to the OPC tool. Since the drawn layout is not affected, the optimization does not introduce design rule violations and does not affect physical verification in any way. The use of OPC to implement gate-length biasing also allows the assignment of different biases to different neighboring transistors within a single cell. As the gate-length biasing does not change the drawn-chip GDSII in any way, no placement or routing changes result, and sign-off analysis can simply be updated with existing golden parasitics. Other degrees of freedom that do not change parasitics—notably, V_{th} reassignment—can be exploited using the same tool.

Intuitively, the Blaze tool is able to make very fine grain optimization choices that are meaningless to attempt before detailed routing is fixed. For example, before placement and routing, logic synthesis can only choose between 1×, 2×, 4×, etc. cell sizing variants. After routing is completed, any resizing based on golden parasitic data will disturb both placement and routing and lead to a nonconvergent iteration. On the other hand, Blaze MO is able to make the equivalent of very small sizing moves (say, 2×, 2.1×, 2.2×, 2.4×) without disturbing either placement or routing; thus, the tool finds opportunities that when accumulated over millions of transistors lead to a substantial chip-level leakage power benefit.

7.2.1 Library Preparation Methodology

Blaze MO requires the preparation of *virtual variants* for those standard cells whose transistors it is allowed to bias. These variants are called virtual because they do not change the physical layout of the cells. Instead, they model the timing and power of the cell instance to reflect the transistor biasing. However, since the biasing is implemented by the OPC tool, the drawn layout of a cell does not change. The sole purpose of the virtual variants is to verify timing closure after the optimization is complete.

In production flows, Blaze MO is used to generate the virtual variant library automatically. As detailed by Shah et al. [11], it is useful to generate several classes of variants to achieve maximum leakage optimization. For example, a variant type (C in ref. 11) that maximally biases all transistors in a cell will achieve maximum leakage savings when there is plenty of available positive timing slack that the optimizer can "convert" into leakage reduction. On the other hand, other variant types (e.g., A, R, F in ref. 11) may be handy in contexts where there is limited timing slack, when the slack is only on rise (fall) arcs, and so on. To gain the most leverage from such "smart" variants, state dependency, transistor stacking, and drive strength must all be considered when determining the transistor-level biasing. A built-in fast SPICE-like engine is used to evaluate and optimize the timing and power impact of transistor-level changes.

As an example, consider the two-input AND gate, implemented as a two-input NAND gate with an inverter, shown in Figure 7.3. In this example, the output of the cell (Z) is 0 in three of the four states, which means that the pMOS transistor P3 is leaking in those states. P3 is typically a large device and is therefore a good candidate for aggressive positive biasing. Also, in three of

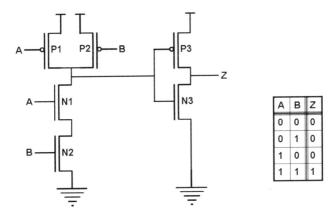

Figure 7.3   Two-input AND gate.

the four states, one or both of the nMOS transistors in the NAND gate (N1 and N2) are zero. Since these transistors are stacked in series, they leak less than they otherwise would, so there may be less benefit in biasing them. Another consideration is the drive strength of the NAND gate relative to the input loading of the inverter and user-specified maximum delay degradation; a stronger output driver can withstand greater biasing. After all virtual variants have been determined, SPICE netlists are extracted for each variant cell, and then golden characterization is used to generate timing and leakage library models (e.g., Liberty) for use during optimization. The automated flow is shown in Figure 7.4.

7.2.2   Leakage Optimization Methodology

During leakage optimization, as gate-length biasing (and optionally, V_t reassignment) is performed, timing closure is achieved automatically without manual ECOs, using a proprietary timing methodology that combines forced correlation with automated multimode/multicorner repair. As shown in Figure 7.5, an internal timing engine monitors and evaluates timing continually during optimization; however, other timers (e.g., Synopsys PrimeTime SI) may be used for golden sign-off. To ensure timing correlation, the first step is to force to Blaze tool's internal timing engine to correlate with the golden analysis. Thus, the optimization begins with a very accurate timing view of the design, and the guardbands (margins) that are typically used by designers to compensate for miscorrelation between timing tools are not needed.

During the course of the optimization, it is possible for timing analysis to diverge from golden results, and hence the Blaze MO tool may create temporary timing violations. However, an automated multimode/multicorner repair capability invokes multiple sessions of the golden timer in parallel to verify timing across all operating modes and process corners, then reverses the

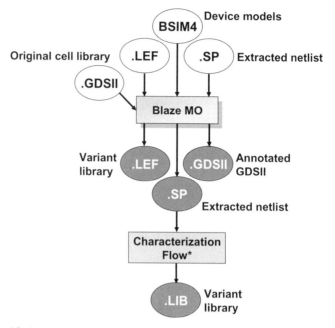

*Only inputs modified by Blaze MO shown

Figure 7.4   Virtual variant library generation and characterization flow.

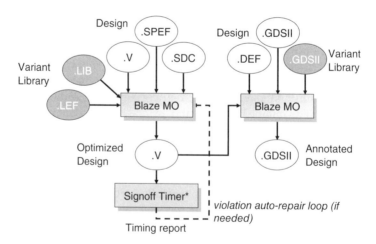

*Only inputs modified by Blaze MO shown

Figure 7.5   Leakage optimization flow.

optimization steps that created any timing violations thus found. If the optimization did not touch the clock network, as in the typical use model, optimization steps can be safely backed out without creating new violations.

7.3 CRITICALITY-AWARE DFM

As highlighted in prior chapters, designs at and below 65-nm node tend to have multiple issues that can cause lower yield or degrade device performance. Designs are no longer entirely free of these issues, but rather, require trade-off among them. What is optimum, therefore, is no longer defined as being "issue-free" but rather as "having an optimal balance among the issues." Further, as scaling continues, both the number of issues and their complexity are exploding.

As highlighted in Section 7.2, for a DFM approach to truly work, it should align itself with existing design environments, bring manufacturing awareness into design, and/or bring design awareness into manufacturing. Another approach to accomplishing DFM while incorporating these mantras beyond what was discussed in Section 7.2 is criticality aware DFM.

The obvious first step in realizing DFM is to understand the causes of manufacturing problems that relate to physical design. These potential causes of problems in physical design properties are what we define as *criticalities*. What is critical depends on a fabricator's manufacturing capability and/or a device's performance requirements. As an example of manufacturability-driven criticality, a fab may choose to reduce stepper illumination coherency to minimize the influence of high-order lens aberrations on printed images while sacrificing the finest image resolution capability at the center of the process window. This may result in reduced aerial image contrast under a defocused condition or a narrower depth of focus (DOF). An example of performance-driven criticality is when one device requires a higher operating frequency than another, therefore requiring more severe timing control and thus a higher priority in gate sizing control. In the next few subsections we discuss how criticalities are defined and used to implement DFM.

7.3.1 Definition of Criticalities

Design rules have traditionally been the bridge between fabricators and design houses. Therefore, it might seem natural to use them to define and screen the criticalities. However, there are several major deficiencies in that approach:

- Design rules are literal rules for physical design. They describe what to do or not to do, but do not define the problems themselves. Because of this, they cannot be used easily for more general DFM purposes.

- They are written in Boolean notation, which tends to be extensive and complex and provides no physical meaning. Because of that, they are extremely difficult to maintain.
- Design rules are written in geometric distances between polygon edges. But as an example, one cannot describe all the possible polygon configurations where the image contrast falls below a certain threshold. So they have a tendency to be either incomplete or overkill.
- The rules provide a yes/no judgment with nothing in between. No values are attached. This is very inconvenient, as trade-offs among conflicting rules become very difficult.

These problems can be solved by using a model-based approach for defining criticalities. Based on physical properties of the design, a model-based approach will tell how critical a pattern actually is, and how trade-offs for improvement can be made. Unlike design rules, this approach also lends itself to portability. Very few commercial tools are available in the market place, for reasons related both to technical complexity and business landscape in the EDA relationship.

In the remainder of this section we discuss the implementation of this approach using a well-known commercial tool, Takumi. In Takumi's current implementation, criticalities are defined in its unique Script language. The language provides links to physical properties such as image contrast, light intensity, and critical area analysis (CAA), each of which is calculated by a specific engine that forms a criticality-aware platform (CAP) along with the language. Within a CAP, each criticality is expressed as a yield/performance formula, and regardless of the type of criticality, the formula calculates the failure rate in parts per billion, which facilitates comparison and trade-off among various types of criticalities.

7.3.2 Applications of the Criticality-Aware Approach

The criticality-aware approach can be used in many different applications without requiring application-specific definitions. If definitions of criticalities are shared among the applications, and refer to what is "good or bad" (i.e., how critical it is), compared to what to "avoid," they can be utilized through the entire design and manufacturing flow. As a result, their accumulated benefit can be larger than the traditional point tools.

As an example, Takumi's application software, LA (for "layout analysis"), uses this information and assigns level of criticalities on every polygon edge in input layouts. The software can then be used to decide which OPC strategy to use among rule-based, model-based, or no OPC, and what tolerance setting to use edge by edge. In this way, the software can minimize the use of more costly and time-consuming model based OPC (MOPC) without compromising on device yield or device performance. Such an approach resulted in a 50% reduction of total OPC CPU time on a 90-nm production tapeout.

The same criticality definitions can be used for photo-mask inspection purposes. Mask vendors consider their job to be reproducing the mask data precisely on the mask substrates; that is, the entire mask surface must be free of defects, the polygon corners must be (close to) rectangular, the CDs must be within nanometer accuracy regardless of which polygon edges are represented in the design, and so on. Using a criticality-aware approach to mask inspection, decisions as to repairing versus ignoring can be automated. Other Takumi software incorporates image enhancement algorithms and simulation capability as well, so that it can import defect images from a mask inspection tool and predict their impact on wafer printing and provide additional guidelines for decision making. Such a system will save mask inspection and repair time, improve mask yield, and shorten mask manufacturing turn around time (TAT).

The two applications described above utilize knowledge of the criticality in order to find where less costly technologies can be used, so that the total cost and TAT can be reduced drastically. However, as design complexity and density grow, more areas become even more critical. Therefore, another application domain of the criticality-aware approach is to reduce or remove such criticalities as the next step.

In the following sections we discuss primarily a *layout optimization* solution, which focuses on trade-off and optimization among criticalities when multiple types of criticalities are present in a layout. This is a typical application for standard cell and other IP blocks. We also touch upon another solution, referred to as the hotspot fixer, which focuses only on the most critical type of hotspots in order to guarantee manufacturability.

7.3.3 Automatic Layout Optimization Flow

Figure 7.6 shows the flow for one of the most appealing applications of the criticality-aware approach: automatic optimization of the physical layout. The layout modification is able to change GDS2 data for a full layer set simultaneously, taking criticalities into account, which can relate to issues all the way from circuit performance (e.g., timing) to mask making (e.g., fracture count). This is a major step toward integrating design and manufacturing.

Automatic modification of the layout involves analysis of the input layout from a wide variety of criticalities affecting it, combining them into a single number, predicting how this number will change as the layout is changed, and providing the final yield or cost benefit with the modified layout. This task can be broken down into two major categories: analysis of input layout and modification of the layout.

7.3.3.1 *Analysis of Input Layout*

Analysis of an input layout is to check which criticalities are to be optimized. Examples of criticalities include yield-related issues such as sensitivity of the layout patterns for random defects, yield loss due to lithographic issues, such as bad printing of patterns in nonnominal printing conditions; yield loss or

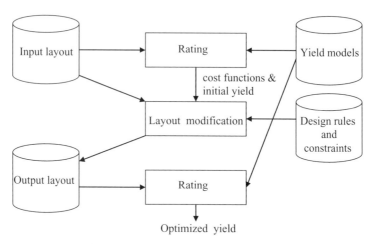

Figure 7.6 Flow of rating and optimization.

device degradation issues due to printing, such as transistor W/L control; and friendliness of the layout for OPC and mask making.

Various yield loss or cost mechanisms have various associated metrics. These various metrics must be consolidated into one before they can be used for optimization. The failure rate (FR) expressed in parts per billion is used as the common metric of the cost (yield loss) of a layout pattern. Although some yield loss mechanisms, such as random defects, are already commonly defined using failure rates, others, such as minimum on-chip clock frequency, need to be converted to a probability of failure.

Defining and calculating yield loss is often done for a layout pattern with a limited size context. Yield loss for different locations in the layout is considered to be statistically independent. This is true for random effects such as particle defects but is not true for systematic issues. For example, printing conditions that degrade a transistor at one location can degrade it equally at another location. For computational efficiency, Takumi treats the various yield loss issues at different locations as being independent. This allows prediction of FR for layout permutations with a local context, rather than recalculation of the total yield iteratively during layout modification. Furthermore, this local modeling matches the way that yield loss is commonly defined.

The total cost of the layout is given by

$$\mathrm{FR}_{\mathrm{layout}} = \sum \mathrm{FR}_{\mathrm{mechanism, pattern}} \qquad (7.1)$$

The cost functions are defined on layout patterns with small perturbations of polygon edges. The process of creating cost functions is called *rating* the layout. The total of the cost functions for a given layout gives the yield loss for that particular layout.

7.3.3.2 Actual Modification of Input Layout

Layout optimization is done by moving individual polygon edges of the shapes in the layout. There is no change in layout topology, such as in a re-route. Further changes are done for all layers simultaneously.

The driver for the change is the cost functions that are dependent on edge positions. The change is limited by design rule constraints, and possibly other constraints, such as keeping transistor W/L fixed or keeping cell boundary or pin access fixed. The key to good optimization is allowing maximum freedom in the layout (given the restriction of topology preservation), which implies that the optimizer be able to switch from one valid pattern to another, where intermediate patterns can be invalid. The result of the optimization is a modified layout, for which the layout rating can be re-run to verify the improvement in cost.

The creation of cost functions as a function of polygon edges or layout rating is best explained using the example shown in Figure 7.7. Assume that we want to calculate cost for CAA (particle defects)-related yield loss. The two metal tracks have a common run length (CRL) and are at a distance D. Now if the particle size distribution model is

$$P(S) = P_0 S^{-2} \tag{7.2}$$

where $P(S)$ is the probability (in particles per area unit) of finding a particle of size S. The probability of a short will be (approximately)

$$P(\text{short}) = \text{CRL} \int_D^\infty P(S)\partial S = \text{CRL}\frac{P_0}{D} \tag{7.3}$$

where CRL and D are the distances as defined in Figure 7.7.

This number would give the yield loss as a simple number for this layout pattern. This or a similar number would be reported by yield analysis tools. Now we need a cost function that predicts the cost (or yield loss) for small layout changes, or small perturbations of the layout. In this example the cost function would look as follows:

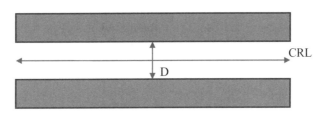

Figure 7.7 Sample layout.

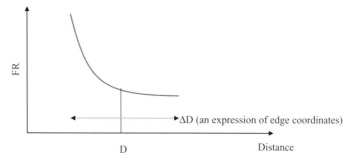

Figure 7.8 Cost function graph.

$$FR = CRL \frac{P_0}{D + \Delta D} \tag{7.4}$$

where ΔD is the change in distance between the two layout edges compared to the distance in input (D), and represents the perturbation of the layout (see Figure 7.8). In summary, the purpose of the rating functions is to calculate the cost for a given layout and to predict or model simply the change in cost if the layout changes.

7.3.4 Optimizing the Layout

7.3.4.1 Design Rule Modeling and Hotspot Fixing

To modify the source layout and change the polygon edge positions in a way that will not introduce design rule error, the design rule constraints need to be captured in a mathematical model. Design rules are not simple anymore and are often defined as a collection of allowed patterns and constraints. For example, most space constraints are defined in combination with the width or common run length of edges of associated shapes. Modeling the constraints must be such that all these combinations can be given and that the optimizer can switch from one option to another. Integer programming is used to model the combinatorial logic in this complicated constraint.

If a designer tries to move one of the polygon edges in an attempt to remove a hotspot, the effect will immediately ripple across the design in all three dimensions. Figures 7.9 and 7.10 compare pre- and post-optimized layouts, respectively, using the Takumi tool, demonstrating the significance of this problem. Figure 7.9 shows a hotspot (highlighted by the white circle) near the upper right corner. Then the automated layout optimization software fixes the hotspot, as shown in Figure 7.10. Almost every edge ended up moving by a few nanometers, some in the x-direction and others in the y-direction, even though the hotspot itself was violating the rule in the x-direction only. Furthermore, the tool rotated some polygons by $90°$ to accommodate the constant footprint constraint.

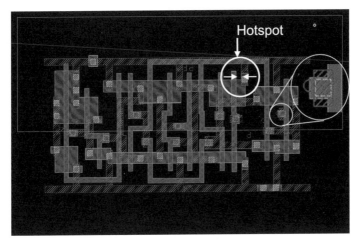

Figure 7.9   Prior to hotspot optimization.

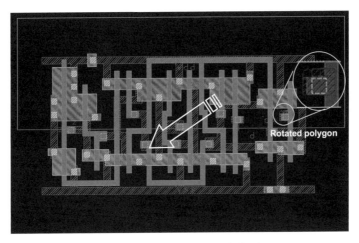

Figure 7.10   Post hotspot optimization.

Considering that modifications need to be made on multiple layers simultaneously and that any edge movement may introduce new hotspots, or worsen existing ones, and that a single standard cell might have multiple hotspots, it is almost impractical to optimize designs manually. Use of automated layout optimization software based on the criticality-aware approach can provide designers with a consistent and quantitative view on priorities of the issues.

7.3.4.2 Cost Functions and Trade-off

As discussed earlier, the driver for change in the layout is the set of cost functions defined on edge positions. Since we assume that there are many competing yield- and performance-driven mechanisms, the trade-off among them becomes very important. Figure 7.11 is an illustrative example of where we assume two distinct yield issues:

1. Transistor device performance changing due to its effective W/L change, due to diffusion rounding, which affects both drive current and capacitance. The yield loss, or failure rate (FR), is dependent on distance A.
2. Possible shorts due to random particle defects during processing. Here FR is dependent on distance B.

If we assume that the left and right sides of the layout are fixed for some reason (e.g., limited cell size), the distances A and B, and therefore the FRs for the two yield hotspots, compete with each other (Figure 7.12). Note that distances A and B are plotted in opposite polarities since they are competing.

Using this cost-based approach to trade off between various yield and performance loss mechanisms can result in very fine tuned tweaks to the layout. This is clearly more precise and flexible than is a more common approach based on preferred distances in design rule sets, also known as *recommended rules* or *DfM rules*. Also, since the generation of cost functions can be model based, the cost functions can take many pattern considerations into account, which is especially important for complicated (very pattern dependent) printing-related yield issues that are getting important for process nodes of 65 nm and below.

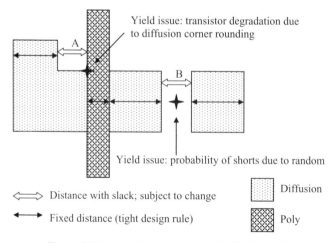

Figure 7.11 Layout example for criticality trade-off.

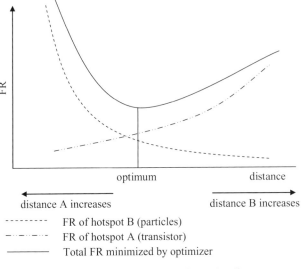

Figure 7.12   Cost function for trade-off.

7.3.5   Examples of Cost Function Trade-offs

7.3.5.1   *Random Defect Sensitivity: Metal Opens Versus Shorts*

In Section 7.3.3.2 we explained yield loss due to random defects to explain the concept of cost functions. The function is fairly simple, with a decreasing FR when width (for particles that cause opens in connections) or space (for particles that cause shorts) increases. We now apply that simple approach to optimizing the critical area for metal shorts versus opens, with the pre- and post-optimization results shown in Figure 7.13. To understand the optimization process, Figure 7.14 shows the FR for the various layouts when trade-offs between metal1-short and metal1-open is done. If P0 for opens is held constant (at some noncalibrated value) and P0 for shorts increases, more focus is given on shorts when comparing layouts. The tool is able to degrade FRshort (at P0-short = 25) in order to make a big improvement for FRopen. But if P0-short is increased to higher values, FRshort gains importance at the expense of a somewhat reduced gain in FRopen.

7.3.5.2   *Image Contrast Versus Contact Hole Doubling*

A higher contrast in the projected aerial image of the mask on the wafer eases dose and focus tolerance in printing, creating a bigger "process window" and resulting in a higher yield. However, increasing contrast might mean sacrificing other criticalities.

Figure 7.15 shows the trade-off between contrast improvement and doubling of contact holes. The layouts and corresponding images from left to right are the original, one with higher priority on doubling, and one with higher

(a)

(b)

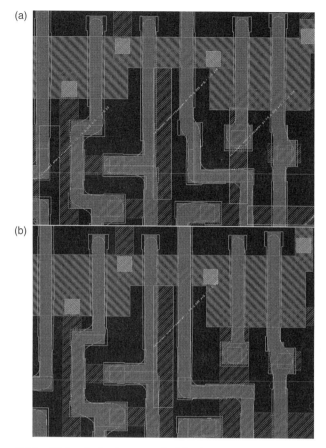

Figure 7.13   CAA optimization example (a) prior to and (b) after optimization.

priority on contrast improvement, respectively. The oval shapes on the layouts highlight regions where contacts are or are not doubled, and those on the aerial images show regions of poor or improved contrast. As expected, when doubling has higher priority (Figure 7.15b), more contact holes were doubled but at the cost of contrast degradation. Note that even then, the fatal contrast degradations identified by the ovals on the original (Figure 7.15a) aerial image were removed. In these areas, severe contrast degradation pushed up their local FR even higher than heavily weighted single contact holes. For the layout or image in Figure 7.15c, only one contact hole was doubled, but improvement in contrast is apparent. Trade-offs like this were made possible by rating different types of criticalities on the same metric.

7.3.5.3 Transistor Gate Variability Optimization
Transistor variability due to variation in poly/diffusion patterns with varying processing conditions across the process window, as well as misalignment

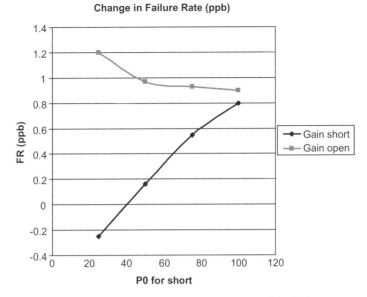

Figure 7.14 Metal open versus short trade-off analysis.

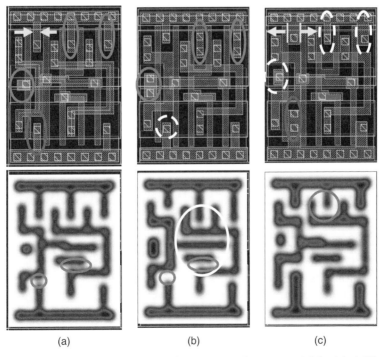

Figure 7.15 Contact hole doubling versus image contrast improvement: (a) original; (b) priority on doubling; (c) priority on contrast.

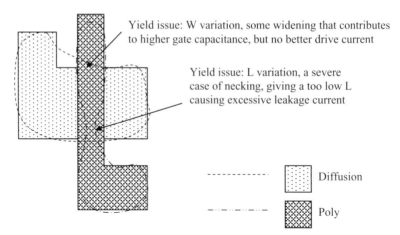

Yield issue: W variation, some widening that contributes to higher gate capacitance, but no better drive current

Yield issue: L variation, a severe case of necking, giving a too low L causing excessive leakage current

- - - - - - - - - - - ⬚ Diffusion

— · · — · · — ▨ Poly

Figure 7.16 Transistor variability.

between the two layers, is a well-known cause of speed, leakage, and power problems at sub 90-nm nodes. Since printing effects can be complicated, the model-based approach has a definitive advantage over rule-based approach. Rating functions in Takumi software handle this by creating cost functions that makes gate size less susceptible to the process variation. Variability is expressed in terms of W_{min}, $W_{average}$, W_{max}, L_{min}, $L_{average}$, and L_{max}, and transistor FR is based on design objectives such as speed or leakage. For example, if leakage is the main concern, as shown in Figure 7.16 for a transistor with contours varying significantly from its layout, it can be checked using L_{min}. A more advanced approach uses models for drive current, leakage current, and gate capacitance (instead of simple values of W and L), possibly in combination with circuit simulation.

Figure 7.17 shows wafer SEM images pre- and post-optimization for transistor gate width variability. This example illustrates the case where the transistor width varies from the target design due to corner rounding in the diffusion layer and misalignment.

7.3.6 Criticality-Aware Layout Optimization Summary

The industry is starting to see the emergence of various layout rating tools. These tools are meant to allow designers to be aware of process- and manufacturability-related problems in their designs. However, there are some challenges that these tools need to solve:

1. Many different criticalities coexist in a design. Without knowing the relative importance among them, fixing one may degrade others. Therefore, tools that can comprehend various criticalities in a single metric are needed.

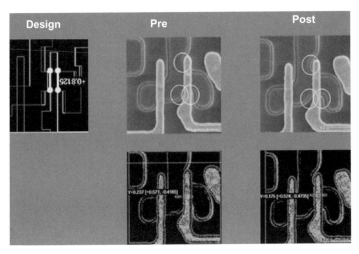

Figure 7.17 Wafer SEM images compares pre- and post-optimization for transistor size variability.

2. Designs are too complex to fix manually. Without knowing how to fix them, being aware of the existence of problems does not really help. This points to the need of an automated tool, which also satisfies the foregoing criteria.

3. A fundamental problem is that there is no easy quantitative way to enforce a certain flavor or priority of issues. Therefore, it is important for the metric mentioned in point 1 to be quantitative.

4. The final challenge, not really discussed in this book because of its nontechnical nature, is the free and timely flow of critical information from the fabricators to the EDA tool vendors or design houses, which is further compounded by criticalities evolving as the process matures. This also points to the fact that solutions in the DFM space will differ by application and market requirements in addition to technical challenges.

The industry's response to some of these challenges has been the introduction of recommended rule decks and restricted design rules (RDRs), which allows for better optimization of the production process to that layout style. The criticality-aware approach to layout optimization using some of the techniques discussed in this section provides a complementary solution to challenges 1 through 3 and enables aggressive density and performance targets.

7.4 ON GUARDBANDS, STATISTICS, AND GAPS

In an ideal future at ≤45 nm, chip designers' power and timing requirements will be used to tailor the manufacturing line for each transistor of each design—without changes or adjustments to the fabricating equipment. One can dream that designers will be able to take advantage of available entitlement or process margin so that the process delivers significantly improved parametric quality of the silicon product, and that designs can be driven to a sweet spot for the process, just as the process is today driven to a sweet spot for the design. The reality is that as physical implementation tools and manufacturing handoffs evolve for 45- and 32-nm nano-CMOS production, the industry must maintain proper expectations regarding the three key issues that we discuss in this section: guardbands and margin reduction, the prospective benefits of statistical design, and the production readiness of process-aware design flows.

7.4.1 Time Constants and the Inevitability of Guardbanding

In the coevolution of silicon technology and silicon products, basic time constants range over nearly three orders of magnitude:

- *O(years):* technology development; application market definition; architectural and front-end design
- *O(quarters):* SPICE model revision; design rule manual revision; library/ IP design; library/IP silicon qualification
- *O(months):* library and IP modeling and characterization; RTL-to-GDS implementation; reliability qualification
- *O(weeks):* fabrication latency (wafer start to wafer out); cycle of yield learning; design re-spin; OPC and mask flow
- *O(days):* process tweak; design ECO

Also, a number of precedence and practical constraints apply; for example, the SPICE model version 1.0 must be fixed before libraries and IPs are fixed; libraries and IPs must be fixed before RTL-to-GDS physical implementation can occur; only limited changes to the SPICE model are permissible after a certain volume of library/IP/chip design activity has taken place; and so on. Furthermore, even though a design change can be made in *O*(days), the latency for assessment in silicon must span the OPC, mask, and foundry flows. Two critical observations follow: (1) the process must continue to adapt to the design, as it does today; and (2) the ability of the foundry to tweak the process even when SPICE and RCX models are fixed implies that significant guardbanding (i.e., overdesign) is inherent in today's design–foundry relationship. This is a fundamental asymmetry between process and design.

In this light, R&D goals for ≤45 nm include:

1. Improved quantification of guardbanding costs and benefits, such as the potential trade-off of guardband reduction and parametric yield loss for faster design closure and improved random defect yield [2]

2. Achieving design robustness to variabilities (cf. self-compensation [3]), which can include intentional (on the part of foundry process engineers or product and yield engineers) model-to-silicon miscorrelations
3. New methods for rapid process adaptation to design (e.g., through improved understanding of how parametric tests in the fab map through SPICE models to design sign-off constraints)*

With respect to the first goal, recent work [2] provides the first-ever quantifications of the impact of modeling guardband reductions on the outcomes of the chip implementation (synthesis, placement, routing) flow in 90- and 65-nm technologies. The results reported show clear potential design quality and turnaround time benefits of model guardband reduction: for example, an average of 13% standard-cell area reduction and 12% routed wirelength reduction as a consequence of the 40% FEOL timing guardband reduction that IBM reported from an iso-dense pitch-aware ("variational") timing analysis methodology [12]. Such data suggest further that there is a potential product "sweet spot" with respect to intentional guardband reduction (with respect to foundry-supplied default values): Reduced guardband sacrifices parametric yield, but wins on raw die per wafer due to the reduced implementation area. (Design cycle time reduction is a separate win.) Figure 7.18 shows level curves of good die per wafer plotted against guardband reduction percentage (i.e., implied parametric yield loss) and area reduction (i.e., raw die per wafer gain)

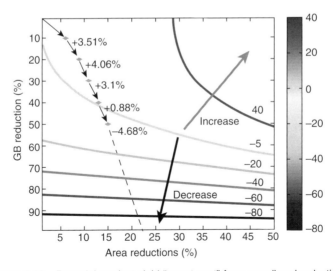

Figure 7.18 Potential product yield "sweet spot" from guardband reduction.

---

*Moore's law value scaling is roughly 1% per week. Even if margin in the form of guardbanded process models is inevitable, big wins are possible from, for example, design-directed yield learning.

[2]. The plot suggests a potential gain of up to 4% good die per wafer at the sweet spot of guardband reduction.

7.4.2 Practicality and the Value of Statistical Design

Despite a great deal of industry attention, deep challenges remain with respect to modeling, characterization, and mitigation of manufacturing variations. Statistical analyses and optimizations have been conceived rapidly but lack consensus on enablement for production flows. We note that interdie [die to die (DTD)] variations are easier to model in the manufacturing process as well as in statistical design techniques. On the other hand, intradie [within-die (WID)] variations have a significant component that is systematic and pattern dependent. With many distinct variability phenomena and length scales in play—from wafer radial bias, reticle bending, lens aberration, CMP planarization length, flare, and so on, down to mask CD and mask error enhancement factor (MEEF), etch, and lithography—modeling of spatial and pattern-dependent correlations is a key challenge to deployment of statistical design flows at ≤45 nm.[*]

An even more basic issue is whether sufficient return on investment (ROI) can be shown for statistical design approaches (e.g., Burns et al. [4] showed limited impact of statistical power optimization). Intuitively, statistical design will have only limited impact with respect to "sum" objectives such as power, as opposed to "max" objectives such as timing. Impact will also be limited for phenomena such as subthreshold leakage, which are exponential in most parameters (L_{eff}, temperature, etc.) and for which sensitivities and variances track nominal values. When statistical optimization drives the design to essentially the same point as deterministic design, as appears to be the case for timing-driven design when spatial correlations are considered for example, the potential differentiated benefits lie mainly in yield prediction.[†]

7.4.3 Gaps in Nascent Flows

As noted in Chapter 6, electrical models of nonrectangular devices and interconnects have enjoyed recent interest as a means of assessing impact of lithographic and CMP errors on power and performance. Such models enable

---

[*]Variability modeling, from easiest to hardest, spans (1) systematic WID (e.g., pattern dependence of lithography and CMP), (2) random DTD (SSTA), (3) random WID, (4) correlated random WID, and (5) systematic DTD. For example, nascent approaches to model 4 still gloss over the question of how to model the fact that BUF = INV + INV or AND = NAND + INV [20]. Because process learning and design rules also address model 1, and because effects of other variation types are substantial, the focus to date on models 1 and 2 may in hindsight constitute "looking under the lamppost."

[†]Business frameworks for statistical design remain unclear. For example, it seems impractical for foundries to deliver the exact process statistics to which a design was optimized. Or if the process evolves during the course of a given design project, optimizations targeted to early process statistics could end up being harmful in the matured process.

process-aware analysis or model-based sign-off, which informs sign-off analyses (e.g., RCX, delay calculation, STA) with results of physical simulations of systematic (deterministic) pattern-dependent variations. Gupta et al. [7] model nonrectangular device channels with comprehension of narrow-width effect and resulting variation of V_t across the gate width. Figure 7.19 illustrates definitions of gate width (W) and edge width (w), with edge regions shown in blue. These concepts are used in Figure 7.20, which shows variation of V_t along the width of the device for different gate widths. Edge width is the width of the region near the boundary between poly and diffusion. Models analogous to those described by Gupta et al. [7] are necessary to capture the electrical impacts of line-edge roughness (LER) and line-width roughness (LWR), which are probably significant contributors to interdevice variation in ≤45-nm nodes. Particularly for analog and mixed-signal circuits, edge roughness can affect matching and delay requirements. While today's design methodologies still model the effects of LER/LWR as random, future electrical DFM flows

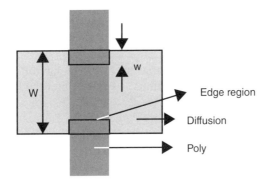

Figure 7.19   Definitions of gate width (W) and edge width (w).

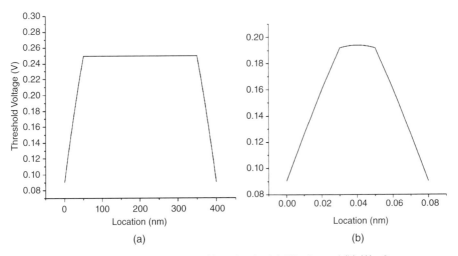

Figure 7.20 V_t as a function of location for (a) $W > 2w$ and (b) $W < 2w$.

require more accurate, model-based accounting for (and bounds on) delay, capacitance, and power variation with LER/LWR.

A critical extension for ≤45 nm is the modeling of diffusion rounding. Poly CD is increasingly well controlled in modern processes, in part due to layout restrictions. However, diffusion is still very irregular, resulting in imperfect printing. Although diffusion patterns have larger CD than poly patterns, corners and jogs are more prevalent in smaller technologies, and process windows are small due to significant corner rounding with defocus. Hence, poly gates placed in close proximity to diffusion edges are more likely to demonstrate larger performance variation than are those away from the edges. Such diffusion patterning issues are likely to become more significant as the average gate width scales down with each technology generation.

Simple models that account for diffusion rounding by adjusting gate width (cf. refs. 6 and 8) have unacceptable errors in I_{off} predictions. Moreover, source-side diffusion rounding and drain-side diffusion rounding behave very differently from an electrical perspective, which strongly suggests that diffusion-rounding modeling must be performed in a design context-aware manner (cf. recent results reported by Gupta et al. [33]). Future flows require new modeling techniques to determine equivalent L and W given both poly and diffusion patterning imperfections.

More generally, process-aware analysis flows for sign-off at ≤45 nm require industry consensus on "deconvolutions" to solve potential blurring of today's analysis capabilities.

- In the FEOL, silicon-calibrated LPE (layout parasitic extraction) rule decks potentially double-count lithography process contour effects (LPC).
- In the BEOL, silicon-calibrated RCX (parasitic extraction) tools potentially double-count post-CMP wafer topography effects.

A final blocker in the electrical DFM roadmap is industry consensus on treatment of sign-off analysis corners in the presence of process-aware electrical model corrections. For example, if a simulator indicates that a device's nominal L_{eff} should be changed from 40 nm to 38 nm due to pattern-specific lithographic variations, it is unclear today how to change the qualified BC/WC SPICE corners for the device. Related issues include (tractable) standardized silicon qualification of process-aware analysis and enablement of full-chip sign-off analyses in cell-based methodologies.

7.5 OPPORTUNISTIC MINDSETS

With industry demand for new design for manufacturability capabilities, the physical verification platform has taken over some functionality (e.g., via doubling, wire spreading and fattening, dummy fill insertion) from the upstream physical implementation [synthesis, place, and route (SP&R)] plat-form. This has led to two distinct mindsets today.

Upstream "prevention," in the sense of correct by construction, focuses on design rules, library design, and manufacturing awareness in the SP&R flow. This is costly in terms of buffer dimensions, modeling margins, and other guardbands; examples include guaranteed standard-cell composability for alternating-aperture PSM, and dummy gate poly at boundaries of standard cells. In some cases, the cost increases with scaling: for example, as the number of pitches in the stepper wavelength increases, the area penalty of buffer dimensions grows rapidly. Solutions can also be too onerous for adoption: for example, a one-pitch-one-orientation poly layout is highly manufacturable but incurs unacceptable area penalties. Finally, prevention may mean attempting to solve problems with too little information, such as trying to solve litho-graphic hotspots that have timing impact during P&R, before golden wire parasitics and signal integrity reports are in hand.

Downstream "cure," in the sense of "construct by correction," is often per-formed at the post-layout handoff between design and manufacturing. This can suffer from shape centricity and loss of design information as well as sepa-ration from implementation flows. Without a grasp of electrical and perfor-mance constraints, timing slacks, slew criticality, and so on, such flows cannot easily determine whether manufacturing nonidealities actually harm the design or how to mitigate such nonidealities to maximize parametric yield. Moreover, any loop back to ECO P&R and re-sign-off is costly, since it has disturbed the "golden" state of the design (with no guarantees of convergence) and affects the tapeout schedule directly.

Where prevention can address manufacturability too early for best results, cure often comes too late in the flow. It is possible simply to bolt manufactur-ing verification and SP&R tools together, but this is not optimal. Rather, optimizations should reach up into the implementation flow to introduce cor-rections at appropriate times. The mantra for ≤45 nm might be: (1) make opportunistic changes that can only help (i.e., "do no harm"), and (2) make changes at appropriate junctures, when enough information is available, and before doing work that will be thrown away. Optimizations should reach up into the implementation flow to introduce corrections at the appropriate times (e.g., the best way to correct lithographic hotspots on poly is after detailed placement and before routing). In the remainder of this section we provide two examples of such opportunism.

7.5.1 The CORR Methodology

With respect to the "mantra" above, the appropriate juncture for correction of poly-layer lithographic hotspots is after detailed placement, but before routing. This is the fundamental insight of the CORR methodology [7], which improves the lithographic process window by removing forbidden pitches for subresolution assist feature (SRAF) and etch dummy insertion.

Adoption of off-axis illumination (OAI) and subresolution assist feature techniques to enhance resolution at specific pitches worsens the printability of patterns at other pitches. These pitches are called *forbidden pitches* because of their lower printability, and designers should avoid such pitches in the layout. Forbidden pitches consist of horizontal and vertical forbidden pitches, depending on whether they are caused by interactions of poly geometries in the same cell row or in different cell rows, respectively. The resulting *forbidden pitch problem* for the manufacturing-critical poly layer must be solved before detailed routing. The *SRAF OPC* technique combines pattern biasing with assist feature insertion to compensate for the deficiencies of bias OPC. The SRAFs are placed adjacent to primary patterns such that a relatively isolated primary line behaves more like a dense line. However, the SRAF technique places more constraints on the spacing between patterns. SRAFs can be added whenever a poly line is sufficiently isolated, but certain minimum assist-to-poly and assist-to-assist spacings are required to prevent SRAFs from printing in the space. The forbidden pitch rule is determined based on CD tolerance and worst defocus level, which, in turn, are dependent on requirements of device performance and yield. SRAF OPC restores printing when there is enough room for one scattering bar. However, larger pitches are forbidden until there is enough room for two scattering bars, as shown in Figure 7.21. Finally, a set of forbidden pitches can be extracted.

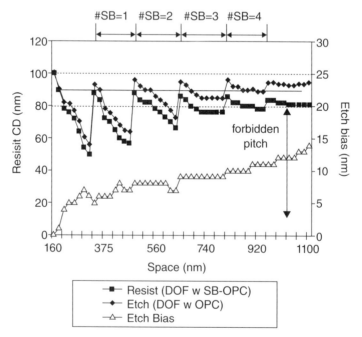

Figure 7.21 Through-pitch proximity plots and etch skew for 90-nm technology: worst defocus with SRAF OPC and worst defocus with etch OPC (left *Y*-axis), along with etch bias (right *Y*-axis), are shown.

On the other hand, the insertion of etch dummy features has been introduced to reduce the CD difference between resist and etch processes for technology nodes 90 nm and below. In dry etch processes such as plasma, ion, and reactive ion etch (RIE), different consumptions of etchants with different pattern density lead to etch skew between dense and isolated patterns. For example, all available etchants in areas with low density are consumed rapidly, and thus the etch rate then drops off significantly. In areas with a high density of patterns, the etchants are not consumed as quickly. As a result, the proximity behavior of the photo process differs from that of an etch process (Figure 7.21). In general, the etch skew of two processes increases as the pitch increases. However, etch dummy rules conflict with SRAF insertion because each technique requires specific design rules.

Since placement of cells can create forbidden pitch violations of the resist process and increase etch skew, the placer must also generate etch dummy-correct placement. This can be solved by assist-feature correctness (AFCORR) and etch-dummy correctness (EtchCORR), which are both techniques for intelligent white-space management for assist-corrected etch dummy-corrected placements. This optimization is achieved by a dynamic programming algorithm that "jiggles" an existing detailed placement to exploit white space while curing any forbidden or weak pitches. The AFCORR and EtchCORR variants are timing and wire-length preserving, in that they can constrain the movement allowed in each cell instance according to timing criticality.

Kahng et al. [9] describe pitch-specific CD variations through OPC and lithographic process windows that are mapped to systematic leakage variations of individual devices. Figure 7.22 illustrates how detailed placement choices (cell ordering, site choice, mirroring) can affect pitches of boundary gates. LeakageCORR then optimizes leakage in a timing- and wire-length-preserving manner. The CORR concept extends to a variety of objectives at the design–manufacturing interface. Examples include phase-shift conflict resolution, lithographic hotspot removal, timing and leakage improvement,

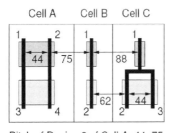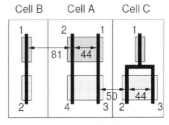

Pitch of Device 2 of Cell A: 44, 75
Pitch of Device 2 of Cell C: 62, 44
Pitch of Device 1 of Cell B: 75, 88

Pitch of Device 2 of Cell A: 81, 44
Pitch of Device 2 of Cell C: 50, 44
Pitch of Device 1 of Cell B: –, 81

Figure 7.22 Leakage reduction through optimization of boundary poly pitches in standard-cell detailed placement.

improvement of recommended rule coverage, proportion of cell-based OPC use (discussed next), and stress exploitation.

7.5.2 Auxiliary Pattern for Cell-Based OPC

The *auxiliary pattern* (AP) methodology of Kahng et al. [9] is motivated by unacceptable scaling of model-based OPC (MBOPC), which is a major bottleneck in the turnaround time of IC data preparation and manufacturing. To address the OPC run-time issue, the cell-based OPC (COPC) approach has been studied by, for example, Gupta et al. [18] and Wang et al. [19]. COPC runs OPC once per each cell definition (i.e., per cell *master*) rather than once per unique instantiation of each cell (i.e., per cell *instance*). Thus, in the COPC approach, master cell layouts in the standard cell library are corrected before the placement step, and then placement and routing steps of IC design flow are completed with the corrected master cells; this achieves significant OPC run-time reduction over MBOPC, which is performed at the full-chip layout level for every design that uses the cells. Unfortunately, optical proximity effects in lithography cause interaction between layout pattern geometries. Since the neighboring environment of a cell in a full-chip layout is different from the environment of an isolated cell, the COPC solution can be incorrect when instantiated in a full-chip layout, and there can be significant CD discrepancy between COPC and MBOPC solutions.

The AP technique of Kahng and Park [16] *opportunistically* shields poly patterns near the cell outline from the proximity effect of neighboring cells. Auxiliary patterns inserted at the cell reduce the discrepancy between isolated and layout-context OPC results for critical CDs of boundary poly features. This allows the substitution of an OPC'd cell with APs directly into the layout[*]; COPC with AP then achieves the same printability as MBOPC, but with greatly reduced OPC run time. Here, opportunism arises in two forms:

1. If the layout context of a standard-cell instance has room to substitute an AP version for the non-AP version, this should always be done, since it reduces OPC cost without affecting OPC quality. Otherwise, if there is no room for AP insertion, we leave the layout as is, and are no worse off than before.
2. The placement of cells in a given standard-cell block might not permit insertion of APs between certain neighboring cell instances. To maximize AP insertion in such cases, the detailed placement can be perturbed using AP-CORR. In other words, we create our own opportunities: an

---

[*]As detailed by Kahng and Park [16], APs consist of vertical (V-AP) and/or horizontal (H-AP) nonfunctional (dummy) poly lines. V-AP features are located within the same cell row and print on the wafer. H-AP features are located in the overlap region between cell rows; their width is comparable to that of subresolution assist features and hence they do not print on the wafer.

efficient, timing-aware dynamic programming code can maximize possible substitutions of AP cell versions and hence the run-time benefits of COPC.

The resulting flow, including AP-CORR, is shown in Figure 7.23. A standard-cell layout is input to an AP generation step and then to an SRAF insertion step. The resulting layout is input to OPC insertion step, which results in a set of OPCed standard-cell layouts corresponding to the master cells. These OPCed cell layouts will be instantiated within the final layout according to the results of post-placement optimization. The AP-correct placement takes the OPCed standard-cell layout as an input. A final cell-based OPC layout is generated from the modified AP-correct placement and the OPCed standard-cell layouts.

We may compare the average CD difference of devices near the cell outline for three cases of COPC with MBOPC. The average CD differences for COPC with no placement optimization, COPC with placement optimization, and COPC with placement optimization and AP over MBOPC are 7.2, 2.5, and 1 nm, respectively. Figure 7.24 shows the actual layouts with various OPC methods. The CD of the COPC with AP can match that of MBOPC within 1%.

Apart from run-time improvement, AP-based OPC benefits process-aware sign-off. Full-chip lithographic simulation is implicit in such a methodology, since two instances of the same standard-cell master can print differently due to context-dependent OPC and lithographic variations. Since an AP version has a predetermined OPC solution and an aerial image in lithographic simulation, the run time of process-aware sign-off can be reduced substantially without loss of accuracy [21].

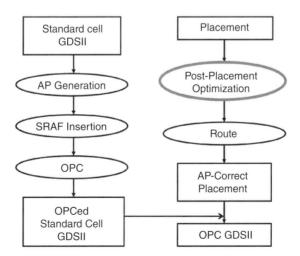

Figure 7.23 System for AP generation and placement perturbation of layout objects.

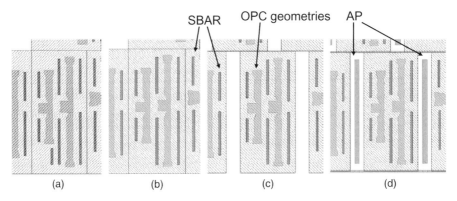

Figure 7.24 Layouts with various OPC methods: (a) MBOPC; (b) COPC with no placement optimization; (c) COPC with placement optimization; (d) COPC with placement optimization and AP.

7.6 FUTURES AT ≤45 NM

In this section we present a sampling of "futures" for nano-CMOS design to address the key challenges described above—variability, power, design productivity, reliability, and so on—with an emphasis on DFM opportunities. We begin, however, with a list of necessary advances in core implementation (place-and-route) tools. The industry cannot afford to leave these gaps unsolved, even as it turns to address emerging challenges at the interface to manufacturing.

7.6.1 Glue Technologies in Place-and-Route Tools

As defined by Smith [22], a *power user*:

- Has a CAD group that develops internal design tools
- Is designing ICs on the latest two process nodes
- Updates the design tool flow at the end of each project
- Uses tools from EDA startups
- Spends at least 33% more on CAD than does a mainstream user
- Has designer productivity in line with an ITRS design cost chart [14]

By contrast, an *upper mainstream user*:

- Has a CAD team but avoids internal tool development when possible
- Designs ICs on processes that have been in production for ≥2 years
- Tends to standardize its tool flow around 1+ major EDA vendors' tools
- Has little or no use of startup tools
- Spends less than 6% of its R&D budget on CAD

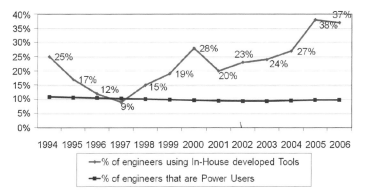

Figure 7.25   Trajectories of power user population, and use of in-house-developed EDA tools. (Courtesy of Gary Smith, EDA.)

- Has designer productivity that lags the ITRS design cost chart by ≥3 years

According to Smith [22], the key to cost control is design productivity. A power user designs the same IC as an upper mainstream user, but at much lower cost. Thus, the power user either prices its competition out of the market or designs a chip with more features in the same amount of time. The catch is that EDA vendors are not supplying the tools necessary for today's designs. Hence, a large percentage of upper mainstream design groups are being forced to develop tools internally (Figure 7.25; note that not all internal tool users satisfy the power user criteria). The resulting trajectory choice [22]: An upper mainstream user must evolve into a power user or else "end up doing RTL handoff." With this preamble, future differentiated, "private label," or internal physical design capabilities take on added significance.

7.6.1.1 Placement Opportunities

Any near-term differentiating capability for the nano-CMOS placement phase of design will address a subset of the following.

1. Clock-to-data spatial management, and more generally, exploitation of correlated variation.
2. Timing path monotonicity. There are many indications that at least one full process node of timing is left on the table by today's SP&R and optimization flows.
3. Demands from 2.5- and three-dimensional integration, including through-silicon via placement, macro and thermal via placement, and other thermal, coefficient of thermal expansion (CTE), and reliability-driven optimizations.
4. Pitch and lithographic hotspot management, invoking methodologies such as the CORR approach discussed earlier, and fast hotspot filtering.

5. Stress mitigation and exploitation.

6. Improved ECO placement—a fix-all knob that addresses timing, SI, power, density, and so on—while remaining closely tied to ECO routing.

With respect to timing path monotonicity, Figure 7.26 traces a typical critical path in a 90-nm implementation of a JPEG encoder. Delay on the critical path is 2.796 ns; delay of the path without an interconnect would be 2.084 ns, which is a difference of 712 ps, or about 25%. Figure 7.27 shows path delays with and without an interconnect for the 5000 most critical paths in the same block implementation. Data such as this, along with ASIC versus custom studies by Chinnery and Keutzer [34,35] and others, suggest that considerable performance—at least a full technology node—is being left on the table by today's chip implementation flows.

7.6.1.2 Routing Opportunities

In a 2003 invited paper on the future of routers [23], a top-10 list of objectives for the industry began with the following items.

1. Sensible unifications to co-optimize global signaling, manufacturability enhancement, and clock/test/power distribution

2. Fundamental new combinatorial optimization technologies (and possibly geometry engines) for future constraint-dominated layout regimes

3. New decomposition schemes for physical design

4. Global routing that is truly path-timing aware, truly combinatorial, and able to invoke "atomistic" interconnect synthesis

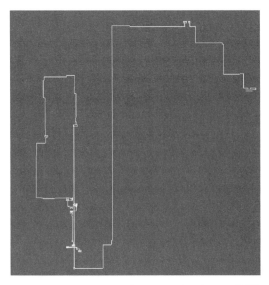

Figure 7.26 Nonmonotonicity of layout for a critical path in a JPEG encoder block.

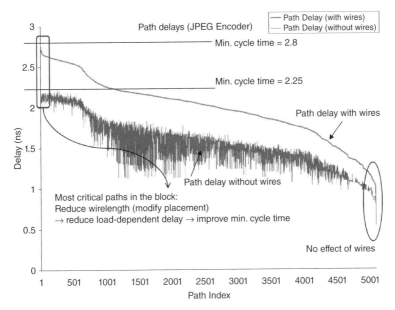

Figure 7.27 Placement-induced delays for the top 5000 critical paths in a small 90-nm block.

5. In-context layout synthesis that maximizes process window while meeting electrical (functional) specifications

Arguably, these are still at the top of today's priority list (see also ref. 24). To highlight the significance of item 3: a DAC-2006 work [25] demonstrates that a simple ECO routing flow can cure "insane" topologies of timing-critical nets so as to improve clock frequency by approximately 5%. Here, an insane topology is one where the rank order of sinks in terms of their slacks or required arrival times is not well correlated with their rank order in terms of source-to-sink delay. Figure 7.28 shows rank correlation of sink slacks versus source-to-sink delays for critical nets in a purportedly timing-optimized block. It is evident that the correlations are not uniformly positive—let alone equal to 1—as we would hope.*

Opportunities for differentiated routing capability also include the following.

- Support for double-patterning lithography (see below), ranging from layout decomposition to autofixing of coloring conflicts, etc. (see below)
- Support for performance-, variability-, and reliability-driven interconnect redundancy, including 100% redundant vias

---

*Following Alpert et al. [25], virtual pin and subnet constructs were used to force the ECO routing to deliver specific timing-driven Steiner topologies.

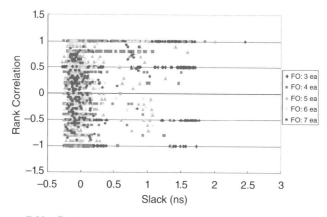

Figure 7.28 Rank correlation of sink slacks versus source-to-sink delays.

• Overhaul of global and detailed routers for restricted pitches, simultaneous performance-aware routing and fill, support for rich library design, and ownership of swapping and final placement
• Overhaul of clock distribution to support nontree network synthesis and adaptive link insertion for variation robustness
• Cleanup of old problems, such as combinatorial timing- and signal integrity–driven routing; and assignment and ownership of pins, buffering resources, and wiring planes in hierarchical design

7.6.2 Double-Patterning Lithography

Double-patterning lithography (DPL) involves partitioning dense circuit patterns into two separate masking steps, so that decreased pattern density can improve resolution and depth of focus (DOF). DPL is a candidate mainstream technology for 32-nm lithography [26]. A key problem in DPL is the decomposition of layout for multiple exposure steps. This recalls strong (alternating-aperture) PSM coloring formulations, along with automatic phase conflict detection and resolution methods [5] (e.g., Kahng et al. [27] gave one of the earliest automated and optimal compaction-based phase conflict resolution techniques). With DPL layout decomposition, two features must be assigned opposite colors if their spacing is less than the minimum color spacing. Fundamental issues for DPL are (1) generation of excess line ends, which cause yield loss due to overlay error in double exposure, as well as line-end shortening under defocus, and (2) resulting requirements for tight overlay control, possibly beyond currently envisioned capabilities. The EDA industry must rapidly bring to market tools for layout perturbation and layout decomposition to minimize the number of line ends created, and for the introduction of layout redundancy that reduces functional failures due to line-end shortening.

Lithographic hotspot finding and fixing with overlay error simulation is another enabler of DPL. Detailed design-stage DPL challenges are as follows.

7.6.2.1 *Layout Decomposition*

As noted above, in DPL layout decomposition, two features must be assigned opposite colors if their spacing is less than the minimum color spacing. In such cases we must split one feature into two parts. It is necessary to develop polygon splitting methods that address the challenges of proximity effect, overlay, and so on. Figure 7.29 shows a typical layout decomposition and the image after the layout that has been reconstituted after patterning.

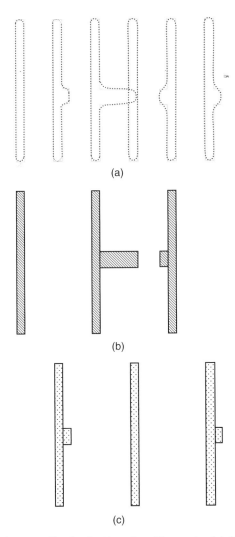

Figure 7.29 Layout decomposition for double-pattern lithography: (a) design intent and reconstituted contour; (b) first decomposed pattern; (c) second decomposed pattern.

Shortening and Rounding Awareness   As a consequence of cutting polygons, new line-end structures are introduced. In an imaging system that operates with 193-nm illumination at 32-nm half-pitch and below, higher-order information about the pattern is lost in diffraction, which results in more line-end shortening (LES) and corner rounding (CR) at the edges of the cutting polygons. Furthermore, under focus and exposure variations, LES and CR may cause a functional failure in circuit operation. To avoid this problem, Drapeau et al. [36] suggest three guidelines for the polygon cutting: (1) avoid creating small stubs and jogging line ends; (2) prefer cuts on landing pads, junctions, and long run; and (3) ensure enough overlap on stitching points.

Overlay Awareness   DPL requires tighter overlay control since one layer is generated by two exposure steps. Providers of lithography systems have their overlay calibration models, for example, expressed by (x,y) coordinates of a point in the exposure field. Based on the overlay distortion measured on-wafer, the lithography system compensates for overlay error by moving the wafer stage for each field. It is necessary to investigate new forms of overlay awareness [e.g., use of a weighted least-squares method (LSM) to provide higher weight in either the x or y directions]. An example context is if we know that the layout is less sensitive to y-direction error, or if the number of cut lines with a small overlap margin in the x-direction is greater than in the y-direction): calibration of the weighted LSM optimization can be changed to reduce residual error in the x-direction at the cost of larger y-direction residual error.

7.6.2.2 Design Compliance for DP Decomposition

Similar to the industry experience with altPSM (alternating phase shift mask), random logic layouts may not be aligned with layout decomposition goals. That is, when complex pattern behavior exists in a layout, there is less available reduction of pitch. For example, in the contact array there may be a $\sqrt{2} \times$ reduction factor, instead of a 2× reduction, since the diagonal pitch remains unchanged. In two-dimensional contexts such as poly and metal layers in random logic, the minimum gap between two line ends is still constrained, even though the pitch of horizontal patterns achieves a 2× reduction factor. It is thus necessary to study design compliance that is aware of the following two issues in addition to pitch doubling.

Grid-Based Regular Design   DPL cannot achieve the best process window with the current irregular design styles. If multiple pitches exist after the layout decomposition, off-axis illumination (OAI) will have a relatively lower process margin at pitches beyond the optimal angle of the OAI aperture. A new design methodology with a grid-based design will be required; that is, all features are placed at a grid and have the same pitch rule, and then all neighbors of a feature are assigned to different masks. With such an approach, all features will have a fixed pitch even after layout decomposition.

SRAF Awareness   Subresolution assist features (SRAF) will probably continue to be a useful technique to enlarge the process window, even with deployment of DPL. SRAFs require specific spacings of poly after the layout decomposition, and cannot be inserted in certain pitches without upstream SRAF awareness during the design stage. Design compliance with respect to DPL layout decomposition thus involves existing RETs as well as pitch doubling.

7.6.3 Hotspot Detection and Fixing

There are layout configurations in actual design where it is not always possible to achieve sufficient overlap between two polygon cuttings, and marginal space between two mask layers. This results in generation of *hotspots* (i.e., actual device patterns with relatively large risk of disconnection, pinching, and bridging). Hotspot detection will be a task within the physical verification sign-off flow in DPL design. Such detection is aimed at confirming the quality of the decomposed layers, with the decomposed layers being simulated through a process window and checked against a variety of DPL process constraints. Hotspot fixing (e.g., with minimum design perturbations around the hotspot pattern) is much more difficult than detection, since the hotspot fixer must preserve design rule compliance in the (decomposed) DPL layer as well as in the upper and lower neighbor layers.

7.6.4 Improvements to Optical Proximity Correction Methodology

OPC treatment of a double-exposure layout is not so different from the existing OPC for a single exposure. However, OPC must achieve several aspects of added awareness, including the following. It is necessary to understand these aspects to determine any need and value of associated OPC methodology changes. Two examples are as follows.

Topography Awareness   A new challenge is caused by the topographic differences between the first and second exposure steps. Whereas the first exposure takes place on a perfect wafer, the second exposure occurs on topography because overetch during the hard mask etch step can generate a height difference. The OPC model must be changed to take into account the different defocus level and resist thickness. Also, the OPC algorithm must allow different OPC retargetings between the first and second layers.

Process-Type Awareness   Most of the known double-patterning techniques have relatively complex process flows, which result in three different formations of process flow, such as the negative PEPE (photo–etch–photo–etch), positive PEPE, and spacer formation (or self-aligned DPL) approaches. According to the tone of the background, the OPC model must consider different polarization effects. In addition, with respect to spacer formation, actual device size after the deposition of oxide and nitride must be used as an OPC input.

7.6.5 Stress Modeling and Exploitation

Engineering of stress and strain is the key means of achieving mobility enhancement, starting with the 65-nm node. Systematic, layout-dependent impacts of stress must be modeled and exploited wherever possible to optimize the performance–power envelope of the design. As an example, Kahng et al. [28] analyze and exploit STI (shallow trench isolation) compressive stress along the device channel, which typically enhances pMOS mobility while degrading nMOS mobility. STI-induced stress on a given device depends on the device location in the diffusion region and the width of the STI on both sides of the diffusion region. The BSIM stress model [29] accounts for STI stress as a function of device location in the diffusion region (cf. the SA, SB, LOD parameters), but not as a function of STI width. On the other hand, TCAD-based studies reported by Kahng et al. [28] show that STI width modeling can affect critical-path timing by up to 6%. Since STI width is determined by cell placement, this naturally recalls the concepts of opportunism and the use of CORR placement to manage deterministic variations.

Indeed, the performance of standard-cell blocks can be enhanced by using detailed placement to modulate STI width and by inserting active-layer dummy shapes [30]. Additional spacing between timing-critical cells (1) increases STI width and hence speed for pMOS devices, and (2) creates space for insertion of active-layer fill next to nMOS diffusion so as to improve nMOS speed. Figure 7.30 shows a standard-cell row before optimization, after placement perturbation, and after fill insertion. In the figure, $STIW^{sat}$ is the STI width beyond which the stress effect saturates. Cells with diagonal-line patterns are timing critical, and "don't-touch" cells with brick pattern cannot move in the placement optimization. As reported by Kahng et al. [28], over 5% in (STI width-aware) reduction in SPICE-computed path delay can be achieved by combined placement and active-layer fill optimization. Figure 7.31 shows delay histograms of the 100 most critical paths in a small test case, before and after the optimization.

7.6.6 Design for Equipment

A wide range of equipment improvements (hooks to "smart inspection", dynamic control of dose [31], various forms of adaptive process control, etc.) continually afford opportunities to leverage design information for cost and turnaround-time improvements. We discuss two examples.

7.6.6.1 Design-Aware Mask Inspection and Defect Disposition
Mask vendors consider their job to be the precise reproduction of mask data onto mask substrates; that is, the entire mask surface must be free of any

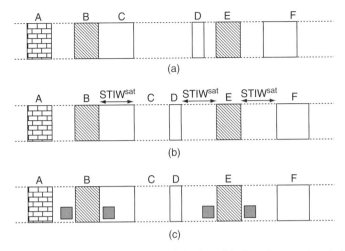

Figure 7.30 Standard-cell row (a) before optimization, (b) after placement perturbation, and (c) after fill insertion.

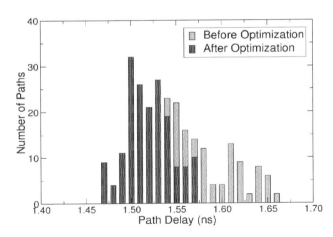

Figure 7.31 Delay histograms of the top 100 paths before and after placement and fill-based timing optimization with respect to STI width-induced stress.

defects, polygon corners have to be (close to) rectangular, and critical dimensions must be within nanometer accuracy regardless of which polygon edges are represented in the design. Such extreme accuracy requirements are artificial and make photomasks very expensive. In a regime of rising mask costs, design-aware modulation of mask complexity [32], and design-aware inspection and defect disposition, can, respectively, reduce mask write times and increase tolerances for functionally insignificant features. In other words, a concept of *design criticality* can potentially be applied in the photomask inspection context.

Recently, Dai Nippon Printing (DNP) and Takumi have announced a collaboration that applies such a design criticality-aware framework to mask inspection such that decisions as to repair versus ignore can be automated. Software from Takumi incorporates image enhancement algorithms along with simulation capability, so that it can import defect images from a mask inspection tool and predict their impact on wafer printing, providing additional guidelines for the defect disposition decision. The resulting mask inspection and repair time savings are anticipated not only to improve yields and shorten turnaround times, but also to improve the overall picture of capital expenditure and equipment depreciation for the merchant mask company.

7.6.6.2 Design-Driven Dose Map Optimization

As another example of passing design intent directly to equipment, we consider ASML's DoseMapper technology, which has been used extensively within the automatic process control context to improve global CD uniformity. The "design for equipment" question here is whether DoseMapper can be used—either for a fixed tapeout, or in synergy with floorplanning and place-and-route—to improve design parametric yield.

Currently, the DoseMapper capability within the step-and-scan tool is used solely to reduce across-chip line-width variation (ACLV) and across-wafer line-width variation (AWLV) metrics for a given integrated circuit during the manufacturing process. At the same time, to achieve optimum device performance (e.g., clock frequency) or parametric yield (e.g., total chip leakage power), not all transistor gate CD values should necessarily be the same. For devices on setup timing-critical paths in a given design, a larger than nominal dose (causing a smaller than nominal gate CD) will be desirable, since this creates a faster-switching transistor. On the other hand, for devices that are on hold-timing-critical paths, or in general that are not setup-critical, a smaller than nominal dose (causing a larger than nominal gate CD) will be desirable, since this creates a less leaky (although slower-switching) transistor. Therefore, a design-aware dose map, that is, a DoseMapper solution that is aware of which transistors in the IC product have critical setup or hold times, can potentially improve circuit performance and leakage power simultaneously.

The ASML DoseMapper concept is implemented by means of Unicom-XL and Dosicom, which change dose profiles in the across-slit (x-dimension) and scan (y-dimension) directions, respectively (Figure 7.32). *Dose sensitivity* is the relation between dose and critical dimension, measured as CD (nm) per percentage change in dose. Increasing the dose decreases the CD; that is, the dose sensitivity has a negative value (in a typical 90-nm process, dose sensitivity is $-2\,\text{nm}/\%$).

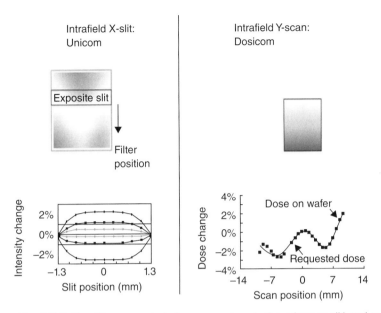

Figure 7.32 ASML DoseMapper control of exposure dose in the *x* (across slit) and *y* (along scan) directions.

For the objective of timing yield and leakage power, the design-aware dose map problem can be stated as follows. *Given an integrated-circuit placement with timing analysis results, determine a dose map to optimize timing yield as well as total device leakage.* To solve this problem, the exposure field can be partitioned into a set of rectangular grids, with the exposure doses in the grids corresponding to optimization variables which affect transistor gate lengths within the grids according to the dose sensitivity. With, for example, an assumption that gate delay increases linearly and leakage power decreases quadratically with increasing gate length, we may represent gate delay and leakage as functions of the dose variables. For IC design, a standard network representation of the circuit timing graph can be used. Then the design-aware dose map problem can be formulated as one of tuning the dose map to adjust channel lengths of gates to minimize a weighted sum of circuit delay in the timing graph and total leakage power, subject to upper and lower bounds on the dose values in the grids, and a dose map smoothness bound to reflect limits of the Unicom-XL and Dosicom capabilities. This problem formulation can be solved efficiently using classic quadratic programming methods, to yield significant timing improvements without worsening leakage [37]. Further optimizations are possible at the placement level; that is, given a dose map, dose map–aware placement can be found to further improve parametric (timing and leakage) yield.

7.7 SUMMARY

It will be evident by the 45-nm node that nano-CMOS technology drives the need for fast, model-based, accurate analysis and correction tools. Most of these tools are still emerging; only a few are production worthy at this time. We are certain that many more will be available as the need becomes compelling. We have described in some detail some of the autocorrection tools that embed the knowledge base into the tool so that designers deploying such a tool in the design process will have the master designer behind the optimizations.

Statistical design has been promoted as the solution for variability, but the practicality is still being debated. The process control of an immature process would result in a large sigma value; at the same time, the data available to support the statistics are limited; therefore, the statistics provided by foundries would at best be a projection. When the process finally matures, there are sufficient data to make the statistics meaningful, but at that point the sigma value has improved significantly, limiting the usefulness of the statistical design.

We have introduced some new concepts for opportunistic DFM and the CORR methodology to achieve improvements with little or no cost. We have made some projections into the future for the 45-nm nodes and beyond where no one can claim a clear vision with a fast-forward view of things. Certainly, the challenges in design would be mounting, and some preparation now would alleviate the need for many cycles of design re-work to achieve a manufacturable, robust, high-performance product, introduced within the market window. We certainly hope that this book has provided some food for thought in your preparation for what is coming.

REFERENCES

1. A. B. Kahng, Key directions and a roadmap for electrical design for Manufacturability, *Proc. European Solid-State Circuits Conference/European Solid-State Device Research Conference*, Munich, Germany, Sept. 2007.

2. K. Jeong et al., Impact of guardband reduction on design process outcomes, *Proc. IEEE International Symposium on Quality Electronic Design*, San Jose, CA, Mar. 2008.

3. P. Gupta et al., Self-compensating design for focus variation, *Proc. ACM/IEEE Design Automation Conference*, Anaheim, CA, 2005, pp. 365–368.

4. S. M. Burns et al., Comparative analysis of conventional and statistical design techniques, *Proc. ACM/IEEE Design Automation Conference*, San Diego, CA, 2007, pp. 238–243.

5. C. Chiang et al., Fast and efficient phase conflict detection and correction in standard-cell layouts, *Proc. ACM/IEEE International Conference on Computer-Aided Design*, San Jose, CA, 2005, pp. 149–156.

6. P. Gupta et al., Lithography simulation-based full-chip design analyses, *Proc. SPIE Conference on Design and Process Integration for Microelectronic Manufacturing*, Santa Clara, CA, 2006, vol. 6156, paper 61560T.

7. P. Gupta et al., Detailed placement for improved depth of focus and CD control, *Proc. Asia and South Pacific Design Automation Conference*, Shanghai, 2005, pp. 343–348.

8. P. Gupta et al., Modeling of non-uniform device geometries for post-lithography circuit analysis, *Proc. SPIE Conference on Design and Process Integration for Microelectronic Manufacturing*, Santa Clara, CA, 2006, vol. 6156, paper 61560U.

9. A. B. Kahng et al., Detailed placement for leakage reduction using systematic through-pitch variation, *Proc. International Symposium on Low Power Electronics and Design*, Portland, OR, 2007.

10. P. Gupta et al., Method for correcting a mask layout, U.S. patent 7,149,999, Dec. 2006.

11. S. Shah et al., Standard cell library optimization for leakage reduction, *Proc. ACM/IEEE Design Automation Conference*, San Francisco, CA, 2006, pp. 983–986.

12. P. Gupta and F.-L. Heng, Toward a systematic-variation aware timing methodology, *Proc. ACM/IEEE Design Automation Conference*, San Diego, CA, 2004, pp. 321–326.

13. http://www.blaze-dfm.com/products/products.html.

14. Y. Chen et al., Performance-impact limited area fill synthesis, *Proc. ACM/IEEE Design Automation Conference*, Anaheim, CA, 2003, pp. 22–27.

15. P. Gupta et al., Topography-aware optical proximity correction for better DOF margin and CD control, *Proc. SPIE*, Santa Clara, CA, 2005, vol. 5853, pp. 844–854.

16. A. B. Kahng and C.-H. Park, Auxiliary pattern for cell-based OPC, *Proc. 27th Bay Area Chrome Users Group Symposium on Photomask Technology and Management, Proc. SPIE*, Monterey, CA, 2006, vol. 6349, paper 63494S.

17. G. Baccarani et al., Generalized scaling theory and its application to a micrometer MOSFET design, *IEEE Trans. Electron Devices*, vol. 31, pp. 452–462, 1984.

18. P. Gupta, et al., Merits of cellwise model-based OPC, *Proc. SPIE Conference on Design and Process Integration for Microelectronic Manufacturing*, Santa Clara, CA, 2004, vol. 5379, pp. 182–189.

19. X. Wang et al., Exploiting hierarchical structure to enhance cell-based RET with localized OPC reconfiguration, *Proc. SPIE Conference on Design and Process Integration for Microelectronic Manufacturing*, Santa Clara, CA, 2005, vol. 5756, pp. 361–367.

20. P. Gupta, personal communication, Jan. 2007.

21. A. B. Kahng et al., Auxiliary pattern-based OPC for better printability, timing and leakage control, *SPIE J. Microlithogr. Microfabrication Microsyst.*, July 2008, pp. 013002-1–013002-13.

22. G. Smith, personal communication, Aug. 2007. See http://garysmitheda.com/.

23. A. B. Kahng, Research directions for coevolution of rules and routers, *Proc. ACM International Symposium on Physical Design*, Monterey, CA, 2003, pp. 122–125.

24. R. Brashears and A. B. Kahng, Advanced routing for deep submicron technologies, *Comput. Des.*, May 1997.

25. C. J. Alpert et al., Timing-driven steiner trees are (practically) free, *Proc. ACM/ IEEE Design Automation Conference*, San Francisco, CA, 2006, pp. 389–392.

26. http://www.edn.com/blog/450000245/post/70007207.html?nid=3389.

27. A. B. Kahng et al., Automated layout and phase assignment techniques for dark field alternating PSM, *Proc. Bay Area Chrome Users Group Symposium on Photomask Technology and Management*, Monterey, CA, 1998, pp. 222–231.

28. A. B. Kahng et al., Exploiting STI stress for performance, *Proc. IEEE International Conference on Computer-Aided Design*, San Jose, CA, 2007, pp. 83–90.

29. http://www-device.eecs.berkeley.edu/~bsim3/.

30. http://imec.be/wwwinter/mediacenter/en/SR2006/681406.html.

31. N. Jeewakhan et al., Application of dosemapper for 65-nm gate CD control: strategies and results, *Proc. SPIE*, Santa Clara, CA, 2006, vol. 6349, paper 6349G.

32. P. Gupta et al., A cost-driven lithographic correction methodology based on off-the-shelf sizing tools, *Proc. ACM/IEEE Design Automation Conference*, Anaheim, CA, June 2003, pp. 16–21.

33. P. Gupta et al., Investigation of diffusion rounding for post-lithography analysis, *Proc. ACM/IEEE Asia and South Pacific Design Automation Conference*, Seoul, Korea, Jan. 2008, pp. 480–485.

34. D. Chinnery and K. Keutzer, *Closing the Gap Between ASIC and Custom: Tools and Techniques for High-Performance ASIC Design*, Kluwer Academic Publishers, New York, 2002.

35. D. Chinnery and K. Keutzer, Closing the power gap between ASIC and custom: an ASIC perspective, *Proc. ACM/IEEE Design Automation Conference*, Anaheim, CA, 2005, pp. 275–280.

36. M. Drapeau et al., Double patterning design split implementation and validation for the 32 nm node, *Proc. SPIE Design for Manufacturability Through Design-Process Integration*, Santa Clara, 2007, vol. 6521, pp. 652109-1 to 652109-15.

37. K. Jeong et al., Dose map and placement co-optimization for timing yield enhancement and leakage power reduction, manuscript, 2007.

INDEX

Nano-CMOS Design for Manufacturability: Robust Circuit and Physical Design for Sub-65 nm Technology Nodes
By Ban Wong, Franz Zach, Victor Moroz, Anurag Mittal, Greg Starr, and Andrew Kahng
Copyright © 2009 John Wiley & Sons, Inc.